Models for Computer Aided Tolerancing in Design and Manufacturing

# Models for Computer Aided Tolerancing in Design and Manufacturing

Selected Conference Papers from the 9th CIRP
International Seminar on Computer-Aided Tolerancing,
held at Arizona State University, Tempe, Arizona, USA,
10-12 April, 2005


*Edited by*

Joseph K. Davidson

*Design Automation Laboratory*
*Department of Mechanical and Aerospace Engineering*
*Ira A. Fulton School of Engineering*
*Arizona State University*
*Tempe, Arizona, USA*

 Springer

A C.I.P. Catalogue record for this book is available from the Library of Congress.

ISBN-10 1-4020-5437-8 (HB)
ISBN-13 978-1-4020-5437-2 (HB)
ISBN-10 1-4020-5438-6 (e-book)
ISBN-13 978-1-4020-5438-9 ( e-book)

---

Published by Springer,
P.O. Box 17, 3300 AA Dordrecht, The Netherlands.

*www.springer.com*

*Printed on acid-free paper*

**The 9th CIRP International Seminar on Computer-Aided Tolerancing**

Arizona State University, Tempe, Arizona 85287-6016, USA
April 10-12, 2005

**Sponsored by**
CIRP (International Institution for Production Engineering Research)
Arizona State University

**International Program Committee**
P. Bourdet (France)
K. Chase (USA)
A. Clément (France)
J. K. Davidson (USA)
A. Desrochers (Canada)
H. Elmaraghy (Canada)
C. Fortin (Canada)
F.J.A.M. van Houten (Netherlands)
L. Joskowicz (Israel)
L. Laperrière (Canada)
L. Mathieu (France)
E. Morse (USA)
J. Shah (USA)
V. Srinivasan (USA)
J.-C. Tsai (Taiwan)
A. Weckenmann (Germany)
R.G. Wilhelm (USA)
E. Zussman (Israel)

**National Organizing Committee**
J. K. Davidson (Chair)
L. Mathieu (Co-chair)
J. Shah (Co-chair)
R.G. Wilhelm (Co-chair)
G. Ameta
Z. Shen
R. Salinas – C. Standiford – L. Mata-Hauksson


**Additional Reviewers**
M. Giordano
T. Kurfess
J.-M. Linares
E. Pairel
A. Rivière
R. Söderberg
J.-M. Sprauel
F. Villeneuve

# Preface

Computer Aided Tolerancing (CAT) is an important topic in any field of design and production where parts move relative to one another and/or are assembled together. Geometric variations from specified dimensions and form always occur when parts are manufactured. Improvements in production systems only mean that the amounts of the variations become smaller, but their presence does not disappear. Clearances nearly always are the measures used to predict how consistently the relative motion will occur and also to ensure that parts will assemble together in a mass-production environment. But clearances, which are attributes of pairs of parts, must be converted to tolerances, which are attributes of individual parts and dimensions. It is this process that CAT is intended to fulfill, all the while being sensitive to the impact of manufacturability, interchangeability, and maintainability on parts and product being designed.

To shorten the time from concept to market of a product, it has been increasingly important to take clearances and the tolerancing of manufacturing variations into consideration right from the beginning, at the stage of design. Hence, geometric models are defined that represent both the complete array of geometric variations possible during manufacture and also the influence of geometry on the function of individual parts and on assemblies of them. The subject of this book, Models for Computer Aided Tolerancing in Design and Manufacturing, focuses on mathematical and computer models and their application to the design and manufacture of machinery and other products. Current CAT systems work for simple problems in which the geometry is not very elaborate. However, for many practical applications, consistent results can be obtained only by those who are expert in both solid modeling and tolerance standards, and sometimes also expert in the computational methods and machinery for quality control measurements. The software for these three fields are not well integrated, and the standards for each were derived from different communities and different premises. Some time ago ISO experts around the world realized how important the integration of these three fields is. Each new standard in these fields improves the integration. The ultimate goal, and the motivation for much of the work described in this book, is to provide CAT systems which are comprehensive enough that ordinary designers can achieve acceptable and consistent results when assigning tolerances to dimensions and features.

In 1996, ISO created a new technical committee, the ISO/TC213, in charge of the standards on geometrical product specification. This committee followed the JHG ISO/TC 3-10-57 which coordinated the technical committees ISO/TC3 "adjustment", ISO/TC57 "Metrology and properties of surfaces" and the subcommittee ISO/TC10/SC5 "Specification and tolerance". The aim of this committee is to provide industrialists with complete and coherent standards in the field of specification and geometrical inspection of products (GPS). In 1996, ISO also published the technical report FD CR ISO/TR 14638 to establish the array of standards to be created or to revised in the future.

In order to achieve this ambitious goal, it is no longer possible to create or develop standards based only on engineers' and experts' knowledge and practice. Now it is necessary to have a global and theoretical approach of the geometrical specification and verification problem. On the basis of French research results, called GEOSPELLING, a model to describe the micro and macro geometry has been developed. The basic concepts are described in the document ISO/TR 17450-1 that was published in 2005. The model is distinct because of its declarative method describing the process of tolerance specification and the process of tolerance verification in inspection.

The contents of this book originate from a collection of selected papers presented at the 9th CIRP International Seminar on Computer Aided Tolerancing (CAT), organized by the Design Automation Laboratory and the Department of Mechanical and Aerospace Engineering of the Ira A. Fulton School of Engineering, Arizona State University, USA from April 10-12, 2005. The CIRP (Collège International pour la Recherche en Production or International Institution for Production Engineering Research) plans this seminar every two years.

The Seminar presentations in Arizona in 2005 began with the keynote address (page 1) by Daniel Whitney from the Massachusetts Institute of Technology. In it he points out specific needs for improvement for CAD and CAT software. Current design practice does not make a clear distinction between creation of a competent nominal design (that is, one that is as close as practical to properly constrained or one in which the designer deliberately inserts desired over-constraint and takes it into account) and performance of a variation analysis (too often called tolerance analysis). Existing CAD systems do not support this distinction or yet provide sufficiently adequate tools for addressing each kind of design. Since the content of his address was taken from the references on page 2, only the abstract of it appears in this book.

This book focuses in particular on the Models for Computer Aided Tolerancing in Design and Manufacturing since accurate and comprehensive models are the basis of the algorithms in software for CAT. Also included are other developments in the field and present applications.

Models for Computer Aided Tolerancing in Design and Manufacturing provides an excellent resource to anyone interested in computer aided tolerancing. The book is intended for a wide audience including:

- Researchers in the fields of product design, Computer Aided Process Planning (CAPP), precision engineering, inspection, quality control, and dimensional and geometrical tolerancing,

- Technicians of standardization who are interested in the evolving ISO standards for tolerancing in mechanical design, manufacturing, and inspection,

- Practitioners of design, design engineers, manufacturing engineers, staff in R&D and production departments at industries that make mechanical components and machines,

- Software developers for CAD/CAM/CAX and computer aided tolerancing (CAT) application packages,

- Instructors and students of courses in design that are offered either for degrees by universities and technical schools, or for professional development through commercial short-courses, and

- Individuals interested in design, assembly, manufacturing, precision engineering, inspection, and CAD/CAM/CAQ.

The book is organized into seven parts, the papers in each one corresponding to a principal topic. The first, Models for Tolerance Representation and Specification, deals with the role that models play in overcoming some of the shortcomings to CAT. Part 2, Tolerance Analysis, is a traditional activity of tolerancing in which values for geometric functional conditions are computationally simulated from tolerance-values that are imposed on dimensions and features of the parts of a mechanism or assembly. Part 3, Tolerance Synthesis, is about determining the part specifications required to comply with the geometric functional conditions of the mechanism. Part 4, Computational Metrology and Verification, concerns the measurements of features and the computational reduction of these to simple values for comparison to the specified tolerances. Part 5, Tolerances in Manufacturing, deals with the quality of part handling within the processes of manufacturing and the simulation of manufacturing processes. Part 6, Applications to Machinery, is about the use of new models and methods to solve specific problems of tolerancing in mechanical systems. Part 7, Incorporating Elasticity in Tolerance Models, shows progress for interfacing models for computer-aided tolerancing in design and manufacturing with the mechanics of elastic deformation. Several of these papers highlight applications for parts made from sheet metal, such as automotive panels and aircraft components.

As Editor, I wish to express my sincere thanks to the authors for their contributions, to the members of the international program committee and the organizing committee, to the additional reviewers, and in particular to Mr. G. Ameta, Mr. N. Joshi, Mr. R. Salinas, Ms. C. Standiford, and Ms. L. Mata-Hauksson for their efforts in getting this book published.

Joseph DAVIDSON

# Table of Contents

## Tolerance Synthesis



## Computational Metrology and Verification



## Tolerances in Manufacturing



## Applications to Machinery

## Incorporating Elasticity in Tolerance Models

# A Unified Approach to Design of Assemblies Integrating Nominal and Variation Design

***D. E. Whitney***
*Massachusetts Institute of Technology,*
*Cambridge MA 02139*
*dwhitney@mit.edu*

Keynote Presentation to the 9th CIRP Seminar on Computer-Aided Tolerancing

In this presentation I sketch out a model of mechanical assemblies that uses the same underlying mathematics, namely Screw Theory, to model both the nominal and varied condition of an assembly. This model is fleshed out in detail in [1] with preliminary presentations in [2-7]. The model represents assemblies as kinematic mechanisms which may or may not be capable of motion by intent. Reference [1] contains references to many papers by other researchers upon whom I relied and who have built up this field.

Assemblies are designed with the intent of achieving one or more Key Characteristics, that is, specifications on relative position and orientation between features on possibly non-adjacent parts, and specifications on allowed variation of the Key Characteristics. Paths called Datum Flow Chains are established by the designer to carry relative position and orientation from part to part in order to establish nominal achievement of each Key Characteristic. Ideally, each Key Characteristic has its own Datum Flow Chain independent of the others, but in practice this goal is often impractical or impossible to attain. The Datum Flow Chain runs from part to part through the joints between them.

Parts are joined by one or more assembly features that are modelled as sets of elementary surface contacts. These features instantiate the part-to-part constraint goals established when each Datum Flow Chain was declared. Screw Theory is used to determine the state of constraint inside each feature and between features in order to characterize the state of constraint of the entire assembly. Variation analysis is conducted by assuming that one or more of the surfaces within a feature may move within its tolerance zone in ways that the tolerance specification allows. Screw Theory is then used to propagate the effect of this variation onto the assembly to see the effects on the Key Characteristics.

Only properly constrained assemblies can be correctly analyzed for the effects of variation at the feature or part level. In the case of over-constraint, a stress analysis is needed. Without taking stress and strain into account, a unique Datum Flow Chain does not exist. In the case of under-constraint, there is no particular nominal condition, requiring the addition of an artificial constraint.

Designers often make constraint mistakes. [8] CAD systems do not offer much help in this regard. CAD systems check for geometric compatibility but do not detect situations

1

*J.K. Davidson (ed.), Models for Computer Aided Tolerancing in Design and Manufacturing, 1–2.*
© 2007 *Springer.*

where locked-in stress could exist under conditions of variation.  Screw Theory permits us to fill this gap.

Current design practice does not make a clear distinction between creation of a competent nominal design (that is, one that is as close as practical to properly constrained or one in which the designer deliberately inserts desired over-constraint and takes it into account) and performance of a variation analysis (too often called tolerance analysis).  Similarly, CAD current systems do not support this distinction or provide adequate tools for addressing each kind of design.  Commercial tolerance analysis software also often fails to notify the user if the assembly is over-constrained.  Finally, some data models of assembly fail to address this distinction and again fail to provide support for improved design methodologies.  The work summarized here is intended to address these issues.

Keywords: assembly, constraint, Screw Theory, tolerance, variation, Datum Flow Chain, Key Characteristic

References

[1] Whitney, D. E., *Mechanical Assemblies: Their Design, Manufacture, and Role in Product Development*, Oxford University Press, 2004.

[2] Mantripragada, R. and Whitney, D. E., "The Datum Flow Chain," *Research in Engineering Design*, v 10, 1998, pp 150-165.

[3] Whitney, D E, Mantripragada, R., Adams, J. D., and Rhee, S. J., "Designing Assemblies," *Research in Engineering Design*, (1999) 11:229-253.

[4] Whitney, D E and Adams, J D, "Application of Screw Theory to Constraint Analysis of Assemblies Joined by Features," *ASME J of Mech Design*, v 123 no 1, March, 2001, pp 26-32

[5] Shukla, G., and Whitney, D. E., "Application of Screw Theory to Analysis of Mobility and Constraint of Mechanisms," to appear in *IEEE Trans on Automation Systems Engineering*, April 2005

[6] Whitney, D. E., and 10 co-authors, "A Prototype for Feature-Based Design for Assembly," *ASME J. Mech. Design*, v 115 no 4, Dec, 1993, pp 723-34.

[7] Whitney, D. E., Gilbert, O., and Jastrzebski, M., "Representation of Geometric Variations Using Matrix Transforms for Statistical Tolerance Analysis in Assemblies, *Research in Engineering Design*, (1994) 6: pp 191-210

[8] Kriegel, J. M., "Exact Constraint Design," *Mechanical Engineering*, pp 88-90, May 1995

# Virtual Gauge Representation for Geometric Tolerances in CAD-CAM Systems

**E. Pairel, P. Hernandez, M. Giordano**
*LMECA (Laboratoire de Mécanique Appliquée),
École Supérieure d'Ingénieurs d'Annecy - Université de Savoie,
BP 806, 74016 ANNECY Cedex - FRANCE
eric.pairel@univ-savoie.fr*

**Abstract:** The CAD software seeks to represent the syntax of the geometric tolerances, i.e. their writing on the drawings. We propose to represent their semantics, i.e. their meaning with respect to the part. We show that the meaning of the geometric tolerances can be defined thanks to a model of virtual gauges. These gauges concern geometrical entities of the part which are represented on the three-dimensional geometrical model of the part (CAD model). The topology of a gauge is related to that of the part. Recording these attributes is sufficient. The advantages of this representation are its simplicity, the semantic coherence which can be guaranteed, the independence from the standards, their limits and their evolutions, and the extension of the tolerancing possibilities for the designer.
**Key words:** Tolerancing, Virtual gauge, CAD-CAM.

## 1.  INTRODUCTION

The subject of this paper is to present the bases of a data-processing representation of the geometric tolerances. The tolerances which are considered are those which are allowed by the ISO and ASME standards. Nevertheless, we will show that the suggested representation allows to specify functional tolerances which are difficult and even impossible to express with the writing rules of the standards. Indeed it is necessary to distinguish the **syntax** of a geometric tolerance, i.e. its writing on the technical drawing, and its **semantics**, i.e. its meaning with regard to the part. Whereas the CAD software packages try to represent the syntax of the geometric tolerances, we propose to represent their semantics. The task is then much simpler because, while the syntactic rules of the standardized tolerances are many and are badly formalized, we think that their semantics can always be interpreted in the form of a virtual gauge.

Several authors have already shown this geometrical interpretation of tolerances [Jayaraman *et al.*, 1989], [Etesami, 1991], [Nigam *et al.*, 1993]. Some of them have tried to model it [Ballu *et al.*, 1997], [Dantan *et al.*, 1999]. This geometrical interpretation is also found in the American standard [ASME, 1994].



*J.K. Davidson (ed.), Models for Computer Aided Tolerancing in Design and Manufacturing,* 3–12.
© 2007 *Springer.*

However we think that none of these contributions has brought as complete and simple a model as the ***fitting gauge model*** which we have developed since 1995 for the three-dimensional metrology [Pairel *et al.*, 1995]. Here we propose using this model to represent the geometric tolerances in the CAD-CAM systems. We will show that it enables to model a multitude of geometric tolerances very simply.

This semantic representation of the tolerances must be accompanied by a checking of the degrees of freedom removed by references [Kandikjan *et al..*, 2001] and left at the tolerance zones [Hernandez *et al..*, 2002] to guarantee the full semantic coherence of the tolerances. This checking will not be detailed here.

From this tolerance representation, it becomes simpler and more direct to generate the domains of the geometrical variations allowed to the faces of the part [Giordano *et al.*, 1999] [Roy *et al.*, 1999] [Davidson *et al.*, 2002], which is necessary to the analysis and synthesis tolerance processes of a mechanism [Giordano *et al.*, 2001].

## 2. INTERPRETATION BY VIRTUAL GAUGE OF THE STANDARDIZED GEOMETRIC TOLERANCES

In order to present the "fitting gauge model" and its use for the geometric tolerance representation, the technical drawing given on figure 1 will be used.

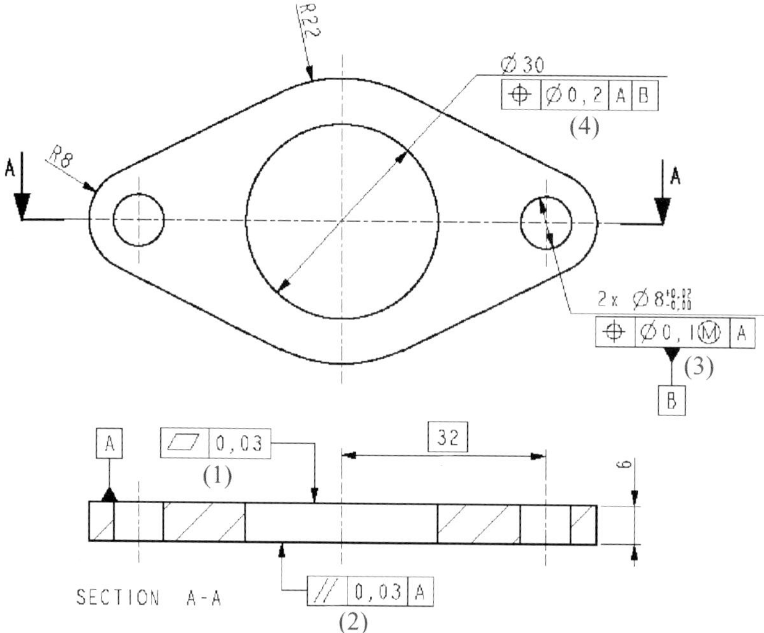


*Figure 1; Examples of geometric tolerances.*

This drawing reveals a broad panel of geometric tolerances: a tolerance of form (No 1 on figure 1), of orientation (No 2), and of position of a single feature (No 4) or of a group of features (No 3), as well as more or less complex ***datum systems*** (tolerances No 3 and 4). The maximum material condition (MMC) is also considered. The other categories of tolerance - run-out, minimum material condition, projected tolerance - will not be presented here but can also be represented. Only, the complementary indications, often added in the form of notes near the geometric tolerances, cannot be directly represented by the model presented here. The case of the dimensional tolerances is not mentioned here either.

### 2.1  Form Tolerance (tolerance No. 1) : the *zone-gauge*

The zone of a form tolerance constitutes a virtual gauge for the toleranced face or line. This gauge is completely free in displacement compared to the part :

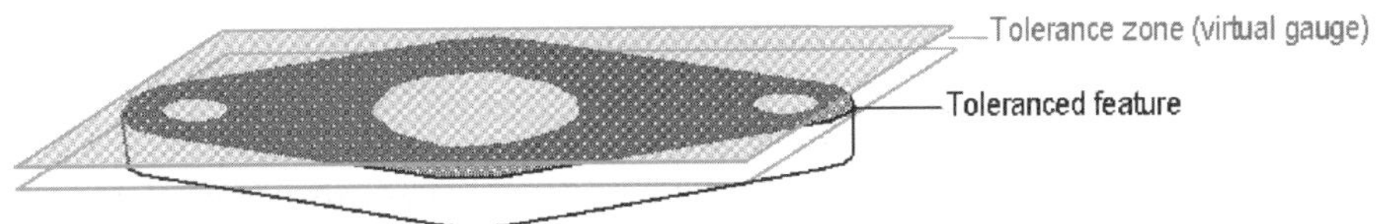


*Figure 2; Interpretation of a form tolerance (flatness).*

### 2.2  Orientation tolerance (tolerance No. 2): *Surface-gauge* - Degree of freedom of a gauge

Now the tolerance zone is "dependent" in orientation on a theoretical datum surface (here a plane). This datum surface is like a perfect plane which must be fitted with the "bottom" face of the part.

The orientation tolerance can be interpreted as a virtual gauge, composed of one plane (surface-gauge) and a tolerance zone (zone-gauge). The zone-gauge is linked to the plane-gauge but can move in translation in the three directions of the space. The Surface-gauge plane is fitted with the "bottom" face of the part. Then the zone-gauge can move in translation to try to contain the "top" face :

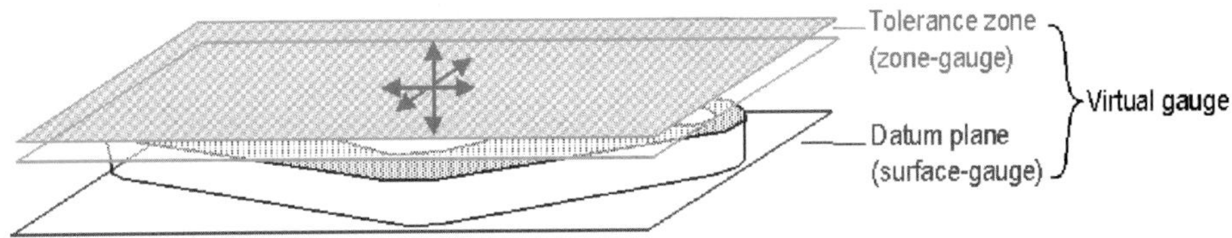


*Figure 3; Interpretation of an orientation tolerance (parallelism).*

### 2.3  Position tolerance for a pattern of features (tolerance No. 3) and maximum material requirement: surface-gauge whose size is fixed

The maximum material requirement defines a theoretical surfaces (border surfaces) that the toleranced faces of the part do not have to cross. These theoretical surfaces are cylindrical surface-gauges of diameter equal to 7.9 mm. They are in theoretical positions between each another and with regard to the surface-gauge plane used as datum :

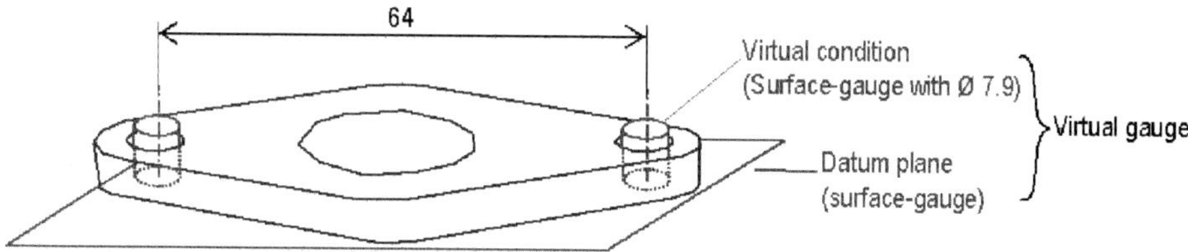


*Figure 4; Interpretation of the maximum material requirement applied to a pattern of holes.*

When the plane is fitted to the bottom face of the part, the virtual gauge, composed of the plane and the two cylinders, has three degrees of freedom corresponding to the established planar joint, which enables it to assemble the two cylinder-gauges inside the holes.

## 2.4 Datum system and pattern of features taken as a datum (tolerance No. 4)

The virtual gauge is composed of four gauges: a plane surface-gauge, two cylindrical surface-gauges and one cylindrical zone-gauge. These gauges are linked together. The plane is fitted first on the bottom face of the part. Then the two cylindrical surface-gauges are fitted simultaneously inside the two small holes by increasing their diameters to a maximum. Then the virtual gauge does not have any degree of freedom with regard to the part. The zone-gauge must contain the axis of the large hole:

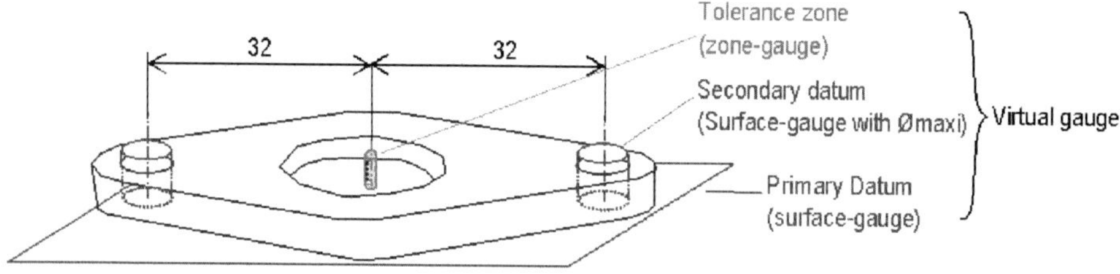


*Figure 5; Interpretation of a position tolerance with a datum system.*

## 3.  PRESENTATION OF THE "FITTING GAUGE" MODEL

The gauges are theoretical constructions of elementary gauges (cylinder, plane, ...), each one being in relation to a geometric feature of the part. These elementary gauges are either the zone-gauges, or surface-gauges.

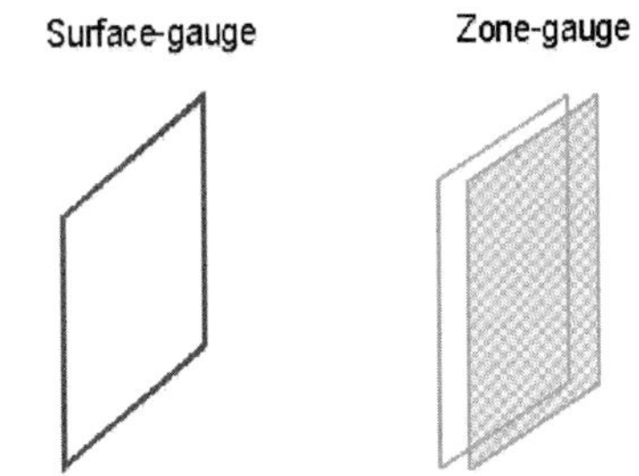


*Figure 6; Surface-gauge and Zone-gauge.*

The zone-gauges "materialize" the tolerance zones whereas surface-gauges "materialize", either the datums, or the virtual conditions.

The topology of a gauge is directly related to the one of the part: a surface-gauge has the nominal shape of the face with which it is in relation. A zone-gauge has the shape generated by the displacement of a sphere - of diameter equal to the tolerance - on the geometric feature with which the zone-gauge is in relation. Thus it is not useful to model the topology of the gauge in the data structure: topology is given by the CAD model. It is sufficient to record the type of gauge and its attributes.

### 3.1 Attributes of a surface-gauge

A surface-gauge can have two types of behavior with regard to the geometric feature of the part: either it seeks to be fitted to the geometric feature of the part, or it only acts as border for it. It seeks to be fitted when it is used as datum and it only acts as border when it represents the virtual condition of the feature.

Levels of priority must be given to allow to define a chronological order for the fitting of the elementary gauges to the faces of the part. We will speak about ***primary-fitting***, ***secondary-fitting*** and ***tertiary-fitting***. We thus define a first attribute for the surface-gauge, which will be called "behavior", and which will be able to take four values:

- ***Behavior = PrimFit, SecondFit, TertiaryFit,* or *Border***

If a surface-gauge has one or more intrinsic dimensions ("sizes"), those are free if the surface-gauge has a fitting behavior. They are fixed with a given value if it is a border. For example, the cylinder must have a variable diameter to be fitted and to be used as a datum on the part and a fixed diameter when it represents a virtual condition of a feature. So we define one or more attributes "size" for the surface-gauges:

- ***Size*= Positive value if the size is fixed or a negative value if the size is not fixed** and if the surface-gauge is fitting.

### 3.2 Attributes of a zone-gauge

The zone-gauge has only a role of border for the geometric feature of the part with which it is in relation. Its shape is determined by that of the geometric feature and by the value of the tolerance. It is thus sufficient to introduce an attribute giving the value of the tolerance:

- *TolValue* = Positive value

When the zone-gauge represents a zone of an orientation tolerance, it can move in translation in all the directions with regard to the datum. It is thus necessary to introduce an attribute indicating if the zone can move or not with regard to the datum:

- *FreeToTranslate* = TRUE or FALSE


## 4. REPRESENTATION OF THE TOLERANCED FEATURES AND OF THE DATUMS FEATURES ON THE CAD MODEL OF THE PART

The semantic representation proposed here requires the three-dimensional construction of the toleranced and datum features on the part. These features can already exist on the CAD model of the part or will have to be added by the designer. For example, a axis hole "will be materialized" by a segment of straight line inside the hole with a starting point at "the entry" of the hole and a final point at the "exit" of the hole. This segment will have to be a "child" of the cylindrical face representing the hole, in the meaning of mother/child relation used in CAD systems. It could be prolonged or axially moved if it is the prolongation of the hole which is functional. Thus the standardized concept of projected tolerance can easily be represented.

Sometimes the tolerance concerns only a piece of the face. In this case a surface corresponding to this piece will be added on the model of the part.


## 5. ILLUSTRATION OF THE REPRESENTATION OF A PART TOLERANCING SCHEME

To simplify the presentation, we consider that each geometrical feature of the CAD model has a number. On the figure 7, only the numbers of the features affected by a gauge were indicated. They are the planar faces (1) and (2), the cylindrical faces (3) and

(4) and finally the straight line segment (5) representing the axis of the large hole:

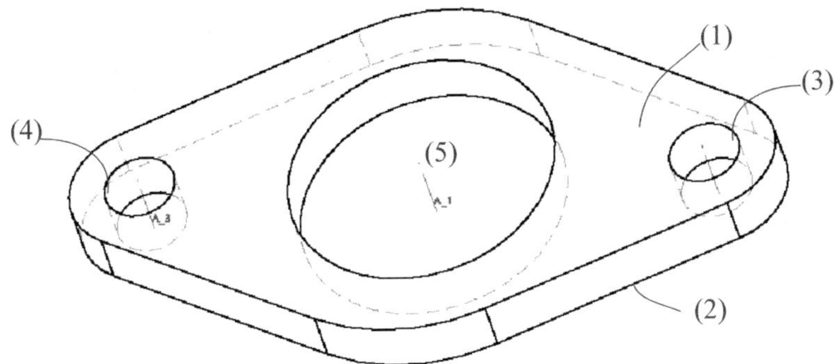



*Figure 7; CAD modeling of the part: each feature has an identifier, represented here by a number.*

The representation of the four geometric tolerances assigned to this part consists in describing the virtual gauges.

Each virtual gauge is a list of elementary gauges, each one in relation to a geometrical feature of the CAD model of the part. So the first attribute of an elementary gauge is a pointer towards the geometrical feature of the part: ***OnFeatureNo***.

According to the type of elementary gauge (***ZoneGauge*** or ***SurfaceGauge***), the values of its attributes (***TolValue*** and ***FreeToTranslate*** or ***Behavior*** and ***Size***) are found. The tolerances of the part represented on figure 1, are defined as follows:

```
ToleranceNo1= { ZoneGauge(OnFeatureNo=1; TolValue=0.03; FreeToTranslate= FALSE) };

ToleranceNo2 = { SurfaceGauge(OnFeatureNo=1; Behaviour=PrimFit);

          ZoneGauge(OnFeatureNo=2; TolValue=0.03; FreeToTranslate= TRUE) };

ToleranceNo3 = { SurfaceGauge(OnFeatureNo=1; Behaviour=PrimFit);

          SurfaceGauge(OnFeatureNo=3; Behaviour=Border; Size=7.9) ;

          SurfaceGauge(OnFeatureNo=4; Behaviour=Border; Size=7.9) };

ToleranceNo4 = { SurfaceGauge(OnFeatureNo=1; Behaviour=PrimFit);

          SurfaceGauge(OnFeatureNo=3; Behaviour=SecondFit; Size= "Free") ;

          SurfaceGauge(OnFeatureNo=4; Behaviour=SecondFit; Size= "Free") };

          ZoneGauge(OnFeatureNo=5; TolValue=0.2; FreeToTranslate= FALSE) }
```

*Figure 8; Illustration of the representation of the geometric tolerances by description of the virtual gauges.*

## 6. PROSPECTS FOR THE USE OF THE REPRESENTATION OF TOLERANCES BY VIRTUAL GAUGES

At present, in the majority of the CAD software packages, the tolerance frames are directly created by the user. The compliance with the standardized rules of syntax is in party ensured by the software which limits the possibilities of writing. The semantic coherence of the tolerances, with respect to the parts, is not verified and depends entirely on the expertise of the user.

To our knowledge, at the moment, two software packages are able to generate the tolerance frames, in a quasi-automatic way, starting from the selection, by the user, of the toleranced features and the datum features on the 3D model of the part. Nevertheless the development and the updating of those software packages are delicate because the writing rules of the standardized tolerances are badly formalized and change regularly.

The "virtual tolerancing gauges" could be generated in the same manner: by selecting the toleranced features and the datum features on the 3D model of the part, the user will define the inspecting gauge corresponding to each functional geometrical requirement of the product. These gauges will be displayed in 3D on the model of part (as shown in figures 2 to 4), which will enable the user to directly visualize the meaning of the geometric tolerances, or could be expressed in the form of tolerance frames on the technical drawings according to the standardized graphic languages (ISO or ANSI). Thus the semantic representation of the tolerances gives more possibilities of tolerancing to the user and releases him from the constraints of standardized writing rules. The tolerancing will be more functional and faster to realize.

The representation by gauges allows to define geometric tolerances impossible to express by the today standardized syntax. For example it is impossible to specify a "self-parallelism" tolerance for the two plane faces (1) and (2) of the part (figure 7) with the standardized graphic language. However the gauge corresponding to this requirement could be defined. It will consist of two plane zones each being able to translate in reference to the other. In this case the algorithm for the writing of the gauges in the form of tolerance frames should propose various solutions to the user: either a standardized tolerance which "degrades" the desired tolerance - it is the parallelism tolerance indicated on the drawing of figure 1 - or a less "standardized tolerance" which expresses the gauge as well as possible such as the one proposed on the figure below.

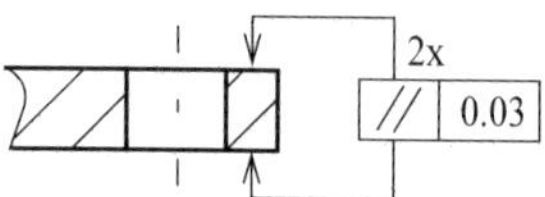

*Figure 9; Instance of non standardized tolerance defining two parallel plane zones.*

Nevertheless we think that the gauge representation could allow to do without the writing of the tolerance frames on drawings in a CAD-CAM environment. Indeed it would be even simpler for the manufacturing designer to see the tolerancing gauges directly in 3D rather than to have "to decode" tolerance frames on drawings. Many mistakes in the interpretation of tolerances would be thus avoided, during their writing and their reading.

In production, we think that the tolerancing by zones is unsuited because it does give the separation of the form, orientation and position defects which is necessary to allow to correctly adjust the production process.

Lastly, the prototype software of three-dimensional metrology that we have already developed, shows that it is possible to directly use the fitting gauge model for the verification of the manufactured parts [Pairel, 1997].

## 7. CONCLUSION

The "fitting gauge model" enables to represent, in an extremely simple way, the near total of the standardized geometric tolerances as well as tolerances by zone which are impossible to express in the standardized graphic language.

This representation guarantees the semantic coherence of the tolerances and can be directly used for the dimensional verification of conformity of the products and also by the manufacturing designer.

We think that this model could represent any geometric tolerance by zone. It could be achieved by improving the model and the data-structure. The case of the geometric tolerances of lines (circularity, straightness, profile tolerance of any line) can easily be modelled by creating a line on the surface of the CAD model of the part. The gauge-zone will be related to this line and not to the surface. The most difficult case is the particular case of the zone having a shape different from the one of the toleranced feature. For example it is the case of a cylinder axis having to be contained within a planar zone. In this case, a solution could consist in creating a plane passing through the axis and directed with regard to another geometric feature of the part defining the secondary datum reference. The zone-gauge would be related to this plane.

The other study to be carried out relates to the representation of the dimensional tolerances with or without envelope condition. For the moment we think that the dimensional tolerances can be "carried" by the CAD model of the part, which is already possible with several software packages. We are currently considering testing a such semantic representation in a CAD software package.

## 8. REFERENCES

**[Jayaraman *et al.*, 1989]** R. Jayaraman, V. Srinivasan, "Geometric tolerancing: 1. Virtual boundary requirements", IBM Journal of Research and Development, Vol. 33, No. 2, March 1989.

**[Etesami, 1991]** Etesami F.; "Position Tolerance Verification Using Simulated Gaging", The International Journal of Robotics Research, Vol. 10, No. 4, 1991.

**[Nigam *et al.*, 1993]** S.D. Nigam, J.D Guilford, J.U. Turner, "Derivation of generalized datum frames for geometric tolerance analysis", ASME Design Automation Conference, Albuquerque, Sept. 1993.

**[Ballu *et al.*, 1997]** A. Ballu and L. Mathieu, "Virtual gauge with internal mobilities for verification of functional specifications"; Proceeding of the 5rd CIRP Seminar on Computer Aided Tolerancing, Toronto, Canada, April 27-29; 1997.

**[Dantan *et al.*, 1999]** J.Y. Dantan, A. Ballu, "Functional and product specification by Gauge with Internal Mobilities", CIRP Seminar on Computer Aided Tolerancing, University of Twente, Netherlands, March 1999.

**[ASME, 1994]** ASME Y14.5M-1994, "Dimensioning and Tolerancing".

**[Pairel *et al.*, 1995]** E. Pairel, D. Duret, M. Giordano, "Verification of a group of functional surfaces on Coordinate Measuring Machine", Proceedings of the XIII IMEKO World congress, torino, Italy, Sept. 1995, pp 1670-1675.

**[Kandikjan *et al.*, 2001]** T. Kandikjan, J.J. Shah, J.K. Davidson, "A mechanism for validating dimensioning and tolerancing schemes in CAD systems", Computer Aided Design, Volume 33, 2001, pp. 721-737.

**[Hernandez *et al.*, 2002]** Hernandez P.; Giordano M., "Outil analytique d'aide au tolérancement géométrique en vue d'intégration en C.A.O.", IDMME 2002, Clermont-Ferrand, France, May 14-16, 2002.

**[Giordano *et al.*, 1999]** Giordano M., Pairel E., Samper S., "Mathematical representation of Tolerance Zones", 6th CIRP Inter. seminar on Computer-Aided Tolerancing, Univ. of Twente, Enschede, The Netherlands, 1999.

**[Roy *et al.*, 1999]** U. Roy and B. Li, "Representation and interpretation of geometric tolerances for polyhedral objects - II. Size, orientation and position tolerances", Computer-Aided Design, Volume 31, No 4, pp. 273-285, 1999.

**[Davidson *et al.*, 2002]** J. K. Davidson, A. Mujezinovic; "A new mathematical model for geometric tolerances as applied to round faces"; Journal of Mechanical Design, Volume 124, Dec. 2002; pp. 609--622.

**[Giordano *et al.*, 2001]** M. Giordano, B. Kataya, E. Pairel; "Tolerance analysis and synthesis by means of clearance and deviation spaces", 7th CIRP International Seminar on Computer-Aided Tolerancing; ENS de Cachan, France, 2001, pp. 345-354.

**[Pairel, 1997]** E. Pairel; "The "Gauge model": a new approach for coordinate measurement"; Proc. of the XIV IMEKO World Congress, Tampere, Finland, June 1997, pp. 278-283.

# Modal Expression of Form Defects

**F. Formosa, S. Samper**
*LMECA ESIA BP 806,*
*74016 ANNECY Cedex FRANCE*
*fabien.formosa@esia.univ-savoie.fr*

**Abstract:** In the fields where the form is functional, the qualification of shape defects is essential to optimize the functional conditions. By using the geometrical tolerances as defined in the standards, the behavior of surfaces inside this space may not be defined (or characterized). To meet specific needs, one of the research orientations consists in correlating the measured defects with the origin of these defects, by generating a basis of characteristic defects. After a brief review of the methods of shape defects analysis, a decomposition based on the dynamic natural modes of their ideal surface is suggested. The method is explained, then applied on a simple example, and compared to the decomposition in the basis of the Zernike polynomials. The article concludes on an opening from the possibilities and the research orientations to consider.
**Keywords**: shape defects, modal tolerancing, dynamic natural modes

## 1. CONTEXT

The control of shape defects is important for the geometric quality of a part, especially when the form has a function in an assembly, or when the shape is functional (optic, perceived quality…). Nowadays, precisions in parts are growing so that size and position deviations becomes in the same order than surface form deviations. It is then necessary to be able to characterize any surface defects by using a minimum of parameters, and to compare it to the customer needs (specifications). We have then to characterize the variations by a new geometrical language. A non-exhaustive assessment of the methods of modal tolerancing that allows surface characterizations is presented. In the following, we show a short view of literature that gives solutions in form characterization.

### 1.1 Medical 3D form analysis

Many works dealing with the active recognition of forms was published and applied since the beginning of the years 1990 [Cootes et al, 2004]. Their purpose is to identify a form compared to a type, then to classify it, and finally to qualify it by using a basis. The pattern of recognition relates to the strategy of sampling and to the rebuilding of the borders criteria [Staib et al, 1992], [Reissel, 1996], [Staib et al, 1996], [Davatzikos et al, 2003]. These operations are carried out on 2D images, a 3D form.



*J.K. Davidson (ed.), Models for Computer Aided Tolerancing in Design and Manufacturing,* 13–22.

### 1.2  Optical Field

Optical elements are parts for which the form of surface is characteristic of the function. The most quoted basis in optics is one made up of the Zernike polynomials [Wyant et al, 1992]. These polynomials have the property to perfectly translate specific defects for the quality of optics such as the astigmatism, the tilt, defocusing, etc. The basis of the decomposition has two entries and characterizes the defects of disc. The method is detailed in the following and is used as confrontation with the suggested method.

### 1.3  Metrological Field

In the field of metrology, the definition of surfaces and the specification of the defects of form relate to three large axes:

- Optimizing the sequences of points acquisition during the control of surfaces,
- Minimizing the error of measurement by choosing the best criterion of association between the measured surface and the nominal surface,
- Comparing the measured defects to the origin of these defects by generating a basis of standard defects.

One of the oldest work seems the minimization of the variation of form, as recalled by K. Carr [Carr et al, 1995] in the introduction of its bearing article on the control of the form defect. Then, the decomposition of the defect in the form of periodic functions as introduced by A. Pandit [Pandit et al, 1981] is pointed (or studied). The third axis simply implies the creation of a basis of the defects in the fields of the medical imagery and optics. Various methods are suggested.

E. Capello, from the Polytechnic School of Milan, presents an approach by "harmonic association" in [Capello et al, 2000]  in order to define the number and the position of the points of sampling during the control of a plane surface.

Kim Summerhays, from the University of San Francisco, presents in [Summerhays et al, 2002] a modeling of a systematic variation of form, by treating each measured surface according to a decomposition in two basis or both. A first basis may consist of the analytical functions, whose  angular dependence is modelled by Fourier series, and axial dependence by Chebyshev polynomials of the second type. Another basis is constituted of the eigenvectors of the measured surface. After comparison, it appears that the basis of the eigenvectors makes it possible to model the abrupt changes of slope with modes of smaller order.

The team of Darius Ceglarek suggests a decomposition of the defect of form by using a discrete transformation into a cosine [Huang et al, 2003]. The field of the variations is then divided into a series of modes with independent defects. This independence allows a statistical distribution of the coefficients for each mode, for a given manufacturing process, and an identity card of each mode may be constituted in order to carry out a follow-up.

### 1.4  GPS Concept

To unify the current and future standards, and to guarantee the coherence of the system under construction, the ISO standard suggests a step in the classification of the standards, according to two axes:

- Geometrical characteristics of a geometrical standard, that (or which) allow the characterization of the face, the form, the orientation, the position, etc...
- Steps of the geometrical and dimensional quotation of a product, that allow the characterization of geometrical standards, the evaluation of the dimensional deviations, and the definition and control of the measuring instruments.

This whole process fits in a two-dimensional board, known as "GPS matrix". Thanks to this matrix one can draw up an assessment of the current standards, locate the redundancies and the deficiencies, and even design (or plan out) the development of the standards to come.

In the area of the control of the defect of form, various basis have been developed in order to be able to identify and qualify the defect. Although a generalized and standardized system to carry out this decomposition does not exist yet, one perfectly realizes that the need is real. In many cases, the definition of a standard zone of tolerance is insufficient to specify a functional requirement in terms of form. Once the customer need has been identified, the standard must give the necessary and sufficient tools to express this functional condition in terms of tolerances. This is  our suggested approach if the form is a functional condition.

## 2.  SUGGESTED METHOD

First, the surface is discretized using a finite element approach. The chosen element type is a shell element. The measured surface is expressed by a displacements vector $(V)$. Then we make a modal analysis in order to obtain the modal basis $(Q_i)$. This one will be used to decompose $V$ and calculate the $\lambda_i$ coefficients. Those coefficients will represent the form deviation in the basis of modal shapes. We use $P$ modes for this decomposition with $\varepsilon_i$ the corresponding residue.

We propose to work on a generic language that permits at the same time:

- To mathematically express the defect of form in the direction "to specify a functional need",
- To simulate this defect,
- To give a direction to measurements, which  is, as of today, achievable by using the means of 3D measurement.

The principle of our modeling is based on a decomposition of the defect of form in an associated modal basis. After reviewing the modal analysis, a step of construction is suggested.

In linear dynamics, the discretized equations of the movement for a conservative, discrete or continuous system, can be written in the general form:

$$M\ddot{q}+Kq=0 \qquad (1)$$

with $M$ the matrix of generalized mass, $K$ the matrix of generalized stiffness generalize, $Q_i$ a vector of amplitudes. $N$ is the number of degrees of freedom of the system.

The mode of a structure is then generally written in the form:

$$q_i=Q_i.cos(\omega_i\, t) \qquad (2)$$

where $Q_i$ is the vector of the amplitudes, and $\omega$ are pulsations in rad.s$^{-1}$.

By taking into account their definition, the modes are solution of the equation:

$$(K - \omega_i^2\, M)Q_i=0 \qquad (3)$$

which admits $N$ eigen solutions, the modes of the structure.

The modal pulsations $\omega_i$ of the various modes are the roots of the characteristic polynomial:

$$det(K - \omega_i^2 M)=0 \qquad (4)$$

The vectors of modal deformations $Q_i$ are the eigenvectors associated the pulsations $\omega_i$, which form a basis M and K orthogonal in the vector space to the movements of the structure. One chooses classically vectors $Q_i$ such as $Q_i^t \cdot M \cdot Q_i =1$. One defines the scalar product $<A.B>=A^t M B$ and the standard $\|A\| = \sqrt{A^t MA}$. The orthogonality of the eigen modes expresses that the inertias and stiffness developed in a mode do not work in the movements of the other modes. There is mechanical interdependence of the modes.

For a continuous system, the number of eigen modes thus obtained is theoretically infinite. Practically, a truncation of the modes is carried out. As $i$ is high, the corresponding form $Q_i$ is more complex. Like in the Fourier Transform analysis, this analysis will stop at an $P$ number of modes with a given residue $\varepsilon$.

## 3. APPLICATION OF THE METHOD ON A SIMPLE CASE

In a first example, a case of disc flatness , we will implement  with details each step of the decomposition of a measured defect of form, and we will compare this decomposition with the one used in optics, calling upon the polynomials of Zernike.

### 3.1 Modal Decomposition

At first, we define a mechanical entity associated to the kind of specified surface. That entity has an elastic behavior. In the case of complex geometries, only a discrete approach leads to approximate solutions. Therefore we naturally choose the context of finite elements computation. The material coefficients (Young modulus, density) is chosen in order to obtain a satisfactory numerical conditioning, as well as a direct geometrical interpretation of the frequency of the modes, in terms of wavelength of defects.

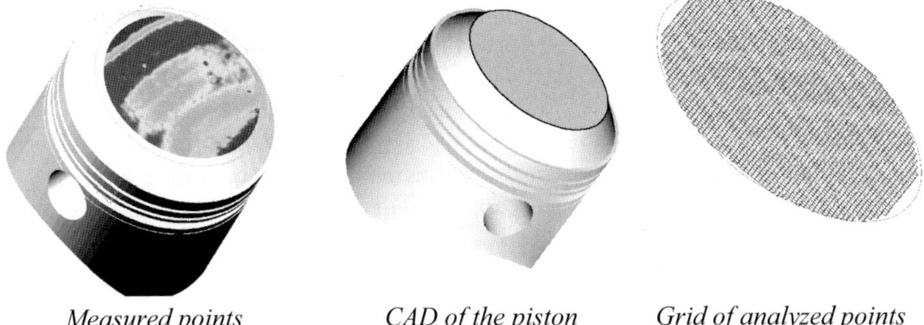



*Figure 1; From the measured to the discretized surface*

The analyzed surface is a disc with a diameter of 63 mm. It is possible to recover a surface representation thanks to groups of measured points, whose distribution follows the grid of measurement. The theoretical surface will be a mesh using shell type elements, and of low thickness (equal to the hundredth of the value of the diameter). The modal basis is then given by using free-free boundary conditions in the general case. Thanks to the presence of rigid body eigen mode in the basis of decomposition, the body motion of the surface may be taken into account.

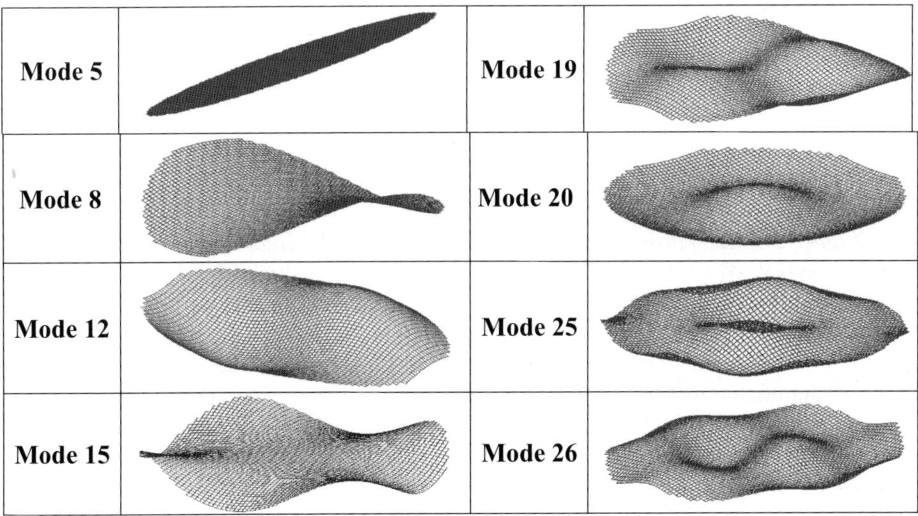



*Figure 2;* Examples of Eigen Modes Obtained of a Disc.

The figure 2 shows some eigen modes of the surface. The basis of the eigen modes have been defined, and one now needs a method of projection of the surface, measured on the modes. The measured defect is represented by a displacement vector $V$. Its decomposition is carried out using the following vectorial projection:

$$\lambda_i = <V.Qi> = V^t \, M_r. \, Q_i \tag{5}$$

$\lambda_i$ is the coefficient associated to the mode $Q_i$. It will be called the modal coefficient. We define the residue vector $\varepsilon$ as the difference between the vector displacement and it's projection in the reduced basis. The expression of this vector residue $\varepsilon$ is as follows:

$$\varepsilon = V - \sum_{i=1}^{P} \lambda_i \cdot Q_i \tag{6}$$

Where $P$ is the number of retained eigenvectors. We define the total residue $E$ as the mass normalization of $\varepsilon$.

$$E = \sqrt{\varepsilon_t \cdot M \cdot \varepsilon} \tag{7}$$

Then, we may consider various significant modes to represent the defect observed.

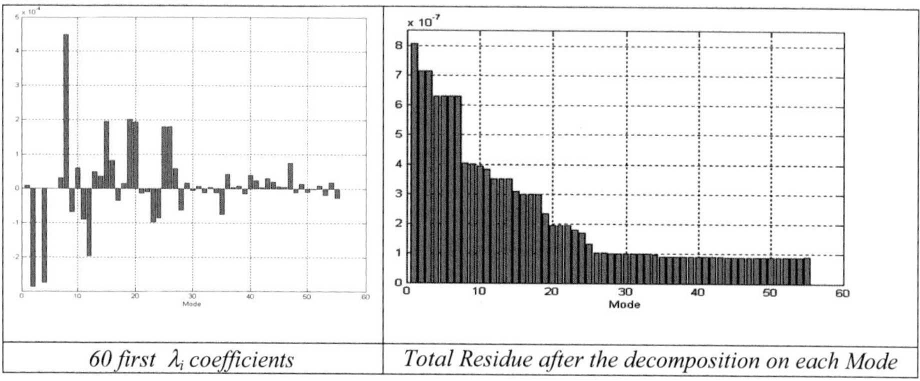

| 60 first $\lambda_i$ coefficients | Total Residue after the decomposition on each Mode |
|---|---|

*Figure 3; Decomposition of the Defect of Form of the Head of the Piston.*

We can see in the figure 3 above that some modes have significant contribution. If we use the 35th first ones we observe that the total residue is small. We can see in figure 4 the comparison between the measured surface and the rebuilt surfaces with 35th modes and the 7th significant modes. The measured surface has positive and negative peaks that cannot be rebuilt with few modes. This phenomenon is the same than the Dirac function in Fourier transform. The modal tolerancing method with mode truncating makes a geometric filter.

We can notice that the first 6 modes give (6 rigid body modes in the case of free boundary conditions) the position of the surface. Then the following modes do not contain the position defect.

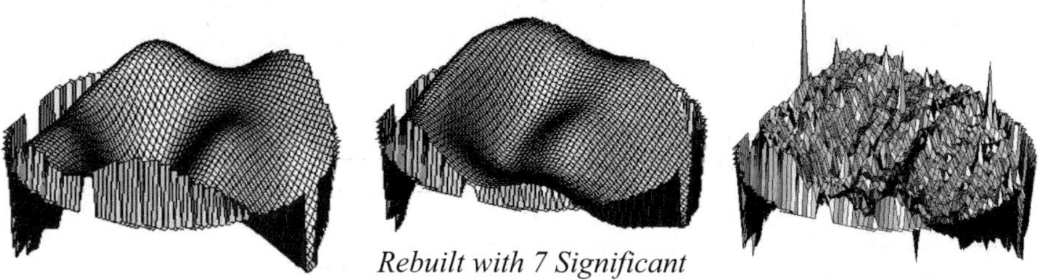



Figure 4; Comparisons between Rebuilt an Measured Surfaces

**Remarks:** -The edge of the figures above us has fringes only due to the shape viewer.
- The measured and the discretized grids must not be the same.

### 3.2  Comparison with the Polynomials Zernike Method

Zernike polynomials are generally defined in polar coordinates $(\rho,\ \theta)$, where $\rho$ represents the radial coordinate  ranging from 0 to 1, and $\theta$ represents the component of azimuth, which varies between 0 and $2\pi$. Each polynomial consists of three components: a normalized factor, a component with radial dependence, and a component with azimuth dependence. The radial component is polynomial whereas the azimuth component is sinusoidal. A double entry diagram is necessary to describe this function without ambiguity, with index $N$ indicating the highest order of the radial polynomial and the index $m$ indicating the azimuth frequency of the sinusoidal component.

According to this diagram, Zernike polynomials are:

$$Z_m^n(\rho,\theta) = \left\{ \begin{array}{lll} N_n^m R_n^{|m|}(\rho) \cos m\theta & ; \quad for & m \geq 0 \\ - N_n^m R_n^{|m|}(\rho) \sin m\theta & ; \quad for & m < 0 \end{array} \right\} \quad (8)$$

where

$$N_n^m = \sqrt{\frac{2(n+1)}{1+\delta_{m0}}} \quad\quad (9)$$

$\delta_{m0}$ is related to Kroneker notation.

$$R_n^{|m|}(\rho) = \sum_{s=0}^{\frac{n-|m|}{2}} \frac{(-1)^s\,(n-s)\,!}{s!\,[0.5\,(n+|m|-s)]!\,.[0.5\,(n-|m|-s)]!}\,\rho^{n-2s} \quad (10)$$

                          F. Formosa and S. Samper

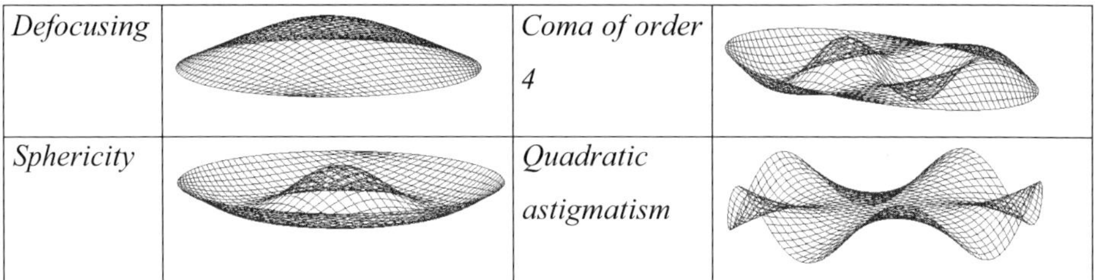

| Defocusing | | Coma of order 4 | |
|---|---|---|---|
| Sphericity | | Quadratic astigmatism | |

*Figure 5; Examples of 4 named Zernike Polynomials*

The figure 5 shows four named Zernike polynomials.

The Zernike polynomials are quoted in the standard [ISO 10110-5] as constituting a simplifying tool for the analysis of the interferograms. The properties of orthogonality of the made up basis allow a fast decomposition of the defect of form of the surface, which is numerically deduced from the interferogram. Moreover, the coefficients resulting from this decomposition are independent concerning the expansion, which is not the case with the other sets of functions.

Since we have a basis of polynomials, it is possible to make a decomposition of the measured defect in this basis, and to compare the obtained results with the modal basis. The first polynomials are named with specific optic defects.

*Figure 6; Decomposition of the Defect in the Two Basis*

The figure 6 shows a decomposition of the defect in the eigen modes and the Zernike polynomials basis.

Note that the modal factors decrease faster than the Zernike coefficients. We verify here that to be effective, the decomposition of a defect in the basis of the polynomials of Zernike must be applied to defects with circular symmetry. The decomposition of an unspecified defect requires much more polynomials than the one for of modes. Whereas one could stop with 35 modes to reach a total residue $E$ lower than $10^{-7}$, we do not reach this level with 55 polynomials.

## 4. CONCLUSION

In this paper we set the basis of a new language for the characterization of shape defects. This modal tolerancing language can express the complexity of any shape. The method, based on the modal analysis uses a finite element solver, linked to a CAD software.

We showed, here, an example with a measured disk. In this example, we can see that we can keep few modes in order to rebuild the measured surface (geometric filtering). Then it is possible to express the specifications in terms of modal coefficients and give them a sense like some Zernike coefficients for optic needs. Our method is not a specific one (like Zernike polynomial), but a general one. We also used it on several industrial complex examples (car fender, plastic hood) with the same kind of results.

The language gives sense to specific forms defects and makes possible to control them. This method is useful where form is function such as geometric perceived quality, high precision parts, quality of complex forms.

## REFERENCES

[Cootes, 2004] COOTES T., "Timeline off developments in algorithms for finding correspondences across sets off shapes and images", www.isbe.man.ac.uk/~bim/Papers/correspondence_timeline.pdf, 2004

[Staib et al., 1992] STAIB L., DUNCAN J., "deformable Boundary finding with parametrically models", IEEE Transactions one Pattern Analysis and Machine Intelligence, vol. 14 (11), pp. 1061-1075, 1992

[Reissel, 1996] REISSEL L. Mr., "Wavelet multiresolution representation off curves and surfaces", Graphical Models and Processing Image, vol. 58 (3), pp. 198-217, 1996

[Staib et al., 1996] STAIB L., DUNCAN J., "Model based deformable surface finding for medical images", IEEE Transactions one Medical Imaging, vol. 15 (5), pp. 720-731, 1996

[Davatzikos et al., 2003] DAVATZIKOS C., CAT X., SHEN D. "Hierarchical activate shape models, using the wavelet transform", IEEE Transactions one Medical Imaging, vol. 22 (3), 2003

**[Wyant et al., 1992]** WYANT J.C., CREATH K., "BASIC will wavefront aberration theory for optical metrology", Applied Optics and Optical Engineering, vol. 11, pp. 2-53, 1992

**[Carr et al., 1995]** CARR K., FERREIRA P., "Checking off form tolerances. I leaves: BASIC exits, flatness and straightness", Precision Engineering, vol. 17 (2), pp. 131-143, 1995

**[Pandit et al., 1981]** PANDIT S. Mr., SHAW Mr. C., "Characteristic shapes and wave length decomposition off surfaces in machining", Annals off CIRP, pp. 487-492, 1981

**[Capello et al., 2000]** CAPELLO E., SEMERARO., "Harmonic fitting approach for planes geometry measurements", The International Newspaper off Advanced Manufacturing Technology, vol. 16, pp. 250-258, 2000

**[Summerhays et al., 2002]** SUMMERHAYS K.D., HENKE R.P., BALDWIN J. Mr., CASSOU R.M., BROWN C.W., "Optimizing discrete not sample patterns and measurement dated analysis one internal cylindrical surfaces with systematic form deviations", Journal off the International Societies for Precision Engineering, vol. 26 (1), pp. 105-121, 2002

**[Huang et al., 2003]** HUANG W., CEGLAREK D., "decomposition Mode-based off share form error by discrete-cosine-transform with implementation to assembly and stamping system with compliant shares", Annals off CIRP, vol. 51 (1), pp. 21-26, 2002

**Standards**
**[ISO 10110-5]** *"Optics and optical instruments -- Preparation of drawings for optical elements and systems -- Part 5: Surface form tolerances"*

# Dependence and Independence of Variations of a Geometric Object

**P. Serré*, A. Rivière*, A. Clément****
**Laboratoire d'Ingénierie des Systèmes Mécaniques et des MAtériaux,*
*3, rue Fernand Hainaut, 93407 Saint-Ouen Cedex, France*
***Dassault Systèmes,*
*9, quai Marcel Dassault, 92156 Suresnes Cedex, France*
*philippe.serre@supmeca.fr*
*ariviere@supmeca.fr*
*andre_clement@ds-fr.com*


**Abstract**:   The general objective of this paper is to analyse the dependency relations that may exist between several specifications of a geometric object. This object will be modelled on the basis of the TTRS concept and vector modelling synthesised by the metric tensor of a sheaf of vectors.  The first problem that we propose to solve therefore is the formal expression of these relations. After that, the dependency will be analysed and illustrated considering a 3D object.
Keywords: Geometric Perturbation, Metric Tensor, Geometrical Specification.


## 1.  INTRODUCTION

ISO dimensioning and tolerancing standards for mechanical parts are not geometric specification standards but standards that specify the metrological verification procedures for a part that already exists, with all the attendant advantages and disadvantages of this restriction. Proof of this assertion can be found in the actual wording of the basic principle, called the "independence principle" expressed as follows:
*Every dimensional or geometric requirement specified on a drawing must be individually (independently) complied with, unless a relation is specified. (ISO 8015)*

Anyone who has even briefly used the 2D or 3D sketcher of a CAD system is pertinently aware that the slightest variation of only one dimension very often leads to drastic changes in the form of the object or even, on occasions, its non-existence, and you are immediately convinced that all the dimensions are interdependent.

The strong point of standardised, zone-based ISO tolerancing is its transactional nature: it enables exchanges to be made between a project manager and subcontractors with the utmost safety. It constitutes a major technological advance over the previous state of the art situation. Its main weakness resides in its non-applicability to dimensioning and tolerancing at the preliminary project stage (conceptual design) of a mechanical system. The possibility of specifying the parameters of a mechanical part in

23

*J.K. Davidson (ed.), Models for Computer Aided Tolerancing in Design and Manufacturing*, 23–34.
© 2007 *Springer*.

a non-nominal state by indicating a variation interval for certain of these parameters (parametric tolerancing) is not totally ruled out. On the other hand, it is impossible to obtain an instance of this part in its final state subsequent to parametric differentiation. Due to this inadequacy, justified mistrust has arisen from an industrial standpoint regarding the use of 3D dimension chains resulting from a parametric differential. Hence the current desire by manufacturers to make sure of the existence of the final object for any combination of tolerances. The explicit values of the dimensions of the final object – based on parametric tolerances – actually result from a computation which calls for extensive expertise, except in simple cases, and which does not necessarily imply the existence of the object.

In order to explain the problem in greater detail from a mathematical point of view, you simply need to imagine a certain nominal dimension $f_j\left(x^1, x^2, \ldots, x^n\right)$, dependent on n parameters, the variation of which must be specified by the designer. All current parametric technology is based on analysis of the differential form:

$$df_j = \frac{\partial f_j}{\partial x^i} . dx^i$$ .[Chase *et al.*, 1997], [Laperrière *et al.*, 2003] and [Serré *et al.*, 2003]. The designer is not only obliged to specify all the $dx^i$ with values that depend on the manufacturing process (which he does not know), for values of coefficients of influence (partial local derivatives) that he is not even aware of, but, above all, nothing goes to prove the existence of the object after a variation of this kind, however small it may be. In other words, there is no biunivocal relation in manufacturing reality between the nominal state and evolution of a system. On a mathematical level, this stems from a possible change in rank of the Jacobian $\left\|\frac{\partial f_j}{\partial x^i}\right\|$ in the neighbourhood of the nominal point. The "nominal value" and, therefore, this Jacobian do not have any precise meaning for an instance made of the object. This is the normal case for mechanical parts where the relative arrangements of points, straight lines and planes are specific (for example, in a combustion engine, the cylinder axis intersects the crankshaft axis at right angles).

When a designer has spent days or even weeks constructing a complex object using CAD, he cannot devote the same time to exploring the dimension variations of the same object. In an endeavour to meet this need for manufacturing security at the design stage, the normal differential parametric approach is completely abandoned in this study in order to adopt the concept of perturbation of the initial state. It is a method of investigation of the final real form of a geometric object with a known nominal form when it is subjected to independent, finished variations (small or otherwise).

*NB. This perturbation concept is a generalisation of the "small displacement torsor", which represented a variation of the position of a solid body [Ballot et al., 2001]; here a perturbation is finished or infinitely small and, above all, the object is subjected to deformation.*

## 2.  TENSOR MODELLING OF GEOMETRIC OBJECTS

In modern technology, all the "finished" variation possibilities of an object must be modelled and not simply the variations considered to be "infinitely small".  The connections between the elements of this object, seen as a system, must be essentially maintained, in order to preserve its identity through its transformations.  From this point, there will no longer be any difference between the dimensional analysis of the object and the tolerance analysis: the same mathematical and computer tool will be used and differential analysis completely rejected.

In analytical geometry, every element of which a complex geometric object is composed is in a unique position relation with the Cartesian reference point (the complexity is $O(n)$). There is no direct topological connection between 2 adjoining elements, except in the eyes of the designer.  The extraordinary advantage of the Cartesian reference point is the small number of independent relations to be processed. The major disadvantage is, in point of fact, that there is no relation between 2 adjoining objects, it is up to the user to "propagate" any modifications.  It is simple for the machine but difficult for the designer. Therefore, the problem is not properly posed!

In declarative geometry, a complex geometric object is perceived as a system of specifications between geometric objects. Each object is potentially in relation with all the others (the complexity is $O(n^2)$). The vector modelling based on the TTRS concept, already presented throughout precedent seminars [Clément *et al.*, 1999], [Serré *et al.*, 2001] and in a PhD [Duong, 2003], and the tensor representation will enable the complexity to be reduced to $O(n)$, that is to say, almost to the level of analytic geometry, at the same time explicitly preserving the relations between objects.

Historically speaking, the inventor of the application of tensor computation to technology was G. Kron, who used the application for electrical networks and rotating machinery.  In particular, he represented Kirchhoff's laws (mesh and node laws) using $n$-dimension, rank 2 tensors, the latter being the electrical currents of active or inactive components of this network.

In physics we normally only use rank 3, 3-Dimension metric tensors since these geometric objects need to be made in our Euclidian 3-Dimension space.  G. Kron [Kron, 1942], however, the first to discover that this relation of equality between the dimension and rank of the metric tensor is not necessarily subject to certain manipulation restrictions.  This tensor concept can be extended to any number ($n$) of vectors forming a geometric object; however, for it to be possible in $R^3$ space, the metric tensor must be restricted to rank 3 and each 3*3 sub-tensor must be defined as positive.  Moreover, not all the usual mathematical operations are valid on these tensors. Inversion, in particular, is an impossible transformation whereas transposition is still valid.

Numerical construction of the metric tensor associated with a valid geometric object in our 3D space composed of hundreds of vectors presents no difficultly. However, the terms of this tensor are not independent and are composed of thousands of implicit relations: as a result, the slightest variation of one of them invalidates the object. From a mathematical standpoint, it seems that expressing all these terms according to a certain number of independent parameters would be all it takes to vary the object in a straightforward manner. Unfortunately, this is very difficult in practice for a number of reasons, the main one being that you cannot explicitly determine all the Euclidian geometry theorems that apply to the object under consideration and it is therefore almost impossible to know all the constraints linking the geometric elements.


## 3. NEW MODELLING: PERTURBATION OF A KNOWN OBJECT

The solution adopted is not to use the geometric parameters of the object but the parameters of a $\Omega$ perturbation of the object and analyse the existence of the "perturbed" object. The perturbation model then becomes independent from the specification model.

Definition of the $\Omega$ perturbation

The $\Delta E$ perturbation of the sheaf of vectors $E$ of an object forming a rank 3 sub-space

$$E_{init} = \left( \vec{e_1}, \vec{e_2}, \vec{e_3}, \ldots, \vec{e_n} \right)$$

will create a valid sheaf of vectors for the $R^3$ space, written as follows:

$$\Delta E_{init} = \left( \overrightarrow{\Delta e_1}, \overrightarrow{\Delta e_2}, \overrightarrow{\Delta e_3}, \ldots, \overrightarrow{\Delta e_n} \right)$$

if, and only if, each perturbation is a linear combination of primitive vectors. Expressed as:

$$\overrightarrow{\Delta e_i} = \sum_k \omega^k . \vec{e_k}$$

The $\vec{e_k}$ vectors are not independent but any linear combination gives a valid and unique vector $\overrightarrow{\Delta e_i}$

Globally, the $\Omega$ perturbation of all these vectors will be expressed as:

$$\begin{pmatrix} \overrightarrow{\Delta e_1} \\ \overrightarrow{\Delta e_2} \\ \vdots \\ \overrightarrow{\Delta e_n} \end{pmatrix} = \Omega \cdot \begin{pmatrix} \vec{e_1} \\ \vec{e_2} \\ \vdots \\ \vec{e_n} \end{pmatrix}$$

The final object will thus be composed of the list of vectors:

$$E_{final} = E_{init} + \Delta E_{init} \quad \ldots\ldots\ldots\ldots Eq1$$

The perturbation defined in this way will give a unique vectoral object, valid only if all the coefficients $\omega_i^k$ are real numbers.

### 3.1. Computation of the metric tensor of the final object

The initial metric tensor is written: $G_{init} = E_{init} \otimes {}^t E_{init}$

The new $E_{final}$ vectors represent the vectors of the final object.

We deduce the expression of the final tensor from the equation (Eq1) via the tensor product

$$G_{final} = E_{final} \otimes {}^t E_{final}$$

i.e. by substituting the value of $E_{final}$

$$G_{final} = \left( E_{init} + \Delta E_{init} \right) \otimes {}^t \left( E_{init} + \Delta E_{init} \right)$$

then by developing the equation

$$G_{final} = \left( E_{init} + \Omega \cdot E_{init} \right) \otimes {}^t \left( E_{init} + \Omega \cdot E_{init} \right)$$

i.e. by expressing the unit matrix of dimension $n$ as $I_n$:

$$G_{final} = \left( I_n + \Omega \right) \cdot \left( E_{init} \otimes {}^t E_{init} \right) \cdot {}^t \left( I_n + \Omega \right)$$

Finally

$$G_{final} = \left( I_n + \Omega \right) \cdot G_{init} \cdot {}^t \left( I_n + \Omega \right) \qquad \text{Eq2}$$

This is the basic formula for the dimensional variations of a geometric object. This tensor equation mathematically defines a differentiable parametric manifold, the $\omega_i^k$ parameters of which are real numbers. In this way, a topological connection is made between 3 models: the $G_{init}$ nominal model, the $\Omega$ perturbation model and the final specified model, $G_{final}$.

*NB. The factor $\left( I_n + \Omega \right)$ demonstrates that a $\Omega$ perturbation has definitely been added to unit $I_n$ since, for a nil $\Omega$ perturbation, we find $G_{final} = G_{init}$*

### 3.2. Example of $G_{final}$ parametric manifold

$G_{final}$ parametric manifold is visualised by a 2D surface in our 3D space for a particular case, too simple to be of practical use, but one that allows the general case of a dimension $n$ rank 3 manifold to be pictured.

Given the $\Omega$ perturbation as the most common for an object, reflected by a $G_{init}$ metric tensor with a dimension $n = 2$ and rank 1 ($b^2 - a \cdot c = 0$), with

$$G_{init} = \begin{pmatrix} a & b \\ b & c \end{pmatrix}$$

28                           P. Serré, A. Rivière and A. Clément

the $G_{final}$ manifold will then represent all 2-dimension, rank 1 tensors.

$$G_{final} = \begin{pmatrix} X & Y \\ Y & Z \end{pmatrix} = \begin{pmatrix} 1+q_1 & q_2 \\ q_3 & 1+q_4 \end{pmatrix} \cdot \begin{pmatrix} a & b \\ b & c \end{pmatrix} \cdot \begin{pmatrix} 1+q_1 & q_3 \\ q_2 & 1+q_4 \end{pmatrix}$$

By identifying the 2 members of the equation, we obtain a parametric representation (see Figure 1) with a 4-parameter manifold:

$$\begin{cases} X = a + 2 \cdot a \cdot q_1 + 2 \cdot b \cdot q_2 + 2 \cdot b \cdot q_1 \cdot q_2 + a \cdot q_1^2 + c \cdot q_2^2 \\ Y = b + a \cdot q_3 + b \cdot q_1 + b \cdot q_4 + c \cdot q_2 + a \cdot q_1 \cdot q_3 + b \cdot q_1 \cdot q_4 + b \cdot q_2 \cdot q_3 + c \cdot q_2 \cdot q_4 \\ Z = c + 2 \cdot b \cdot q_3 + 2 \cdot c \cdot q_4 + 2 \cdot b \cdot q_3 \cdot q_4 + a \cdot q_3^2 + c \cdot q_4^2 \end{cases}$$

This is the only $G_{final}$ manifold that can be visualised in 3D

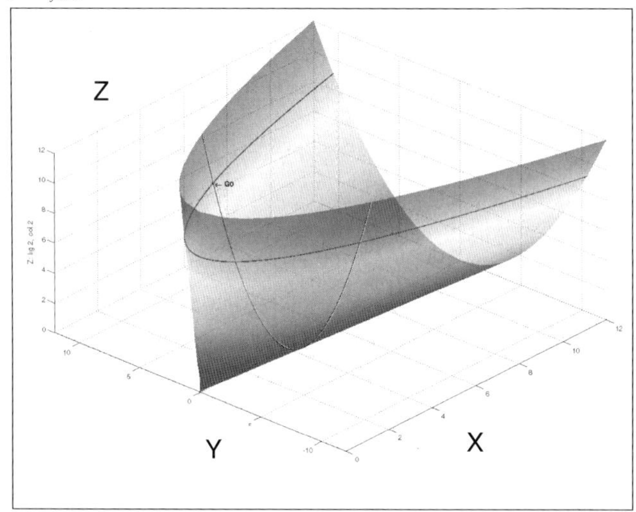


*Figure 1; Parametric representation of the 2-dimension, rank 1 $G_{final}$ manifold*

Any point on this manifold represents a valid object.

### 3.3. Fundamental property

The theorem for the product of determinants immediately shows that if the rank of $G_{init}$ is 3, then $G_{final}$ is also rank 3, even if rank $r$ of $\Omega$ is $3 < r < n$.

In other words, the Eq2 formula always gives a valid object for the 3D space, irrespective of the real $\Omega$ perturbation.

With this tensor equation, we are faced with the usual 2 problems, as follows: one, called the **"direct problem"** which, since $\Omega$ and $G_{init}$ are known, consists of calculating $G_{final}$, the other, called the **"reverse problem"**, which since $G_{final}$ is partially known, consists of determining $\Omega$, followed by the complete $G_{final}$, and then returning to the direct problem.

### 3.4. Direct problem

For example, let us consider the perturbation of the vector $\vec{e_i}$ :

$$\Delta \vec{e_i} = \omega_i^1 .\vec{e_1} + \omega_i^2 .\vec{e_2} + \omega_i^3 .\vec{e_3} + \omega_i^4 .\vec{e_4} + ...... + \omega_i^n .\vec{e_n}$$

The designer specifies the desired perturbation by giving values to the $\omega_i^k$ coefficients of the $\Omega$ matrix. This specification method gives him all the flexibility required to "sculpt" the final object since he has $\dfrac{n!}{6 \cdot (n-3)!}$ independent ways of indicating it for each vector.

For example:

$\left( \omega_i^1, 0, \omega_i^3, 0, \omega_i^5, 0, 0, ..., 0 \right)$ specifies that the $\vec{e_i}$ vector is subjected to a variation in 3D space, the perturbations of which are imposed on the triplet: $\left( \vec{e_1}, \vec{e_3}, \vec{e_5} \right)$.

*NB. It is obviously possible to specify more than 3 components, provided they are real numbers.*

The designer can thus specify a variation by adding perturbations in relation to numerous successive references. The final vector itself will naturally have a position resulting from the combination of these perturbations in the 3D space.

### 3.5. Reverse problem

This substantially important advantage in the preliminary project stage thus brings greater security to the resolution of the reverse problem. In fact, over and above 3, the additional $\omega_i^k$ variables allow vector variation redundancy to be ensured by introducing free parameters – in principle serving no purpose – but which will help the solver find solutions in specific cases of poor conditioning. In IT, the limited accuracy of floating numbers very often changes the rank of a matrix during a computation, which numerically becomes rank $n$ although it is theoretically rank 3, for example. It then turns out to be numerically advantageous to carry out intermediary computations in an $n$-dimension, rank $n$ $\Omega$ space to avoid breaking the calculation chain. The actual existence of a 3-D space vector with these $n$ dependent but coherent components must then be ascertained.

*NB. Experiments have broadly confirmed this result in all cases. Moreover, it speeds up convergence of the resolution process.*

### 3.6. Reverse problem resolution strategy

By assumption, the terms of $G_{init}$ are all known. For a size $n$ $G_{final}$ there are $3n-3$ independent terms to be specified at the most but normally the coherence of these specifications is never guaranteed. This is one of the problems that we are going to resolve by showing the 3D solution that "best" verifies the specifications.

The $\Omega$ matrix introduces a maximum of $n^2$ dependent parameters. If only $3n-3$ $\omega_i^k$ variables are introduced, with a maximum of 3 per vector, we are confronted with an iso-constrained problem, which may or may not have a solution. This traditional approach is unsatisfactory for the designer, who receives a "no solutions" type of message giving him no indication of the nature of the specification changes to be made. This is why we use a different strategy that consists of always providing a real object, as close as possible to the designer's specification, which then clearly indicates the specifications that are not met.

For this the problem will always be deliberately underconstrained by systematically introducing more variables than constraints into the $\Omega$ matrix. We will solve this undetermination by seeking the $G_{final}$ tensor which minimises the perturbation of the initial object and by verifying the specifications using the "best" solution. The anticipated benefit of this methodology is to always give a solution to the problem posed, either an accurate solution or the "nearest" solution for the initial object when there is no solution that accurately verifies the constraints.

### 3.7. Algebraic expression of constraints

**Specification of the metric tensor** $\left(G_{final}\right)_{i,j} = SPEC_{i,j}$

Given the list of angle specifications

$$SPEC_{i,j} = \left(\left(I_n+\Omega\right)\cdot G_{init}\cdot{}^t\left(I_n+\Omega\right)\right)_{i,j} \qquad\qquad \text{Eq3}$$

where $G_{init}$ is known and $\Omega$ undetermined.

**Specification of the possible closure constraint**, where $\Delta L$ represents the perturbation of the lengths of vectors of the loop under consideration:

$$\Delta L\cdot\left(I_n+\Omega\right)\cdot G_{init}\cdot{}^t\left(I_n+\Omega\right)\cdot{}^t\Delta L = 0 \qquad\qquad \text{Eq4}$$

Since the problem is still underconstrained, we will seek the $\Omega$ matrix that verifies all the Eq3 and Eq4 type specifications, at the same time minimising the function $\left(\Omega^2+\Delta L^2\right)$. The problem is properly posed: search for the minimum of a convex function subjected to convex constraints. Any off-the-shelf software such as Matlab, automatically provides a valid response, i.e. a certain value for the $\Omega$ matrix. The solution is unique if, and only if, the function to be minimised and the constraints are convex, otherwise a local minimum will be obtained which, in any case, is of interest to the designer.

This valid response offers two possibilities:

- The constraints Eq3 and Eq4 are accurately verified: this is the solution sought. The $G_{final}$ tensor is then recalculated using the basic Eq2 formula.

- One or several constraints are not verified; however, the designer is shown the $G_{final}$ object obtained by applying the basic Eq2 formula. This view will enable the designer to understand the modifications he needs to make to the specifications.


## 4. CASE STUDY

To illustrate the approach proposed, the study of the perturbation of the geometric object presented in Figure 2, is examined. This object is composed of seven planes, called n1, n2 ,.. n7, and a cylinder called l1.
Without losing its general nature, the study only covers angle specifications and, as a result, the generation of vectoral closure equations is not presented in the following explanation.

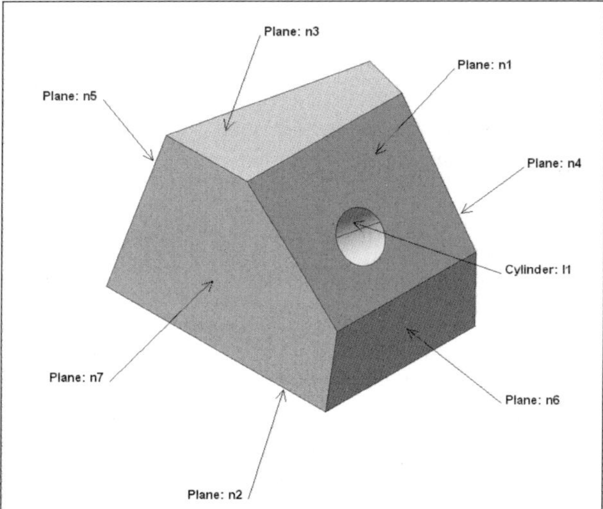


*Figure 2; Visualisation of the geometric object studied*

### 4.1. Internal representation of the object

Tensor modelling applied to the initial object enables the angle specification model to be obtained, which is the initial metric tensor, called $G_{init}$. This defines a vectoral space for the sheaf of vectors $E$ useful in the representation of the specifications covered in this paragraph.

Here, $E = \left( \vec{n_1}, \vec{n_2}, \vec{n_3}, \vec{n_4}, \vec{n_5}, \vec{n_6}, \vec{n_7}, \vec{l_1}, \right)$ with the following convention: the normal to plane $n_i$ is called $\vec{n_i}$ and the director vector of the axis of the $l_1$ cylinder is called $\vec{l_1}$.

The values of the angles between the vectors, measured on the initial object, are noted in the table below.

*Table 1: Initial values of angles (in °)*

| angle | $\vec{n_1}$ | $\vec{n_2}$ | $\vec{n_3}$ | $\vec{n_4}$ | $\vec{n_5}$ | $\vec{n_6}$ | $\vec{n_7}$ | $\vec{l_1}$ |
|---|---|---|---|---|---|---|---|---|
| $\vec{n_1}$ | 0° | 40° | 40° | 83.3239° | 124.2760° | 45° | 77.5901° | 180° |
| $\vec{n_2}$ | sym. | 0° | 0° | 85° | 85° | 85° | 85° | 140° |
| $\vec{n_3}$ | sym. | sym. | 0° | 85° | 85° | 85° | 85° | 140° |
| $\vec{n_4}$ | sym. | sym. | sym. | 0° | 83.7013° | 85.1644° | 159.5953° | 96.6761° |
| $\vec{n_5}$ | sym. | sym. | sym. | sym. | 0° | 165.6258° | 113.0501° | 55.7240° |
| $\vec{n_6}$ | sym. | sym. | sym. | sym. | sym. | 0° | 76.2796° | 135° |
| $\vec{n_7}$ | sym. | sym. | sym. | sym. | sym. | sym. | 0° | 102.4099° |
| $\vec{l_1}$ | sym. | sym. | sym. | sym. | sym. | sym. | sym. | 0° |

## 4.2. Problem posed

Based on an initial object of a known form, the designer wishes to obtain a new object (called the final object) which meets certain angle constraints.

This scenario corresponds to what we call a "reverse problem". The method of resolution consists of determining the $\Omega$ perturbation then the complete metric tensor, with $G_{final}$, representing the final object.

To show the genericity of the model proposed, two specifications are studied. For each of them we have noted the numerical values obtained after computation in one table. A second table shows the angle differences obtained between the initial and final values and, finally, two images show the initial object and the final object after perturbation.

## Specification 1:

Any value has been chosen for the constraints specified (grey boxes in Table 2).

*Table 2: Final angle values (in °) obtained for specification 1*

| | | | | | | | |
|---|---|---|---|---|---|---|---|
| 0° | 40.0098° | 44.8648° | 85.3239° | 125.5864° | 47° | 79.5901° | 178° |
| sym. | 0° | 4.8599° | 86° | 87° | 87° | 84.9059° | 141° |
| sym. | sym. | 0° | 86° | 82.2412° | 91.8491° | 86.0000° | 136.1401° |
| sym. | sym. | sym. | 0° | 77.9964° | 86.1644° | 164.4357° | 92.9796° |
| sym. | sym. | sym. | sym. | 0° | 163.4088° | 114.0501° | 54.9089° |
| sym. | sym. | sym. | sym. | sym. | 0° | 80.7734° | 131.8911° |
| sym. | sym. | sym. | sym. | sym. | sym. | 0° | 102.0285° |
| sym. | sym. | sym. | sym. | sym. | sym. | sym. | 0° |

Despite the low angle differences specified (all between 1° and 2°), we can see in Table 3 that certain differences can be large (see the boxes with bold borders).

*Table 3: Angle differences (in °) for specification 1*

| | | | | | | | |
|---|---|---|---|---|---|---|---|
| 0° | 0.0098° | 2° | 2° | 1.4648° | 2° | 2° | -2° |
| sym. | 0° | 1.9922° | 1° | 2° | 2° | 0.5336° | 1° |
| sym. | sym. | 0° | 1° | 0.0442° | 3.9880° | 1° | -0.9922° |
| sym. | sym. | sym. | 0° | -4.8966° | 1° | 4.5680° | -3.6965° |
| sym. | sym. | sym. | sym. | 0° | -1.4640° | 1° | -0.9365° |
| sym. | sym. | sym. | sym. | sym. | 0° | 3.7706° | -3.1089° |
| sym. | sym. | sym. | sym. | sym. | sym. | 0° | -0.4021° |
| sym. | sym. | sym. | sym. | sym. | sym. | sym. | 0° |

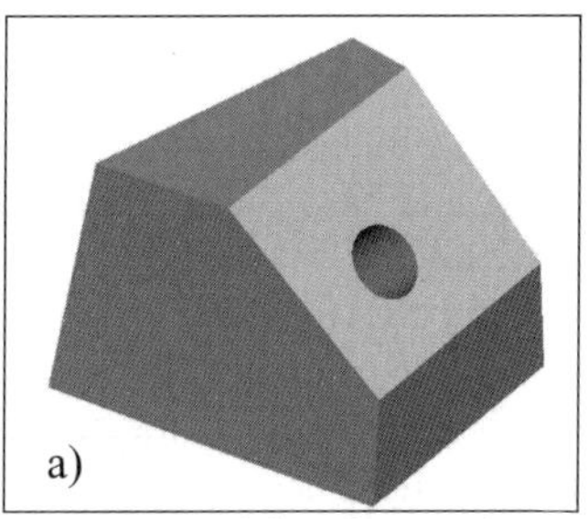
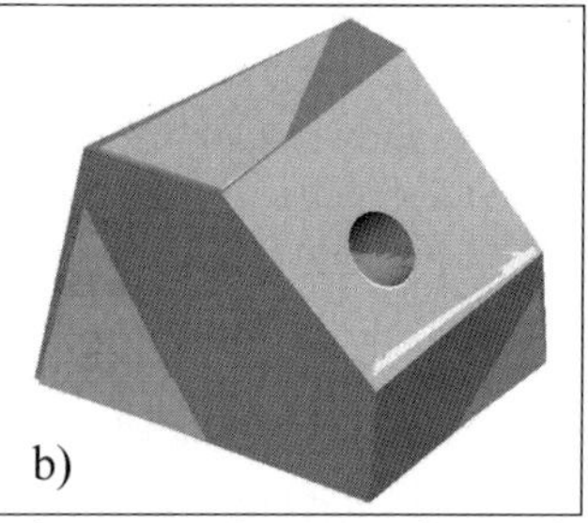



*Figure 3; Images of the object after perturbation in the case of Specification 1*
*a) Final object, b) Final object in grey and initial object in white*

## Specification 2:

The angle constraints specified (grey boxes in Table 4) are placed on the first two diagonals, i.e. the angular position of a vector is defined in relation to the two vectors that precede it.

*Table 4: Final angle values (in °) obtained for Specification 2*

| | | | | | | | |
|---|---|---|---|---|---|---|---|
| 0° | 41° | 42° | 128° | 91.1804° | 85.4806° | 81.2767° | 137.4902° |
| sym. | 0° | 1° | 87° | 87.0895° | 81.8917° | 119.8227° | 128.7360° |
| sym. | Sym. | 0° | 86° | 87° | 81.8504° | 120.7479° | 128.2132° |
| sym. | Sym. | sym. | 0° | 84.7013° | 87.1644° | 145.9142° | 68.2537° |
| sym. | Sym. | sym. | sym. | 0° | 166.6258° | 115.0501° | 48.0625° |
| sym. | Sym. | sym. | sym. | sym. | 0° | 77.2796° | 137° |
| sym. | Sym. | sym. | sym. | sym. | sym. | 0° | 103.4099° |
| sym. | Sym. | sym. | sym. | sym. | sym. | sym. | 0° |

As already noted, despite the low angle differences specified, certain differences may be large. In this instance, they are actually very large (see the boxes bordered in bold in Table 5) and the form of the final object is no longer the same as the initial object.

*Table 5: Angle differences (in °) for Specification 2*

| | | | | | | | |
|---|---|---|---|---|---|---|---|
| 0° | 1° | 2° | 44.6761° | -33.0956° | 40.4806° | 3.6866° | 42.5098° |
| sym. | 0° | 1° | 2° | 2.0895° | -3.1083° | 34.8227° | -11.2640° |
| sym. | Sym. | 0 | 1° | 2° | -3.1496° | 35.7479° | -11.7868° |
| sym. | Sym. | sym. | 0° | 1° | 2° | -13.6811° | -28.4224° |
| sym. | Sym. | sym. | sym. | 0° | 1° | 2° | -7.6615° |
| sym. | Sym. | sym. | sym. | sym. | 0° | 1° | 2° |
| sym. | Sym. | sym. | sym. | sym. | sym. | 0° | 1° |
| sym. | Sym. | sym. | sym. | sym. | sym. | sym. | 0° |

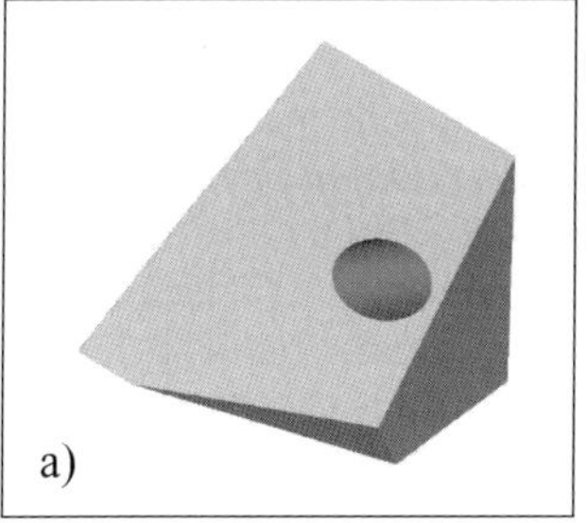
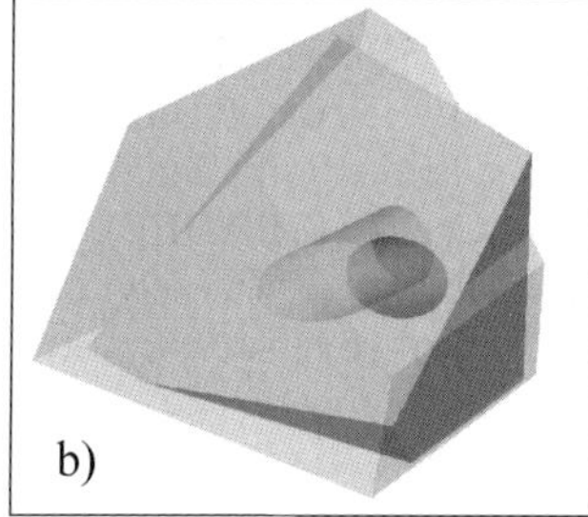



*Figure 4; Images of the object after perturbation in the case of Specification 2*
*a) Final object, b) Final object grey and initial object transparent*

## 5. CONCLUSION

We have initially shown that, if necessary, the dimension variations, specified or executed for a 3D geometric object are dependent, and have then demonstrated the basic formula for dimensional variations of a geometric object.

There is a wide variety of domains of application for this formulation, for example:

- the dimensioning or parameterisation of a part or mechanical assembly for CAD/CAM system sketchers and assembly module applications.
- the geometric tolerancing of mechanical assemblies with the development of analysis and tolerancing synthesis tools,
- the analysis and synthesis of complex engineering problems by associating the geometry equations presented in this article with other engineering equations describing the globally specified problem.


REFERENCES

[**Ballot et al., 2001**] Ballot, E.; Bourdet, P.; Thiébaut, F.; "Determination of Relative Situations of Parts for Tolerance Computation"; In: *7th CIRP International Seminar on Computer Aided Tolerancing*, Cachan (FRANCE), April 24-25, (2001).

[**Chase et al., 1997**] Chase, K.W.; Gao, J.; Magleby, S.P.; "Tolerance Analysis of 2-D and 3-D Mechanical Assemblies with Small Kinematic Adjustments"; In: *Advanced Tolerancing Techniques*, John Wiley, pp. 103-137, (1997).

[**Clément et al., 1999**] Clément, A.; Rivière, A.; Serré, P.; "Global Consistency of Dimensioning and Tolerancing"; *Keynote paper of CIRP Computer Aided Tolerancing, 6th Seminar*, Enschede, The Nederlands, March 22-24, (1999).

[**Duong, 2003**] Duong, A.N.; "Spécification, analyse et résolution de problèmes géométriques 2D et 3D modélisés par contraintes"; Thèse de Doctorat, Ecole Centrale de Paris (2003).

[**Kron, 1942**] Kron, G.; "A short course in tensor analysis for electrical engineers"; Wiley, New York; Chapman & Hall, London, (1942).

[**Laperrière et al., 2003**] Laperrière, L.; Ghie, W.; Desrochers, A.; "Projection of Torsors": a Necessary Step for Tolerance Analysis Using the Unified Jacobian-Torsor Model"; In: *8th CIRP International Seminar on Computer Aided Tolerancing*, Charlotte (USA), April 28-29, (2003).

[**Serré et al., 2001**] Serré P.; Rivière A.; Clément A.; "Analysis of functional geometrical specification"; In: *7th CIRP International Seminar on Computer Aided Tolerancing*, Cachan (FRANCE), April 24-25 (2001).

[**Serré et al., 2003**] Serré, P.; Rivière, A.; Clément, A.; "The clearance effect for assemblability of over-constrained mechanisms"; In: *8th CIRP International Seminar on Computer Aided Tolerancing*, Charlotte (USA), April 28-29, (2003).

# A Model for a Coherent and Complete Tolerancing Process


**L. Mathieu[*], A. Ballu[**]**
[*]*LURPA ENS de Cachan, 61, Av. du Pdt Wilson, 94235 Cachan Cedex, France*
*mathieu@lurpa.ens-cachan.fr*
[**]*LMP – CNRS UMR 5459, 351 Cours de la libération, 33405 Talence cedex, France*
*alex.ballu@u-bordeaux1.fr*



**Abstract**: Few CAD/CAM software in mechanical engineering offer to the designer integrated tools for 3D tolerance analysis and synthesis. For tolerance analysis and synthesis, they have to take into account geometrical specification data,. but each software consider a specific geometrical specification representation. Moreover, this representation is often unknown from the users. Consequently, it is very difficult to understand the models used, the mechanisms implemented and the results provided. GeoSpelling model, a complete and coherent tolerancing process, including tolerance analysis, is presented. GeoSpelling, proposed to ISO for rebuilding standards in the fields of tolerancing and metrology, allows a unified description of the nominal and the non ideal geometry. It allows also a unique mathematical parameterization of the geometry. The proposed approach should help researchers and engineers to better explain the tolerance representation for different activities in a tolerancing process.



**Keywords**: Tolerance representation, Tolerance process, GeoSpelling, Uncertainty.


## 1. INTRODUCTION

Tolerance analysis and synthesis in mechanical engineering remain non solved problems. Just few CAD/CAM software offer to the designer integrated tools for a 3D tolerance analysis and synthesis. The actual solutions are very specific and are described to the users as black boxes. It is very difficult to understand the models implemented and the results provided. The solution is not totally satisfactory for industry. A lot of uncertainty sources between functional needs and results on actual parts and actual assemblies are not controlled.

Salomons et al. in CAT 1997 [Salomons et al., 1997] have proposed a review on "Current status of CAT Systems" for tolerance analysis. The study based on academic papers and also on four commercial systems (CATIA 3D FDT from Dassault Systèmes, TI/TOL3D+ form Texas Instruments, VSA-3D from VSA Inc and Valysis from Tecnomatix) points out a main difference between them, the two first analyze one "sample" of an assembly and are based on a linear algebraic problem, the two second




*J.K. Davidson (ed.), Models for Computer Aided Tolerancing in Design and Manufacturing, 35–44.*
© 2007 *Springer.*

require a large number of "samples" to achieve reasonable accuracy and are based on statistics. The models used within the systems are not clearly presented because it is very difficult to obtain information from CAT system vendors. The authors have distinguished 4 aspects to analyze the CAT systems, **tolerance representation**, **tolerance specification**, **tolerance analysis** and **tolerance synthesis**. These 4 aspects are not at the same level, the first one refers to how tolerances are represented computer internally, this aspect point out directly on the models used for the description of the mechanism without and with geometric variations. Tolerance specification is an important activity for tolerancing. It tries to answer the question: Which tolerance types and values are needed on features to control functional requirements? Tolerance analysis is a method to verify the values of functional requirements after tolerance specification on each isolated part. This method is totally dependent on the models chosen before. A lot of tools are also generally provided to the designer to understand the results. Tolerance synthesis is regarded as a tolerance allocation and a tolerance optimization method taking into account manufacturing and inspection aspects. The conclusions of Salomons paper were on future research directions in CAT. He distinguished: tolerancing for non rigid bodies, clarifying the relation between physics of functioning and tolerances in order to improve tolerance specification, clarifying the relation between physics of manufacturing and tolerances in order to improve tolerance synthesis. Eight years after, the problems are still not solved although a lot of progress in research have been made.

This paper focus on **tolerance representation**, the basis for a coherent and complete tolerancing process. It presents a unified solution to describe ideal and non-ideal geometry of parts and assembly along product lifecycle. After the introduction, section 2 explains what could be a coherent and a complete tolerancing process. Section 3 presents GeoSpelling, the model proposed to ISO experts for a new approach to build specification and verification standards. It explains how GeoSpelling could be a good solution for a coherent and a complete tolerancing process. A short example illustrates the description of ideal and non ideal geometry and a mathematical expression of the specification based on characteristics is given. Finally, section 4 presents a summary and plans for future works.


## 2.   A COHERENT AND COMPLETE TOLERANCING PROCESS

The need in mechanical engineering is to manage geometric variations along the product lifecycle. The relations between client, suppliers and subcontractors speed up the need to have a common view of geometric specification and verification. In particular, product design, process design, inspection design, manufacturing activity and inspection activity are concerned. It is very useful today in the context of concurrent engineering to have a Coherent and Complete Tolerancing Process.

The tolerancing process is defined through all the activities involved by geometric product variations management. Salomons proposed three classical activities: **tolerance**

**specification, tolerance analysis and tolerance synthesis**. Tolerance verification is not included and it is for us a lack.

**Tolerance Verification** defines inspection planning and metrological procedures for functional requirements, functional specifications and manufacturing specifications. It is very important to consider tolerance verification early in the design activities to be able to evaluate uncertainties [ISO/TS 17450-2, 2002], [Mathieu et al., 2003]. Tolerance verification allows to close the process loop, to check the product conformity and to verify assumptions made by the designer (Figure 1).

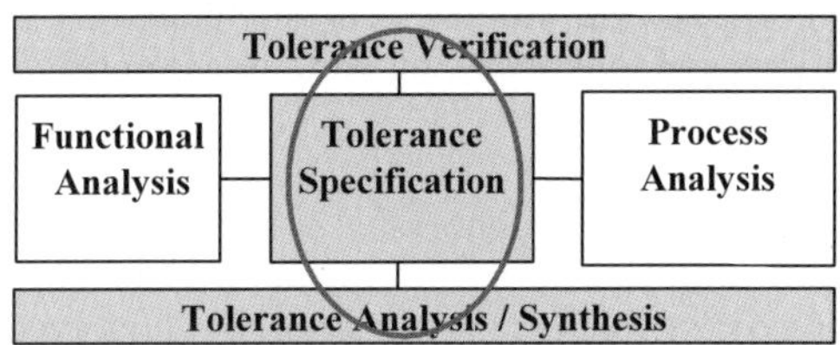


*Figure1 : Tolerancing process*

**Tolerance Specification** is the main activity in the tolerancing process. For each geometric requirement, tolerance specification specifies assemblies and parts involved with a functional view point. During the last 10 years, a lot of academic researches point out the link between functions and specifications [Dantan et al., 2003], [Dufaure et al., 2003]. Functional Analysis, early in the design activities, allows to better analyze the relation between physics of functioning and tolerances. The functional requirements have to be translated into geometric requirements on parts and assembly. Tolerance specification must also take into account data coming from process analysis. Process analysis points out the setting of machines on the shop floor. The dimensional and geometrical variations are analyzed for actions on the machines and also for the feed back to the tolerance analysis activity. The assembly sequence, manufacturing plan and the variations of each process are useful. In particular the link between Statistical Process Control and tolerance specification has to be improved to access statistical data.

**Tolerance Analysis** is totally based on the models used for tolerance specification [Guilford et al., 1993]. There are two main aproaches for tolerance analysis: parametric and geometric analysis.

Parametric analysis (similar to vectorial) views each object as a set of scalar parameters. In a dimension-driven variational geometry system, the parameters are linear and angular parameters. Tolerances are viewed as limits on the parameters values [Chase, 1999]. Parametric tolerancing is based on a different concept than the geometric tolerance and is not supported by standards. A conversion process is needed to derive parametric tolerances introducing uncertainties. The final result of the analysis depends on the parameters chosen [Serre et al., 2003]. The advantages of parametric analysis are that algorithms tend to be relatively simple and the geometry involved is only a perfect geometry.

Geometric analysis define constraints between faces expressing the fact that a geometric element must lie in a zone. This approach is the most important solution adopted by researchers [Ballot et al., 1997], [Giordano et al., 1993], [Desrochers et al., 2003] and CAT developers: Valysis from Tecnomatix, VSA-3D from VSA Inc. Geometric tolerancing tends to be in conformance with standards. Different geometric tolerancing schemes define the zones in different ways, but the significant effect is that the control is directly on the geometry. It reflects more accurately the "real-world" conditions, particularly mating-part relationships. The drawback of geometric analysis is that it tends to be complex and difficult to analyze.

As seen above, the tolerancing process depends totally on the used models. The representation of deviations and tolerances, on parts or assembly, is the key problem of tolerancing.

### 2.1. A coherent tolerancing process

To be coherent, the tolerancing process has to use the same language based on a unified mathematical model to express tolerancing for each one, involved during the process. The mathematical model has to express the "real-world" of the product with the minimum of uncertainty.

An example of a coherent tolerancing process could be based on Prof. Wirtz model, called vectorial tolerancing model. But this model is not complete and for different cases very far from the reality. It implies uncertainties. An other way could be geometric tolerancing but today the standardized possibilities are not apply in manufacturing and are characterized by contradictions.

Today, dimensional and geometric tolerances are expressed with graphical symbols on a 2D drawing. These symbols are defined in different national and international standards. The main existing ISO standards for tolerancing isolated parts, ISO 8015 (1985), ISO 1101(1983), ISO 2692 (1988) and ISO 5459(1981) are not recent and have a lot of gaps and contradictions [Bennich, 1993]. Also, these standards do not cover the need to express geometric requirements on assembly and does not consider statistical aspects. After the creation within ISO organization of Geometrical Product Specification (GPS) committee in 1996, a lot of work has been accomplished, but today, the most useful standards are unfortunately always the same. Probably, in a very near future, engineers may hope to have new ISO standards. This situation put a very important brake on engineering progress.

### 2.2. A complete tolerancing process

To be complete, the tolerancing process has to cover all the aspects of geometric variations during the product life cycle. It must include all the geometric features and all the parameters to describe the functionalities of a product.

For example, the geometric model proposed by Prof. Clement, based on the TTRS [Clement, 1994], model could be a good way to build a complete and coherent tolerancing process. This model is able to describe all the ideal geometry involved in a nominal model for each stage of the tolerancing process. But it must now take into account tolerances and explain how ideal geometry is connected to actual parts.

The geometric variation aspects are not only, functional aspect but also assembly, manufacturing and metrological aspects. The mathematical model used to describe the ideal geometry of the product, the non ideal geometry, the variation and its limits has to cover all the needs for each activity of the tolerancing process.

The following chapter presents GeoSpelling, the first solution for a coherent and complete tolerancing process.

## 3.  THE UNIFIED DESCRIPTION BY GEOSPELLING

### 3.1. GEOSPELLING

GeoSpelling is a model used to describe ideal and non-ideal geometry [Ballu et al., 1995], [ISO/TS 17450-1, 2005]. Indeed, it allows to express the specification from the function to the verification with a common language. This model is based on geometrical operations which are applied not only to ideal features, defined by the geometrical modelers in a CAD system, but also to the non-ideal features which represent a real part. These operations are themselves defined by constraints on the form and position of the features.

GeoSpelling is based on the following basic concept :

- a specification is a **condition** on a **characteristic** defined from **geometric features,**
- these **geometric features** are features created from the **model of the real surface of the part (skin model)** by different **operations**.

A condition defines an interval inside of which the value of a characteristic of geometric features must lie. These geometric features are identified by operations from the skin model. These operations are the operations of partition, extraction, filtration, collection, association and construction.

As it is impossible to completely capture the variation of the real surface of the workpiece, the skin model is an imagined surface, without any representation (drawing, numerical, …), representing all the variations that could be expected on the real surface of the workpiece. It allows to write and read specifications. For tolerance analysis or synthesis, the skin model is reduced to a predictive model, taking into account parametric variations (angles and distances) and sometimes form variation of the workpiece. For coordinate metrology, the skin model is reduced to the points measured on the workpiece.

*Geometric features*

We distinguish different types of elements: ideal features, non-ideal features and limited features.

We name **ideal features,** features such as :

- a plane of the nominal model of the part,
- a cylinder fitted to a real feature nominally cylindrical.

We name **non-ideal features**, features such as:

- a surface portion of the model of the real surface of a part,
- a real axis.

*Operations*

The operations used to define the specifications can be divided into six classes, partition, extraction, filtration, collection, association and construction.

A **partition** is an operation used to identify bounded feature(s) from non-ideal feature(s) or from ideal feature(s)

A **filtration** is an operation used to create a non-ideal feature by reducing the level of information of a non-ideal feature.

An **extraction** is an operation used to identify specific points from a non-ideal feature

A **collection** is an operation used to consider more than one features together.

An **association** is an operation used to fit ideal feature(s) to non-ideal feature(s) according to a criterion.

A **construction** is an operation used to build ideal feature(s) from other ideal features with constraints.

*Characteristics*

Characteristics, useful to the definition of specifications, belong to five families:

- Intrinsic characteristic
- Situation characteristic between ideal features
- Situation characteristic between ideal and limited features
- Situation characteristic between ideal and non-ideal features
- Situation characteristic between non-ideal features

*Summary*

The model is mainly based on operation and characteristic concepts. These concepts are developed to obtain a small set of operations and characteristics to describe the quasi totality of current specifications (standardized or not). This model allows to communicate geometrical information for design, manufacturing or inspection. With the consideration of the non-ideal features and the generic concepts brought forth by the operations and the characteristics, the model is a "company-wide" model. It is a univocal language, common to design, manufacturing and inspection.

Each point of view can be expressed with this unique language. With this common language, the differences between the various existing approaches and their lacks can be pointed out. Moreover, thanks to the generic concepts, new types of specification to express design, manufacturing or inspection intent can be defined [Ballu et al., 2001].

### 3.2. Why GeoSpelling could be a good solution for a coherent and complete tolerancing process ?

The main originality of GeoSpelling is to build geometric models for tolerance not from the nominal model but from the Skin model. This new approach will permit in the future to ask the good questions on the mechanical behavior and after that to put the best assumptions for modeling. With nominal model, all the variations cannot be imagined, and their influences on the functioning cannot either. Before managing variations on parts and product, it is useful to express mathematically the different parameters involved in the behavior of the mechanisms.

With the different operations and in particular the filtration operation, it will be possible to have a good link between micro and macro geometry. In geometrical specifications, micro geometry generally concerns roughness and waviness, macro geometry concerns form, orientation and location. Until now, the standards are not very clear on this subject and there are also different explanations considering the micro or the macro point of view. This aspect is an important point for a unified approach. With GeoSpelling concepts it is possible to describe micro specifications in the same way as macro specifications.

Only six operations are useful to create all the geometric features involved in micro and macro geometry. Only fives classes of features are distinguished, integral, smoothed, substitute, limited and nominal. In particular for the nominal features, all the seven classes of elements are considered (TTRS classes). GeoSpelling offers to the user the possibility to distinguish clearly the geometric elements manipulated during tolerance specification, tolerance inspection and tolerance analysis activities.

The next important point of GeoSpelling is to provide univocal expressions of tolerances on geometry thanks to the characteristic concept. A characteristic represents a linear or angular quantity. Taken into account directly in the expression of the specification, the meaning, based on a mathematical expression, is unique and clearly described for everyone. There is no more interpretation for the designer, the manufacturer and the metrologist. Only five classes of characteristics are useful to express all the parameters needed through the tolerancing process. The characteristics exist, on a feature or between features on a part, or, between features on two different parts. They permit to express geometric requirements on an assembly. They permit also to express constraints between features. They support finally a statistical approach for tolerancing because they are only based on measurable quantities.

In summary, GeoSpelling for the product geometry integers :

- the description of the nominal geometry,
- the design specifications to express the various functions of the part or between parts,
- the manufacturing specifications to express the various manufacturing processes,
- the inspection specifications to express the various inspection processes.

GeoSpelling seems to be able to support the complete and coherent tolerancing process.

### 3.3. Illustration of GeoSpelling model

Let us consider the geometrical tolerance defined in figure 2.

*Standardized meaning*

The toleranced feature shall be contained in a cylinder at the maximum material condition. The cylinder axis must be perpendicular to the datum plane A. The datum plane A "shall be arranged in such a way that the maximum distance between it and the simulated datum feature has the least possible value". This association criterion is not expressed clearly, mathematically. it may have different interpretations.

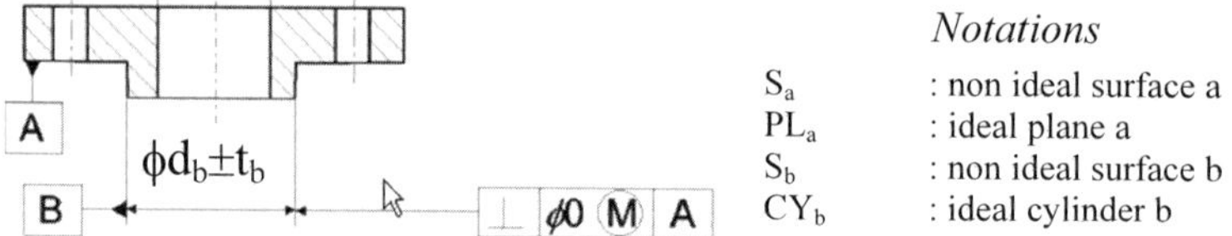


*Figure2 : Perpendicularity with maximum material requirement*

*GeoSpelling  expression*

The list of operations to describe the specification is called an operator. The translation of the standardized specification in GeoSpelling language is given in table I.

| SPECIFICATION OPERATOR | |
|---|---|
| **1) Partition** $S_a$, non ideal feature, from the skin model<br><br>**2) Association** $PL_a$, ideal feature, type plane<br>*Constraints*:<br>    minimum signed distance $(S_a, PL_a) \geq 0$<br>*Objective* to minimize :<br>    maximum signed distance $(S_a, PL_a)$ | **3) Partition** $S_b$, non ideal feature, from the skin model<br><br>**4) Association** $CY_b$, ideal feature, type cylinder<br>*Constraints*:<br>    minimum signed distance $(S_b, CY_b) \geq 0$<br>    angle $(axis(Cy_b), Pl_a)= 90°$<br><br>*Objective* to minimize :<br>    diameter $(CY_b)$ |
| **Characteristic** C1, defined from an intrinsic characteristic of an ideal feature<br>    C1 = diameter $(CY_b)$ - $(d_b + t_b)$ | **Condition** $C1 \leq 0$ |

*Table I: Specification operator*

## 4. SUMMARY AND CONCLUDING REMARKS

Tolerance analysis or tolerance synthesis are based on tolerance representation. Given a particular tolerance representation, efficient and accurate algorithms are needed to actually perform the tolerance analysis. Unfortunately, as the geometric tolerances are complex, it is worse for the algorithms using these tolerances. In reality, the tolerance representation and the analysis algorithms are chosen for the convenience of the developers rather than the user. It is why the user has great difficulties to understand what happens when he uses commercial packages for tolerance analysis. For the majority of CAT systems, the language used for tolerance specification is a national or international standardized language with its lacks and contradictions. This language is also the mean to communicate between each people involved by tolerance activities.

To manage efficiently geometrical variations all along product lifecycle and also to answer to the problems mentioned above, we propose a tolerance representation for the tolerancing process including tolerance specification, tolerance analysis or synthesis and tolerance verification.

GeoSpelling, a unified model, is presented as the basis of a complete and coherent tolerancing process. This model allows to communicate geometrical information between design, manufacturing and inspection. With the consideration of the non-ideal features and the generic concepts brought forth by the operations and the characteristics, the model is a "company-wide" model. It is a univocal language, common to design, manufacturing and inspection.

REFERENCES

**[ISO/TS 17450-1, 2005]** Geometric Product Specification (GPS) – General concepts – Part 1: Model for geometrical specification and verification.

**[ISO/TS 17450-2, 2002]** Geometric Product Specification (GPS) – General concepts – Part 2: Operators and uncertainties.

**[Ballot et al., 1997]** Ballot E., Bourdet P. ; "A Computation Method for the Consequences of Geometric Errors in Mechanisms", *Geometric Design Tolerancing: Theories, Standards and Applications*; pp.197-207; 1998; ISBN O-412-83000-0.

**[Ballu et al., 1995]** Ballu A. and Mathieu L., "Univocal expression of functional and geometrical tolerances for design, manufacturing and inspection", *Computer Aided Tolerancing*; pp. 31-46;1995; ISBN O-412-72740-4.

**[Ballu et al., 2001]** Ballu A., Mathieu L., Dantan J.Y. "Global view of geometrical specifications", *Geometric Product Specification and Verification: Integration of Functionality*; pp. 13-24; 2001; ISBN 1-4020-1423-6.

**[Bennich, 1993]** Bennich P. "Chain of standards: a new concept in tolerancing and engineering drawing GPS-Standards – Geometric Product specification standards",

*Proc. Of the 1993 International Forum on Dimensional Tolerancing and Metrology*; pp. 269-278;  1993; ISBN 0-7918-0697-9.

**[Chase, 1999]** Chase K. "Multi-Dimensional Tolerance Analysis", *Dimensioning and Tolerancing Handbook*; pp. 13-1, 13-27;  1999; ISBN 0-07—018131-4.

**[Clement, 1994]** Desrochers A., Clement A., "A dimensioning and tolerancing assistance model for CAD/CAM systems", *International Journal of Advanced Manufacturing Technology,* pp. 352-361 ; Vol. 9 ; 1994.

**[Dantan et al., 2003]** Dantan J.Y., Anwer N., Mathieu L.; « Integrated Tolerancing Process for conceptual design »; pp.-, *Annals of the CIRP* - Vol. 52/1/03, 2003.

**[Dantan et al., 2002]** Dantan J.Y., Ballu A.; "Assembly specification by Gauge with Internal Mobilities (G.I.M.) : a specification semantics deduced from tolerance synthesis"; *Journal of manufacturing systems*, vol. 21, N°3, pp 218-235, 2002

**[Desrochers et al., 2003]** Desrochers A., Ghie W., Laperrière L.; "Application of a unified Jacobian-Torsor Model for Tolerance Analysis"; *Journal of computing and Information science in engineering;* vol.3, N°1, pp 2-14, 2003

**[Dufaure et al., 2003]** Dufaure J. and Teissandier D, "Geometric tolerancing from conceptual to detail design", *Proc. of 8th CIRP Seminar on Computer Aided Tolerancing*; pp.176-186; The University of Charlotte, North Carolina, USA; April 28-29, 2003.

**[Giordano et al., 1993]** Giordano M., Duret D., "Clearence space and deviation space, application to three-dimensional chain of dimensions and positions", *CIRP Seminar on Computer Aided Tolerancing*, ENS Cachan, May 1993 .

**[Guilford et al., 1993]** Guilford J. and Turner J.; "Advanced Tolerance Analysis ans Synthesis for geometric tolerances"; *Proc. Of the 1993 International Forum on Dimensional Tolerancing and Metrology*; pp.187-198; 1993; ISBN 0-7918-0697-9.

**[Mathieu et al., 2003]**. Mathieu L. and Ballu A, "GEOSPELLING: a common language for Specification and Verification to express Method Uncertainty", *Proc. of 8th CIRP Seminar on Computer Aided Tolerancing*, The University of Charlotte, North Carolina, USA; April 28-29, 2003.

**[Mathieu et al., 1997]** Mathieu L., Clement A., Bourdet P.; "Modeling, Representation and Processing of Tolerances, Tolerance Inspection: a Survey of Current Hypothesis"; *Geometric Design Tolerancing: Theories, Standards and Applications*; pp. 1-33; 1998; ISBN O-412-83000-0.

**[Salomons et al., 1997]** Salomons O., van Houten F., Kals H., "Current Status of CAT Systems", *Geometric Design Tolerancing: Theories, Standards and Applications*, pp. 438-452;1998; ISBN O-412-83000-0.

**[Serre et al., 2003]** Serre P., Riviere A., and Clement A, "The clearance effect for assemblability of over constrained mechanisms", *Proc. of 8th CIRP Seminar on Computer Aided Tolerancing*; pp.102-113;  The University of Charlotte, North Carolina, USA; April 28-29, 2003.

# Tolerance-Maps Applied to the Straightness and Orientation of an Axis


*S. Bhide*, G. Ameta**, J. K. Davidson**, J. J. Shah***

**ExperTeam Services,UGS The PLM Company,*
*Onsite at HP-Vancouver,WA, USA*
*** Mechanical and Aerospace Engineering Department,*
*Arizona State University, Tempe, AZ -85287, USA*
ameta@asu.edu**,** j.davidson@asu.edu



**Abstract**. Among the least developed capabilities in well-developed mathematical models for geometric tolerances are the representation of tolerances on form, orientation, and of Rule #1 in the Standards, i.e. the coupling between form and allowable variations for either size or position of a feature. This paper uses Tolerance-Maps$^{®1}$(T-Maps$^{®1}$) to describe these aspects of geometric tolerances for the straightness and orientation of an axis within its tolerance-zone on position. A Tolerance-Map is a hypothetical point-space, the size and shape of which reflect all variational possibilities for a target feature; for an axis, it is constructed in four-dimensional space. The Tolerance-Map for straightness is modeled with a geometrically similar, but smaller-sized, four-dimensional shape to the 4D shape for position; it is a subset within the T-Map for position. Another internal subset describes the displacement possibilities for the subset T-Maps that limits form. The T-Map for orientation and position together is formed most reliably by truncating the T-Map for position alone.


**Keywords:** form tolerance, position tolerance, straightness, axis, Tolerance-Map**.**


## 1. INTRODUCTION AND LITERATURE REVIEW

When a part is to be manufactured, there are allowable geometric variations for its features. As explained in the ASME Y14.5M Standard [ASME,1994] and the ISO 1101 Standard [ISO,1983], the allowable geometric variations are specified using three-dimensional *tolerance-zones* in which the feature is permitted several degrees of freedom for displacement. The tolerance-zone is located with a basic dimension. The shape of the tolerance zone is defined by the type and value of the tolerance and the feature to which it

---

[1] Patent No. 6963824.

45


*J.K. Davidson (ed.), Models for Computer Aided Tolerancing in Design and Manufacturing, 45–54.*
© 2007 *Springer.*

is applied. The Standard defines six-tolerance types; size, orientation, location, profile, form and runout; and their subclasses. The Standards permit as many as three different tolerances (e.g., size, orientation, and form) on the same feature. The interaction of form and size tolerance applied to a feature constitutes Rule #1 of the Standards.

The major bottleneck, in computerizing the tolerance specification and the downstream analysis and allocation, is the lack of a compact math model. One summary of various attempts to model the variations that are described in the Standards appears in [Davidson, *et al.*, 2002], including a comparison with our model. [Pasupathy, *et al.*, 2003] review the literature for ways to construct geometric tolerance-zones, with special attention to singular points at the vertices of polyhedral objects. In summary, the other models contribute substantially towards representing geometric variations and tolerances, but each model either does not represent, or has not been developed to represent, all the variations that are described in the Standards. The aspects of the Standard where most of the other models are having trouble are: form tolerances, floating zones, Rule #1 tradeoff, bonus tolerance arising from material condition, and/or datum precedence. Our model [Davidson, *et al.*, 2002, Mujezinović, *et al.*, 2004, and Davidson and Shah, 2002] of variations is one of several *vector space models* that map all possible variations into a region in parametric space. Of the vector space models [Whitney, *et al.*, 1994], [Giordano, *et al.*, 1999, 2001] and [Roy & Li, 1999] just one [Giordano, *et al.*, 1999, 2001] models the variations of an axis. In these papers the authors create parametric point-spaces from four of the six Plücker coordinates of a line, one form of a torsor. Since only four of these six coordinates are independent [Davidson & Hunt, 2004], these spaces are limited to dimension 4. Our own model for lines [Davidson and Shah, 2002, Bhide, 2002, and Bhide, *et al.*, 2003] is based on areal coordinates and has no dimensional limit. Further, by making appropriate choices when building the model outlined in [Davidson and Shah, 2002], our model reduces to that of [Giordano, *et al.*, 2001] when no material condition is specified.

The purposes for this paper are to review briefly our model for an axis and to describe the aspects of geometric tolerances for the straightness and orientation of the axis within its tolerance-zone on position. We will also demonstrate the modeling of the relationships between tolerances of position, orientation and straightness applied to an axis.


## 2. THE TOLERANCE-MAP (T-MAP) FOR POSITION OF AN AXIS


The entire range of variational possibilities for an axis will be represented by a Tolerance-Map® (T-Map®), a hypothetical Euclidean point-space; its size and shape will reflect the variations for a target feature. It is the range of points resulting from a one-to-one mapping (eqn 2) from all the variational possibilities of a feature within its tolerance-zone. These variations are determined by the tolerances that are specified for size, position, form, and orientation.

The tolerances shown in Fig. 1($a$) establish the variations for sizes and positions of the two holes in the plate.  According to the Standards [ASME, 1994 and ISO, 1983], the feature control frame $\boxed{\oplus \ | \ \varnothing \ t \ | \ A \ | \ B \ | \ C}$ specifies a *tolerance-zone* (Fig. 1($b$)) for each of the two holes.  Each tolerance-zone is a very thin right-circular cylinder of length $\ell$ and defined by circles $C$ and $\overline{C}$ of diameter $t$, which are located in the upper and lower faces of the plate. The tolerance zone is exactly perpendicular to Datum A (Fig. 1($a$)) and exactly located with dimensions a and b (or c) from Datums C and B, respectively.

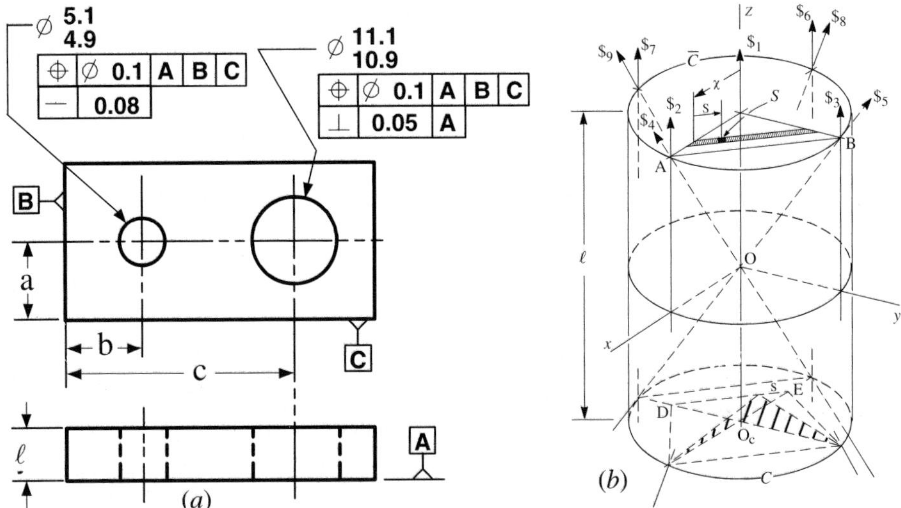

Figure 1. a) Two holes in a plate of thickness $\ell$. Both holes are located with the tolerance t = 0.1 mm. The larger hole is to be held perpendicular to Datum A with the tolerance t″ = 0.5 mm. b) A cylindrical tolerance-zone for the axis of a hole determined by circles $\overline{c}$ and C, each with its diameter equal to the positional tolerance t. Basis-lines $\$_1...\$_5$ are shown, together with four others that complete the symmetry.

To build the Tolerance-Map for the position of the axis of either hole in Fig. 1($a$), choose $\$_1$, $\$_2$, $\$_3$, $\$_4$, and $\$_5$ to be five *basis-lines* that define the space of the four-dimensional set of lines in the tolerance-zone of Fig. 1($b$). These lines are mapped to five corresponding *basis-points* in a hypothetical Euclidean four-dimensional point space which points are arranged to be the vertices $\$_1$, $\$_2$, $\$_3$, $\$_4$, and $\$_5$ of a 4-simplex (simplest polyhedron in a 4-D point-space). One of the basis-lines in the tolerance-zone is assigned to a corresponding vertex in the four-dimensional T-Map point space. Our choices for locating the basis-points are shown in Fig. 2. The geometry for the simplex causes every angle at apex $\$_1$ to be 90° so that $\$_1$ can be regarded as the origin of a 4-D Cartesian frame of reference that is overlain on the T-Map. The four edges of the 4-simplex that are joined at

$\$_1$ become Cartesian axes corresponding to four of the Plücker[2] coordinates $(L,M,N;P,Q,R)$ of a line (see e.g. [Davidson & Hunt, 2004]). Coordinates $P$ and $Q$ are scaled directly on two axes in Fig. 2, but $L$ and $M$ are first multiplied by length $\ell/2$ to give measures $L'$ and $M'$ in units of [length] (see [Davidson and Shah, 2002 or Bhide, *et al.*, 2003] for more detail).

The linear relation

$$\$ = \lambda_1 \$_1 + \lambda_2 \$_2 + \lambda_3 \$_3 + \lambda_4 \$_4 + \lambda_5 \$_5 \tag{1}$$

contains the coordinates $\lambda_1,\ldots,\lambda_5$ [Coxeter, 1969], and its linearity derives from the extremely small ranges for orientational variations which are imposed by tolerance-values in a tolerance-zone. Note that the position of $\$$ depends only on four independent ratios of these magnitudes, thereby requiring one condition among them[2] and confirming 4 as the dimension of the space. When the coordinates $\lambda_1,\ldots,\lambda_5$ are normalized by setting $\Sigma\lambda_i = 1$, they become *areal coordinates* [Coxeter,1969]. To reach all the lines in the tolerance-zone (points in the T-Map), some of the $\lambda_1,\ldots,\lambda_5$ will be negative. Equation (1) can be used to construct the entire boundary of the T-Map (see [Bhide, *et al.*, 2003]); the result is the four-dimensional T-Map that is shown in Fig. 2 with four 3-D hypersections.

Equation (1) implies a one-to-one relationship between the line-segments in the tolerance-zone of Fig. 1(*b*) and the points in the 4D space that is described with areal coordinates. Therefore, it can be used to identify any point in the T-Map of Fig. 2 by interpreting $\$_1$, $\ldots,\$_5$ to represent the five basis-points chosen in Fig. 2. Correspondingly, it can be used to identify any line in the tolerance-zone of Fig. 1(*b*) by assigning $\$_1$, $\ldots,\$_5$ to represent the five basis-lines chosen in Fig. 1(*b*), a suggestion that clearly is not valid in general because linear combinations of two lines yield *screws*, not lines. However, Eq (1) *may* be used for the lines in a tolerance-zone because every one of these lines is rotated only slightly from the theoretical orientation of the feature axis. This assertion was proved in [Davidson & Shah, 2002]. (The constraint of slight rotations causes the quadratic identity to be met with only a residue of order $[t^4]$, yet the terms added together are of order $[t^2]$. Since position tolerance $t$ is two or more orders of magnitude smaller than any dimension on a part, the residue is four or more orders of magnitude smaller than the largest term in the sum.)

Equation (1) can be expanded to

$$\begin{bmatrix} L \\ M \\ N \\ P \\ Q \end{bmatrix} = [X]\begin{bmatrix} \lambda_1 \\ \lambda_2 \\ \lambda_3 \\ \lambda_4 \\ \lambda_5 \end{bmatrix} = \begin{bmatrix} 0 & 0 & 0 & t/\ell & 0 \\ 0 & 0 & 0 & 0 & t/\ell \\ 1 & 1 & 1 & 1 & 1 \\ 0 & 0 & t/2 & 0 & 0 \\ 0 & -t/2 & 0 & 0 & 0 \end{bmatrix}\begin{bmatrix} \lambda_1 \\ \lambda_2 \\ \lambda_3 \\ \lambda_4 \\ \lambda_5 \end{bmatrix} \tag{2}$$

---

[2] Vector $(L,M,N)$ contains the direction cosines of the line, and vector $(P,Q,R)$ is the moment that a unit force along the line would produce about the origin. Consequently, the six Plücker coordinates are constrained by the quadratic identity $LP + MQ + NR = 0$.

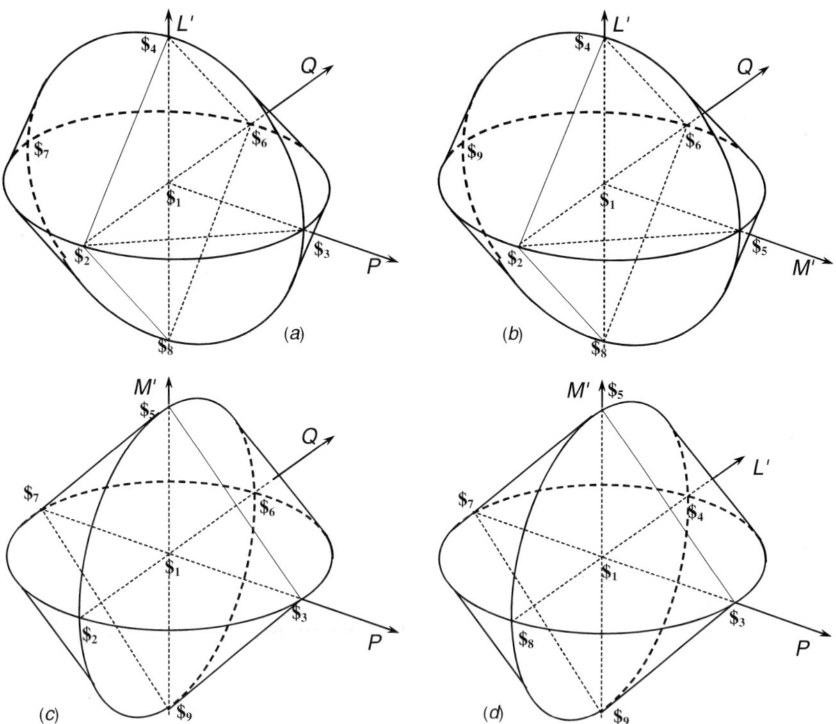

*Figure 2. Four three-dimensional hypersections of the T-Map and its basis 4-simplex for the tolerance-zone in Fig. 1(b) (for position tolerance t only); all circles are of diameter t and all squares have diagonals of length t. (a) The central hyperplane $\lambda_5 = 0$. (b) The hyperplane $\lambda_3 = 0$. (c) The hyperplane $\lambda_4 = 0$. (d) The hyperplane $\lambda_2 = 0$.*

in which each column of matrix $[X]$ represents five Plücker coordinates of one of the basis-lines $\$_1...\$_5$ in the tolerance-zone of Fig. 1(b). Plücker coordinate $R^2$ is omitted in eqs (2) because it is a higher order small quantity (and negligible) for every line in the tolerance-zone. Since the tilt for each line in the tolerance-zone is tiny, coordinate $N$ is unity for every line [Davidson & Shah, 2002]; then the third of eqs (2) provides the normalizing condition $\Sigma\lambda_i = 1$.

From the above definitions, every T-Map for an axis is the range of points that results from the mapping $[X]^{-1}$ applied to every line in a given tolerance-zone.

## 3. TOLERANCE MAP FOR STRAIGHTNESS OF AN AXIS

A straightness tolerance of $t'$=0.08mm is specified on the axis of the smaller hole in Fig.1($a$) with the lower feature control frame $\boxed{-\ |\ t'}$ . This tolerance defines a floating cylindrical tolerance-zone [e.g. ASME, 1994] of diameter $t'$ and length $\ell$, within which all points of the axis must lie. This zone can float/wobble in the cylindrical tolerance-zone for position (diameter $t = 0.1$mm in Fig. 1($a$)) of the axis. Consequently, the Standard [ASME,1994] requires $t'$ to be smaller than the position tolerance on the same axis, indeed smaller than any orientational tolerance also. The T-Map for the position of the axis is still the 4-D point-space defined by the location of basis-points in Fig 2, but now there are two geometrically similar internal subsets whose Minkowski sum add to form the hypersections in Fig 2. The 3-dimensional hypersections of both subsets are shown in Figs 3($b$) and ($c$). The upper hypersection represents the maximum number of lines that are devoted to straightness (the floating zone); its dimension is equal to the diameter of the smallest cylinder that envelopes the imperfect axis. The lower hypersection represents all the locations, i.e. the wobble that the floating zone can occupy, within the position tolerance-zone $t$. Rule #1 of the Standard [ASME, 1994] requires a tradeoff between the diameter of the floating zone (i.e. lines devoted to form) and the range of positions for it. This tradeoff is shown in the Figs 3($a$), ($b$) and ($c$). An axis of perfect form (subset T-Map of size zero) is shown in the Fig 3($a$); its range of position is the entire tolerance-zone for position, and its T-Map is represented by the hypersection of dimension $t$. In Fig 3($b$), the axis has an imperfect form that has an intermediate measure less than $t'$. Fig 3($c$) shows the measure of form equal to $t'$, the limit for the axis of the smaller hole in Fig. 1($a$).

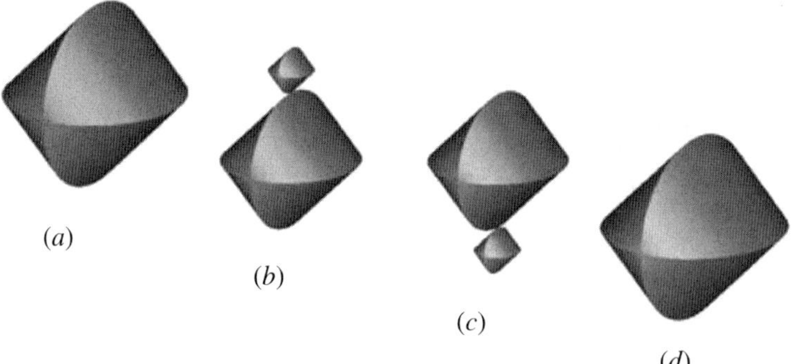



*Figure 3. The tradeoff between internal subsets for straightness and range of positions, which lie within the T-Map for the axis of the smaller hole in Fig 1(a).*

When no tolerance on straightness is specified, Fig 3($d$) provides the upper limit on form variation, as suggested by Rule #1 in the Standards: all the lines in the tolerance-zone on position are devoted to form, the floating zone completely fills the tolerance-zone on position, and its T-Map is the same as that for position, i.e. it has dimension $t$.


## 4. TOLERANCE MAP FOR PERPENDICULARITY OF AN AXIS


A perpendicularity tolerance of $t'' = 0.05$mm is specified on the axis of the larger hole in Fig 1($a$) with the lower feature control frame $\boxed{\perp\ \ |0.05\ \boxed{A}}$. According to the Standards [e.g. ASME, 1994], this tolerance defines a floating cylindrical tolerance-zone of diameter $t''$ and length $\ell$, within which the axis can take any orientation. This cylindrical tolerance-zone is exactly perpendicular to the datum plane A, but it can translate laterally inside the tolerance-zone for position (diameter $t = 0.1$mm in Fig. 1($a$)) of the axis. Therefore, $t''$ must be smaller than the position tolerance on the same axis. The effect of this specification is to limit the variations in orientation of the axis in a tolerance-zone of diameter $t$ to a smaller tolerance-zone of diameter $t''$ but not to limit further its allowed change in position.

The Tolerance-Map for an axis, on which both perpendicularity and position tolerances are specified simultaneously, will be created by truncating the existing hypersections of the Tolerance-Map, those shown in Fig. 2. Since the perpendicularity tolerance limits only the orientational variations of the axis, the T-Map will be truncated only in the directions of the $L'$ and $M'$ axes and will remain unchanged in directions parallel to the axes $P$ and $Q$ that correspond to translational variations; the truncation is accomplished with two parallel planes separated by the dimension $t''$. Two truncated hypersections can represent all four that would result from Fig. 2 because Figs 2($a$) and ($c$) contain only one orientation axis ($L'$ or $M'$) while the sections represented by Figs 2($b$) and ($d$) contain both orientation axes ($L'$ and $M'$) together. When Fig 2($a$) is truncated, the result is the hypersection shown in Fig 4($a$); except for a minor re-labeling, it applies also when Fig 2($c$) is truncated. For the hypersections in Figs 2($b$) and ($d$), the T-Map is truncated by a cylinder of diameter $t''$; the result is shown in Fig 4($b$).


## 5. TOLERANCE MAP FOR AN AXIS TO BE PARALLEL TO A FACE


Suppose now that, instead of the *perpendicularity* tolerance of $t''$ being applied to the axis of the larger hole in Fig 1($a$), a *parallelism* tolerance of $t'' = 0.05$mm ($t'' = t$) were specified in the lower feature control frame, i.e. $\boxed{//\ |0.05\ \boxed{B}}$. According to the Standards [e.g. ASME, 1994], this tolerance defines a floating tolerance-zone consisting of two parallel planes that are separated by distance $t''$ and that are both always parallel to Datum B.   The consequences of this arrangement are that the angle between Datum B and the axis of the hole can never exceed $t''/\ell$, yet all other limits to orientational and translational variations remain unchanged from those required by the position tolerance-zone of diameter $t$.

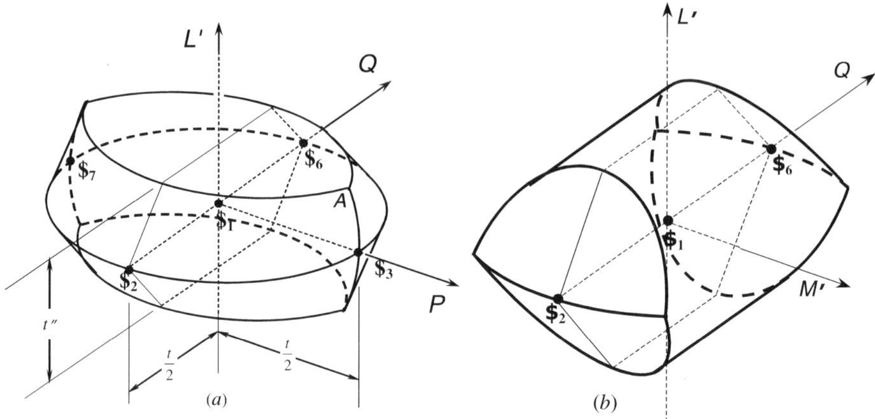


*Figure 4 (a) The hypersection $\lambda5 = 0$ and (b) The hypersection $\lambda3 = 0$ for perpendicularity tolerance applied with position tolerance to an axis.*

To create hypersections for the Tolerance-Map for parallelism and position tolerances that are specified simultaneously, it is necessary to truncate three of the existing hypersections in Fig. 2, those that contain the coordinate $M'$. Presuming the $x$-axis in Fig. 1($b$) to be aligned parallel to Datum B, the feature control frame $\boxed{// \;|\; 0.05 \;|\; B}$ on parallelism limits to 0.05 mm the projection in the $y$-direction of a length $\ell$ along the axis; i.e. coordinate $M'$ is limited to the range $-0.05 \le M' \le 0.05$ mm. The truncated hypersection for Fig 2($b$) is shown in Fig 5. When the hypersections in Figs 2($c$) and ($d$) are truncated along the $M'$ axis, the resulting solids are congruent to the one shown in Fig. 4($a$), although the placement of coordinate axes would be different.

## CONCLUSION

Tolerance-Maps were created for both a perpendicularity tolerance and a parallelism tolerance, relative to a datum plane, each applied to an axis located with a position tolerance. Using areal coordinates and traditional geometry of lines, the T-Maps were created as 4-D objects and visualized using 3-D hypersections. The T-Map for straightness of an axis was developed as an internal subset to the T-Map for position. This concept led to the modeling of Rule#1 and the tradeoff of form and position variations associated with it. Although the T-Maps are modeled in four dimensions and are not so intuitive to visualize, 2-D and 3-D sections through them permit their use in developing stackup conditions.

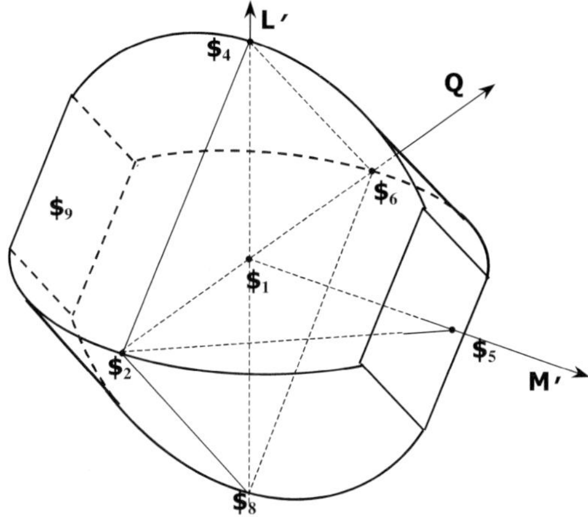



*Figure 5. The hypersection $\lambda_3 = 0$ for parallelism tolerance applied with position tolerance to an axis.*

ACKNOWLEDGEMENT

The authors are grateful for funding provided to this project by National Science Foundation Grants #DMI-9821008 and #DMI-0245422.

REFERENCES

[**ASME Standard, 1994**] ASME Y14.5M.; "Dimensioning and Tolerancing"; The American Society of Mechanical Engineers, NY.

[**Bhide, 2002**] Bhide, S. (2002). *A New Mathematical Model for Geometric Tolerances Applied to Cylindrical Features*, MS Thesis, Arizona State University.

[**Bhide, et al., 2003**] Bhide, S., Davidson, J.K., and Shah, J.J.; "A New Mathematical Model for Geometric Tolerances as Applied to Axes", In: *CD Proc., ASME Des. Technical Conf's.*, #DETC2003/DAC-48736; Chicago, IL.

[**Coxeter, 1969**] Coxeter, H. S. M. (1969). *Introduction to Geometry*, 2nd ed. Wiley.

**[Davidson, *et al.*, 2002]** Davidson, J.K., Mujezinović, A., and Shah, J. J. "A New Mathematical Model for Geometric Tolerances as Applied to Round Faces", *ASME Transactions, J. of Mechanical Design*, **124**, pp. 609-622.

**[Davidson & Shah, 2002]** Davidson, J.K. and Shah, J.J. (2002). "Geometric tolerances: A new application for line geometry and screws." *IMechE J. of Mechanical Eng. Science, Part C*, **216(C1)**, pp. 95-104.

**[Davidson & Hunt, 2004]** Davidson, J.K. and Hunt, K.H. (2004). *Robots and Screw Theory*. Oxford.

**[Giordano, *et al.*, 1999]** Giordano, M., Pairel, E., and Samper, S. (1999). "Mathematical representation of tolerance zones." In *Global Consistency of Tolerances*, Proc., 6th CIRP Int'l Seminar on Computer-Aided Tolerancing, Univ. of Twente, Enschede, Netherlands, March 22-24 (ed. F. vanHouten and H. Kals), pp. 177-86.

**[Giordano, *et al.*, 2001]** Giordano, M., Kataya, B., and Samper, S. "Tolerance analysis and synthesis by means of clearance and deviation spaces." In *Geometric Product Specification and Verification*, Proc., 7th CIRP Int'l Seminar on CAT, Ecole Norm. Superieure, Cachan, France, April 24-25, (eds. P. Bourdet and L. Mathieu), pp. 345-354.

**[ISO 1101, 1983]** "Geometric tolerancing—Tolerancing of form, orientation, location, and run-out—Generalities, definitions, symbols, and indications on drawings"; International Organization for Standardization.

**[Mujezinović, *et al.*, 2004]** Mujezinović, A., Davidson, J.K., and Shah, J. J. "A New Mathematical Model for Geometric Tolerances as Applied to Polygonal Faces", *ASME Trans., J. of Mechanical Design*, **126**, pp. 504-518.

**[Pasupathy, *et al.*, 2003]** Pasupathy, T.M.K., Morse, E.P., and Wilhelm, R.G. "A Survey of Mathematical Methods for the Construction of Geometric Tolerance Zones", *ASME Transactions, J. of Computing & Information Science in Engr.*, **3**, pp. 64-75.

**[Roy and Li, 1999]** Roy, U. and Li, B. (1999). "Representation and interpretation of geometric tolerances for polyhedral objects– I: Form tolerance." *Computer-Aided Design* **30**, pp. 151-161.

**[Whitney, *et al.*, 1994]** Whitney, D. E., Gilbert, O. L., and Jastrzebski, M. (1994). "Representation of geometric variations using matrix transforms for statistical tolerance analysis in assemblies", *Research in Engineering Design*, **6**, pp. 191-210.

# Information Modeling to Manage Tolerances during Product and Process Design

**J.-Y. Dantan, T. Landmann, A. Siadat, P. Martin**

*Laboratoire de Génie Industriel et de Production Mécanique,*

*E.N.S.A.M. de Metz, 4 rue A. Fresnel, 57070 METZ Cedex, France*

*jean-yves.dantan@metz.ensam.fr*

**Abstract:** For car and aircraft industries, the management of tolerances has become an important issue in product and process design. Indeed, designers need to manage the tolerances and to know information that contributed to their determination. To do so, we propose to integrate a qualitative and a quantitative aspect of tolerancing. This quantitative aspect is based on the Key Characteristics approach, the mathematical models of tolerance, and the mathematical tools for tolerance analysis and tolerance synthesis. Our approach uses graphic conventions of UML to represent the information model. A prototype has been realized using an Object Oriented Data Base.

**Keywords**: Tolerancing and life cycle Issues, Tolerancing process, Product and process design, Information model

## 1 INTRODUCTION

Influence of design on manufacturing cost is usually great. Errors made during the early stages of design tend to contribute as much as 70% to the cost of production. It is better to consider manufacturing issues as early as possible in the product design process. However, making sound decisions in the early design phase is rather difficult since it involves many unpredictable factors in manufacturability, quality, etc

For car and aircraft industries, the management of tolerances has become an important issue in product and process design.

- Designers need to manage the tolerances and to know information that contributed to their determination.

- Tolerancing process and Design for Manufacturing are a key activity to evaluate manufacturability, to improve design, …



*J.K. Davidson (ed.), Models for Computer Aided Tolerancing in Design and Manufacturing,* 55–64.

- The inherent imperfections of manufacturing processes and resources involve a degradation of functional part characteristics.

The goal here is to put tolerancing in a concurrent engineering context. There are important questions that would need to be looked upon: How to integrate the tolerancing process in the product and process design? How to ensure the transition from function to manufacturing tolerances? How to evaluate the impact of a tolerance during the product and process design?

Several answers exist today in academic works: [Johannesson et al, 2000], [Roy et al, 2001], [Marguet et al, 2001], [Desrocher et al, 2001], [Dantan et al, 2003], [Dufaure et al, 2004], … But, these different approaches integrate a qualitative aspect.

Evaluating the impacts of manufacturing tolerance on functional characteristics during product and process design is a highly critical problem in industry and it is usually managed by the actors themselves, based on their expertise. A manual resolution of the various impacts is only possible if the product is very simple. For complex product, this evaluation needs to be aided.

In this paper, we propose to integrate a qualitative and a quantitative aspect. This quantitative aspect is based on the Key Characteristics approach [Thornton, 1999], the mathematical models of tolerance, and the mathematical tools for tolerance analysis and tolerance synthesis.

## 2    QUALITATIVE APPROACH TO MANAGE DESIGN TOLERANCES & INTEGRATION OF QUANTITATIVE ASPECT

In order to determine quickly the tolerances of parts or subassembly for complex product like aircraft or car body, we propose a method with a graphical tool (this graphical tool is based on assembly graphs).

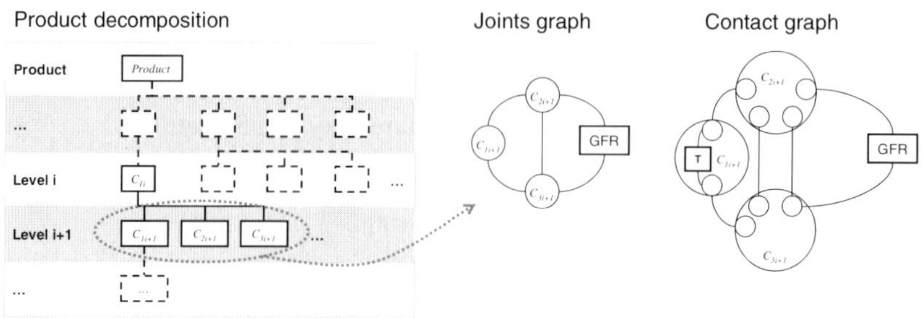


*Figure 1: Structural graphs.*

At each level of the decomposition of the product into subassemblies (Figure 1), for an assembly, by using functional analysis method, designers define major functional requirements and technical requirements. Moreover, the technical functional analysis allows determining the geometrical functional requirements, which limit the functional characteristics of the mechanism. The geometrical functional requirements may be definied at the higher level. To express the geometrical functional requirement and to analyze it by graphs, a joints graph (Figure 1) modelizes the structure of the mechanism; each vertex represents part and each edge between two vertexes represents a cinematic joint. For the needs of tolerancing, the cinematic joint must be decomposed into elementary joints between surfaces [Ballu et al, 1999], [Marguet et al, 2001], [Dantan et al, 2003]. The contact graph (Figure 1) is an extension of the joints graph; each vertex represents a part, each pole of a vertex represents a surface of the corresponding part and each edge between two poles represents an elementary joint. A rectangular vertex represents these geometrical functional requirements.

To determine the influences of the parts, the surfaces or the deviations on the considered functional requirement, the graph analysis method was developed by A. BALLU and L. MATHIEU [Ballu et al, 1999]. Indeed, the key deviations (corresponding to the deviations of surfaces on which the functional requirement is dependent) are determined by using graph simplification rules. Designers study the impact of the deviations and the gaps on the considered functional requirement to define the functional cycles (example – Figure 2). All edges participating to the functional cycles have an impact on the realization of the geometrical functional requirement.

The determination of the tolerances constitutes the last stage of the method. The tolerances corresponding to a requirement are related to all the key surfaces of the key parts and strictly to them and limit the key deviations and strictly them. To determine the functional tolerances of each part, the following criteria are adopted [Ballu et al, 1999]: "the *choice of the datum must be realized according to the type of contact, …*"

The tolerance may be represented into the graph like a geometrical functional requirement (Figure 1) and it becomes a geometrical functional requirement at the lower level.

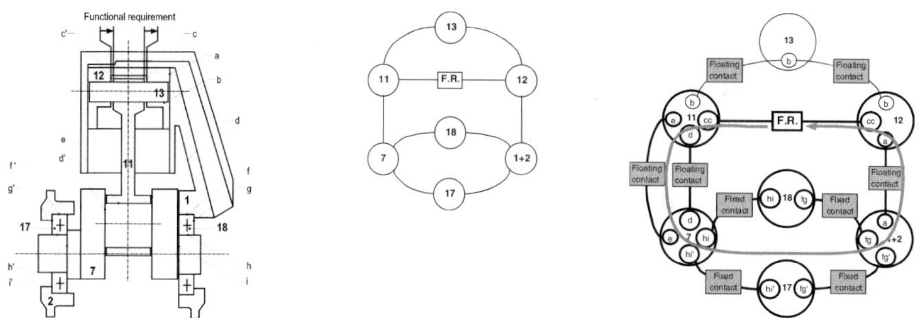

*Figure 2: Example of Graphs.*

Figure 3 shows classes related to Design tolerances:

*Structural entity* is a class that represents a generic structural element which can be specialized: a component like a product, a sub assembly, and a part; a geometrical feature like a functional set of surfaces, a surface, … *Structural entity* represents an element of the structural product decomposition. Decomposition operation is represented in the class diagram using a reflexive aggregation [Dufaure et al, 2004]. This allows to describe a structural tree which represents the product decomposition into assembly, sub assembly, …, part,…, and surface. An object *Structural entity* is a vertex of Joints graph or Contact graph.

*Structural relation* is an association class which represents a generic structural link between generic structural elements, which can be specialized: Cinematic joint, Elementary joint, Topological link, Geometrical requirement, Geometrical specification … The decomposition of a structural relation is represented in the class diagram using a reflexive aggregation, example: cinematic joint break up into elementary joints; a geometrical functional requirement generates tolerances. An object *Structural relation* is edges of Joints graph or Contact graph.

*Function* is a class that represents a product function which can be specialized: Functional requirement, Technical requirement … A function could be mapping to structural entities or/and relations [Dantan et al, 2003], [Dufaure et al, 2004], [Johannesson et al, 2000].

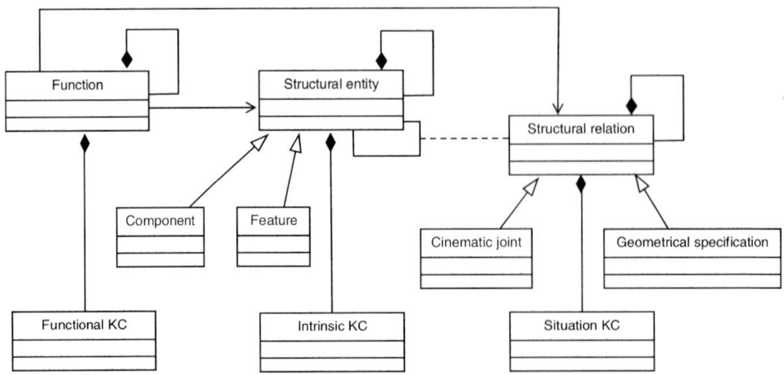


*Figure 3: Class diagram of design point of view.*

To include the quantitative aspect, we add some classes which represent the characteristics. Thornton proposes hybrid definitions of KCs [Thornton, 1999]:

*Key Characteristics (KCs) are the product, sub-assembly, part, process and resource parameters that significantly impact the final cost, performance, or safety of a product when the KCs vary from nominal.*

We specify this definition with GPS definition [ISO 17450-1]:

*A geometrical characteristic is single geometrical property of one or more feature(s). A GPS Characteristic is characteristic describing the micro or macro geometry of one or several features.*

A *Structural entity* could be defined by *Intrinsic key characteristics*. An *Intrinsic key characteristic* is a single Key Characteristic of one product element (example of Intrinsic KC: form deviation). A *Situation key characteristic* is a single Key Characteristic defining a *Structural relation (example of Situation KC: gap of cinematic joint; position deviation of geometrical specification). A Functional key characteristic* represents the product's performance, function, ...


## 3 QUALITATIVE APPROACH TO MANAGE MANUFACTURING TOLERANCES & INTEGRATION OF QUANTITATIVE ASPECT

In the same way of design tolerances, to determine the manufacturing tolerances and to analyze the impact of the process uncertainties on the part characteristics, we propose to use the same graphical tool (manufacturing graph).

Manufacturing graph (Figure 4) modelizes the manufacturing process and the intermediate states of the part. Each vertex represents an intermediate state of the part, each pole of a vertex represents a surface of the corresponding state and each edge between two poles represents an elementary joint. A rectangular vertex represents a manufacturing operation, a fixture or a tolerance.

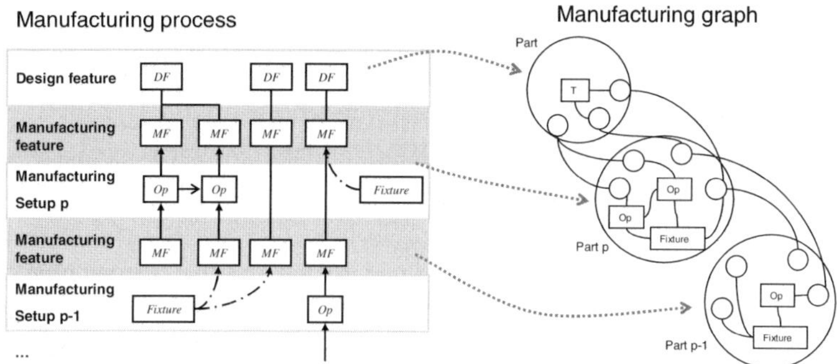


*Figure 4: Manufacturing graph.*

Figure 5 shows the class diagram of Manufacturing Process.

The class *Manufacturing feature* represents a set of features which is obtained by a *Manufacturing Operation* [Shah et al, 1995]. A *Manufacturing operation* is an association class between *Manufacturing features* because a manufacturing operation convert a manufacturing feature at state p-1 into manufacturing features at state p.

The decomposition of the *Manufacturing process* into *Manufacturing setups* and into *Manufacturing operations* is modelized by aggregations between the classes *Manufacturing Process, Manufacturing setup* and *Manufacturing operation* [Halevi et al, 1995]. These Classes represent the manufacturing activities.

These manufacturing activities need some resources. *Manufacturing operation* is related to a *Tool* [Feng et al, 2000]; *Manufacturing setup* to a *fixture* and a *machine. Resource* is a class that represents a physical object that is used in a manufacturing process: *Tool, Fixture* or *Machine.*

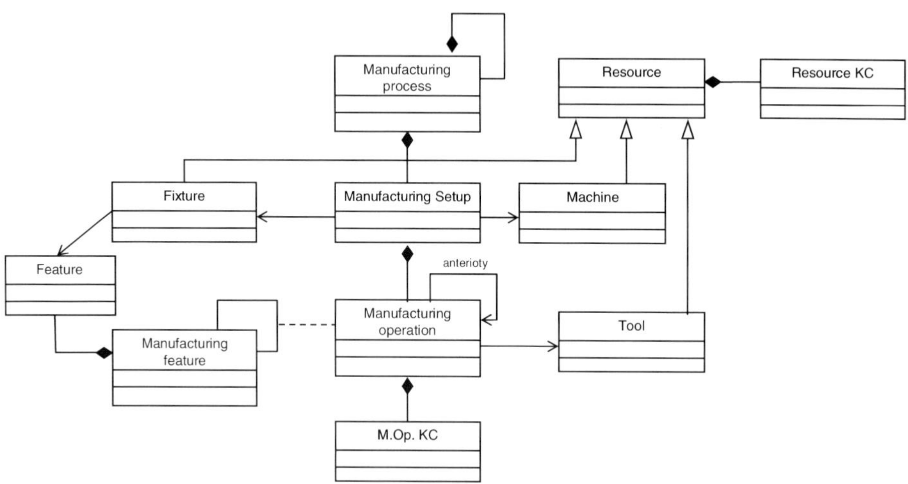


*Figure 5: Class diagram of manufacturing point of view.*

To include the quantitative aspect, we add some classes which represent the characteristics of the process. Manufacturing Operation Key Characteristics (M.Op.KCs) are the manufacturing process parameters (like a cutting speed) and Resource Key Characteristics (Resource KCs) are the resource parameters (like a position uncertainty of fixture) that significantly affect the realization of product key characteristics [Thornton, 1999].

Moreover, an aggregation between Feature and Manufacturing feature represents the manufacturing feature decomposition into features. The manufacturing tolerance is modelized like a design tolerance by a *Structural relation.*

## 4　QUANTITATIVE APPROACH BY KEY CHARACTERISTICS

In a complex product, it is not economically or logistically feasible to control and/or monitor thousands of tolerances and processes. To identify what tolerances and processes to control, many organizations are using a Key Characteristics method (KCs) [Thornton, 1999]. KC methods are used by design to identify and communicate to manufacturing where excess variation will most significantly affect product quality.

Most KCs approaches are based on the concept of a KC flowdown. The KC flowdown prides a system view of potential variation risk factors and captures designers and manufacturers knowledge about variation and its contributors. A hierarchical tree structure is commonly used to describe the Key Characteristics of a product [Thornton, 1999]. A KC flowdown allows for a decomposition of the product into essential features and processes – enabling traceability of cause and effect. Figure 6 shows a KC flowdown. Product level KCs are identified at the highest level of the flowdown and are product requirements. Product KCs are linked to subassembly KCs. These subassembly

KCs are then flowed down to Part-level KCs. Part-level KCs are critical parameters at the lowest level of the product including product features. We add a level: Design feature-level KC. These can be flowed down further to the process KCs : Intermediate Manufacturing feature-levels KCs and Manufacturing setup-levels KCs. Many layers exist within a KC flowdown. The complexity and interrelationships in a tree correspond to the complexity in the product and its manufacturing processes.

Axiomatic design [Suh, 1990] provides a similar systematic method for mapping functional requirements into physical design attributes and mapping physical parameters into process variables but does not specifically focus on variation issues.

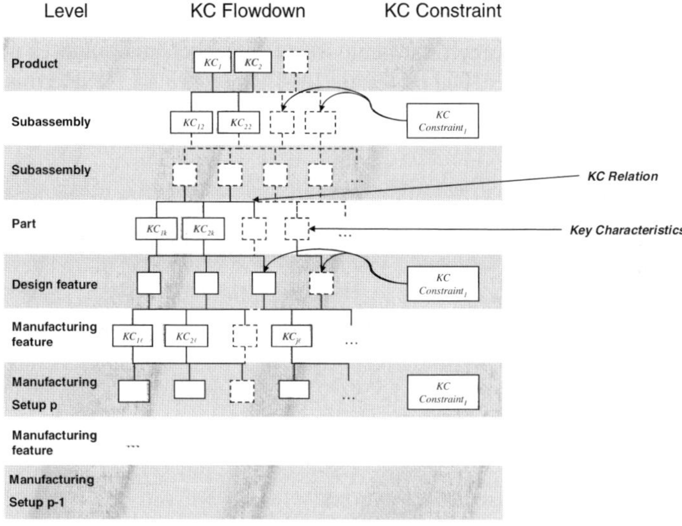


*Figure 6: KC Flowdown.*

In this application, we add the notion of KC level constraint; we specify the KC link by a mathematical aspect and we focus on the geometrical KC:

Indeed, the best way to determine the optimal tolerances is to simulate the influences of deviations on the geometrical behavior of the mechanism. Usually, for mathematical formulation of tolerance analysis or tolerance synthesis, the geometrical behavior is described using different concepts as Variational geometry, Geometrical behavior law, Clearance space and deviation space, Gap space and kinematic models. We principally need a detailed description of each variation to characterize the geometrical behavior. **Each variation is a key characteristic**: surface deviations of each part (situation deviations and intrinsic deviations) and relative displacements between parts according to gap (gaps and functional characteristics) [Dantan et al, 2005]. Example:

$$\alpha_{1a/1}, \beta_{1a/1}, w_{1a/1}, \alpha_{1b/1}, \beta_{1b/1}, u_{1b/1}, v_{1b/1}, \dots$$

And various equations and inequations modelize the geometrical behavior of the mechanism. Composition relations of displacements in the various topological loops of graph (Figure 1) express the geometrical behavior of the mechanism. The composition relations define compatibility equations between deviation, gaps, ... (for design tolerances : [Ballot et al, 1997], for manufacturing tolerances : [Villeneuve et al, 2001]. **A compatibility equation is a KC relation.** Example:

$$-\alpha_{1a/1} + \alpha_{1a/2a} + \alpha_{2a/2} - \alpha_{2b/2} - \alpha_{1b/2b} + \alpha_{1b/1} = 0$$
$$-\beta_{1a/1} + \beta_{1a/2a} + \beta_{2a/2} - \beta_{2b/2} - \beta_{1b/2b} + \beta_{1b/1} = 0$$

*or*

$$Y = f(X)$$

Interface constraints limit the geometrical behavior of the mechanism and characterize non-interference between surfaces [Dantan et al, 2005]; they are **KC level constraints**. Example:

$$\left(v_{ia/ja,O} - z_M \alpha_{ia/ja}\right)^2 + \left(u_{ia/ja,O} - z_M \beta_{ia/ja}\right)^2 \leq \frac{\left(d_{ja} - d_{ia}\right)^2}{4}$$

A geometrical specification is a condition on part deviation (Key Characteristics); it is a **KC level constraints.** Example:

$$\left|x_M \beta_{ia/ib} - y_M \alpha_{ia/ib}\right| \leq \frac{t}{2} \qquad or \qquad x_i \leq t$$

This flowdown with the KC relation and the KC level constraint allows to evaluate the impacts of manufacturing tolerance (KC level constraints) on functional characteristics during product and process design.

Figure 7 shows the class diagram of KCs relations:

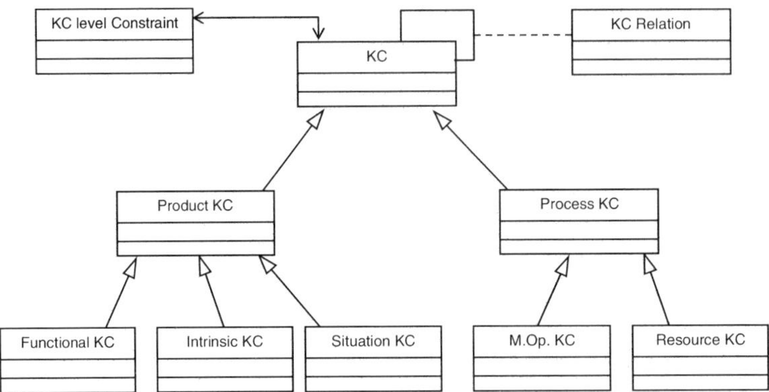


*Figure 7: Class diagram of KC point of view*

*KC* is a class that represents a generic key characteristic which can be specialized: *Product KC, Functional KC, Intrinsic KC, Situation KC, Manufacturing Process KC, Manufacturing operation KC* and *Resource KC*.

*KC relation* is an association class between KCs which represents a generic relation between KCs, which can be specialized: mathematical relations (equation, inequation, and hull), causality … This allows to describe a KC tree.

*KC level constraint* represents a generic constraint between KCs which are defined at the same level.

To demonstrate the efficiency of the proposed approach, a prototype with Ozone (Object Oriented Data Base) has been realized (Figure 8). And the GRT example has been used.

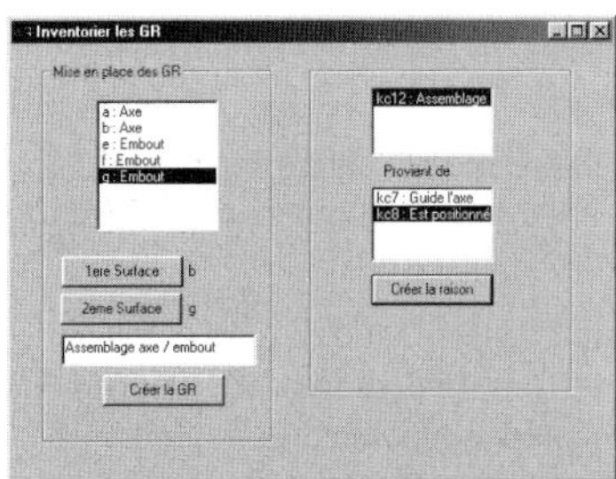
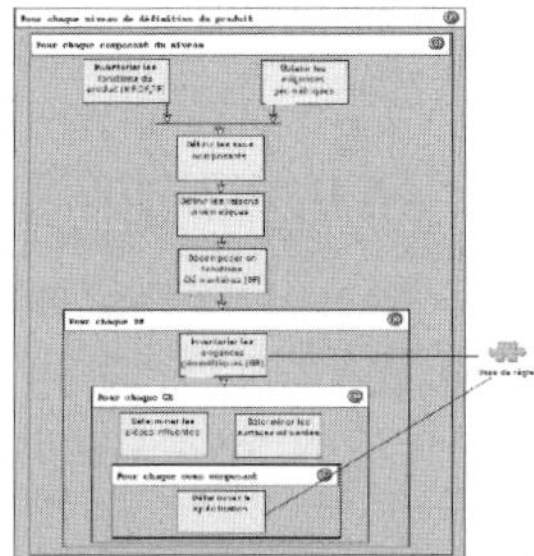

*Figure 8: examples of interface and workflow.*

## 5    CONCLUSION

To manage tolerances during the product and process design, we propose an approach which includes:

- an information representation by graph,

- an information model which stores data related to Product model, Process plan, Manufacturing resources and on the Key Characteristics approach. The extension of the KC approach allows to integrate the quantitative aspect of the tolerances management.

This quantitative aspect is based the mathematical models of tolerance, and the mathematical tools for tolerance analysis and tolerance synthesis and it allows to evaluate the impacts of manufacturing tolerance on functional characteristics during product and process design.

REFERENCES

**[Ballot et al, 1997]** BALLOT E., BOURDET P., 1997, "A Computation Method for the Consequences of Geometric Errors in Mechanisms". Proceedings of CIRP Seminar on Computer Aided Tolerancing, Toronto, Canada.

**[Ballu et al, 1999]** BALLU A., MATHIEU L., 1999, "Choice of functional specifications using graphs within the framework of education", Proc. of the 6th CIRP International Seminar on Computer Aided Tolerancing, Enschede, The Netherlands, pp. 197-206.

**[Dantan et al, 2005]** DANTAN J.Y., MATHIEU L., BALLU A., MARTIN P., 2005, « Tolerance synthesis : quantifier notion and virtual boundary ». Computer Aided Design, Vol.37, n°2, pp.231-240.

**[Dantan et al, 2003]** DANTAN J.Y., NABIL A., MATHIEU L., 2003, "Integrated Tolerancing Process for Conceptual Design". The Annals of CIRP, Vol. 52, n°1, pp. 135-138.

**[Desrochers et al, 2001]** DESROCHERS A., LAPERRIERE L., 2001, "Framework proposal for a modular approach of tolerancing", Proc. of the 7th CIRP International Seminar on Computer Aided Tolerancing, Cachan, France, pp. 93-102.

**[Dufaure et al, 2004]** DUFAURE J., TEISSANDIER D., DEBARBOUILLE G., 2004, "Product model dedicaced to collaborative design: A geometric tolerancing point of view". Proceedings of IDMME, Bath, UK.

**[Feng et al, 2000]** FENG S., SONG E., 2000, "Information Modeling on Conceptual Process Planning Integrated with Conceptual Design". Proceedings of the 5th Design For Manufacturing Conference in the ASME Design Engineering Technical Conferences, USA.

**[Halevi et al, 1995]** HALEVI G., WEILL R.D., 1995, "Principles of Process Planning – A logical approach". (Chapman & Hall, Inc.).

**[Johannesson et al, 2000]** JOHANNESSON H., SODEBERG R., 2000, "Structure and Matrix for Tolerance Analysis from Configuration to Detail Design", Research in Engineering Design, Vol. 12, pp. 112-125.

**[Marguet et al, 2001]** MARGUET B., MATHIEU L., 2001, "Integrated design method to improve productibility based on product key characteristics and assembly sequences", Annals of the CIRP, 50(1), pp. 85-91.

**[Roy et al, 2001]** ROY U., PRAMANIK N., SUDARSAN R., SRIRAM R.D., LYONS K.W., 2001, "Function to form mapping: model, representation and application in design synthesis". Computer Aided Design, Vol.33, pp. 699-719.

**[Shah et al, 1995]** SHAH J.J., MANTYLA M., 1995, "Parametric and Feature-based CAD/CAM". (John Wiley & Sons, New York).

**[Suh, 1990]** SUH N. P., 1990, "The Principles of Design". (Oxford University Press, New York).

**[Thornton, 1999]** THORNTON A. C., 1999, "Variation Risk Management Using Modeling and Simulation". Journal of Mechanical Design, 121, pp. 297–304.

**[Villeneuve et al, 2001]** VILLENEUVE F., LEGOFF O., LANDON Y., 2001, "Tolerancing for manufacturing : a three dimensional model". International Journal of Production Research, Vol. 39, No. 8, pp. 1625-1648.

# Relative Positioning of Planar Parts in Toleranced Assemblies

*Y. Ostrovsky-Berman, L. Joskowicz*
*The Hebrew University of Jerusalem, Jerusalem 91904, Israel,*
*yaronber@cs.huji.ac.il*

**Abstract:** This paper presents a framework for worst case analysis of the relative position variation of toleranced parts in assemblies. The framework is based on our general parametric tolerancing model for planar parts. We present six types of relative position constraints designed to model all types of contact and clearance specifications between features of two parts. To model the relative part position variation in the entire assembly, we introduce the assembly graph, a generalization of Latombe's relation graph that includes cycles, toleranced parts, and three degrees of freedom. We show how to compute the sensitivity matrices of each vertex from the pairwise relative position constraints and the assembly graph. These matrices serve to compute the tolerance envelopes bounding the areas occupied by the parts under all possible assembly instances.
**Keywords**: part models, geometric constraint solving, tolerance envelopes.

## 1. INTRODUCTION

Manufacturing and assembly processes are inherently imprecise, producing parts that vary in size and form. Tolerance specifications allow designers to control the quality of the production and to manufacture parts interchangeably. Tolerancing methods have been developed and incorporated into most modern CAD software. However, these methods are limited in the types of interactions they can model and in the quality of the results they produce.

Determining the variations of the relative positioning of parts with tolerances in an assembly is a key problem in assembly planning [Halperin et al., 2000] and mechanism design [Sacks, 1998]. For example, nearly all assembly planners produce plans for nominal parts. However, because of shape and position variability due to manufacturing imprecision, the relative part positions vary as well. Thus, the nominal assembly plan might not be feasible for certain instances of parts, and a valid plan for one instance might not be suitable for others. In mechanism design, interference between two part instances can occur even when there is no blocking between the nominal parts.

The relative position of imperfect planar parts was studied by Turner [Turner, 1990], who reduces the problem to solving a non-linear system of constraints for a given cost function. Sodhi and Turner [Sodhi, 1994] later extended this work for 3D parts. Li and Roy [Li, 2001] show how to find the relative position of polyhedral parts



*J.K. Davidson (ed.), Models for Computer Aided Tolerancing in Design and Manufacturing*, 65–74.

with mating planes constraints. These methods compute the placement of a single instance of the assembly, and thus cannot be extended to analyze the entire variational class of the assembly. Inui et al. [Inui et al., 1996] propose a method for bounding the volume of the configuration space representing position uncertainties between two parts.    However, their method is only applicable for polygonal parts and is computationally prohibitive. Latombe et al. [Cazals, 1997] present a simple tolerancing model in which polygonal parts vary in the distance of their edges from the part origin, but not in their orientation. They show how to compute the relative position between two parts when the variational parameters span their allowed range, and use it in assembly planning with infinite translations [Latombe et al., 1997]. They acknowledge the limitations of their model and point to the need for developing a more general tolerancing model and for supporting other motion types. This motivated our work.

In this paper, we present a framework for worst case analysis of the relative position variation of toleranced parts in mechanical assemblies. The framework is based on our previously developed general parametric tolerancing model for planar parts. We introduce six types of relative position constraints designed to model all types of contact and clearance specifications between features of two parts. To model the relative part position variation in the entire assembly, we introduce the assembly graph, a generalization of Latombe's relation graph that includes cycles, toleranced parts, and three degrees of freedom. We show how to compute the sensitivity matrices of each vertex from the pairwise relative position constraints and the assembly graph. These matrices serve to compute the tolerance envelopes bounding the areas occupied by the parts under all possible assembly instances. The envelopes provide an accurate characterization of geometric uncertainty that is useful in assembly planning and mechanism design.


## 2.    TOLERANCED ASSEMBLY SPECIFICATION

Assemblies of toleranced parts require a representation that accounts for part variations. The goal is to develop a framework within which part variations can be represented and efficiently computed. Our starting point is the general model of planar toleranced parts whose boundary consists of line and arc segments that we developed in previous research. Throughout the paper, we will use the assembly shown in Figure 1 as an example to illustrate the concepts.

### 2.1. Toleranced parts

We model part variation with the parametric tolerancing model described in [Ostrovsky-Berman, 2004]. In this model, part variation is determined by $m$ parameter values $p=(p_1...p_m)$, specifying lengths, angles, and radii of part features. The parameters have nominal values and can vary along small tolerance intervals. The coordinates of the part vertices are standard elementary functions of a subset of the $m$ parameters. An instance of the parameter values determines the geometry of the part. Figure 2(a) shows the tolerance specification of part $P_3$.

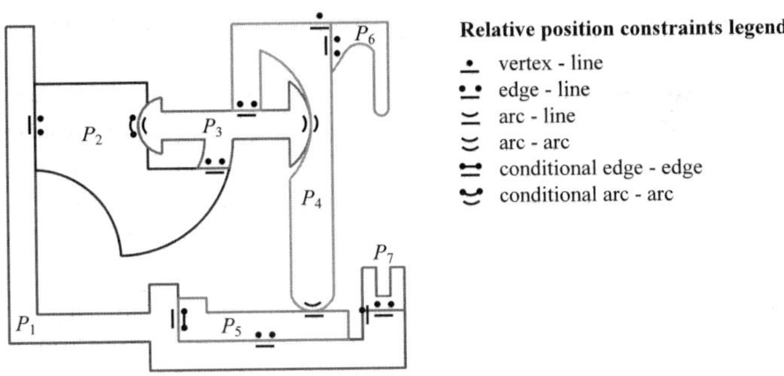



*Fig. 1. Example of a simplified seven part planar mechanism with all types of contacts between parts.*

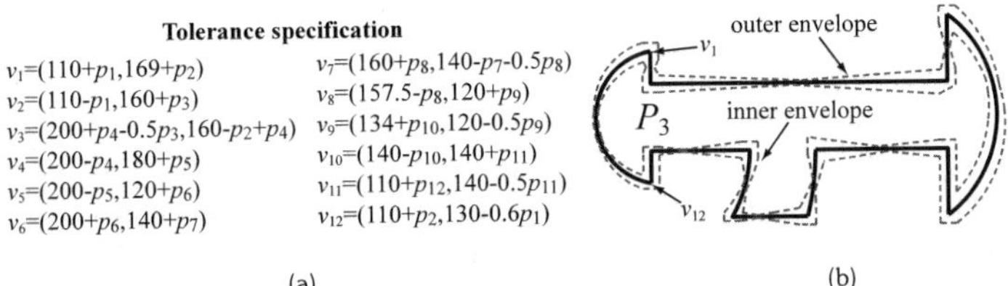



*Fig. 2. (a) Tolerance specification and (b) envelope of part $P_3$. Vertices $v_1$ to $v_{12}$ are ordered clockwise. Parameters $p_1...,p_{12}$ have all nominal values equal to zero.*

In [Ostrovsky-Berman, 2004], we describe an algorithm for computing the outer and inner tolerance envelopes, which are boundaries of the union and the intersection of all possible parts, respectively. The algorithm inputs the partial derivatives of the vertices according to the $m$ variational parameters, and computes the envelopes under the linear approximation of the model. For a part with $n$ vertices, the algorithm computes the most accurate tolerance envelope in $O(nm^2)$ space and $O(nm^2\log m)$ time. Figure 2(b) shows the tolerance envelope of part $P_3$.

### 2.2. Relative position of two parts

The relative position of one part with respect to another is modeled with contact and clearance constraints. These constraints specify the location of the part boundaries with respect to each other, or with respect to a reference datum. For planar parts consisting of line and arc segments, the constraints describe how to position a part feature (vertex, edge, or arc) with respect to another one or with respect to a datum (line). The variability of the feature parameters determines the variability of the relative position of parts in the assembly.

We have identified six types of relative position constraints that describe all types of contact and clearance constraints, including simultaneous contacts. The constraints yield the possible variation in the position of part $B$ (the free part) relative to part $A$ (the fixed part) when $B$ is positioned according to the specification and the variational parameters of both parts span their allowed values. For each vertex $u$ of $B$, the goal is to compute the transformation matrix that describes the sensitivity of the vertex to variations in the parameters of parts $A$ and $B$. The $2 \times m$ sensitivity matrix $S_u$ has one column for each of the variational parameters. We first describe the relative position constraints and their associated equations. We then show how to solve the resulting system of equations and compute the sensitivity matrices of $B$.

### 2.2.1 Relative position constraints

Planar part $B$ has three degrees of freedom, two for translation and one for rotation. Thus, to uniquely determine its position relative to $A$, three independent constraints are required. For each instance of the parts, there is a rigid transformation $T = (t_x, t_y, \theta)$ that positions $B$ relative to $A$ and satisfies the constraints. Since the part variations are typically at least two orders of magnitude smaller than the nominal dimensions, we approximate the transformation angle with $cos(\theta) \approx 1$ and $sin(\theta) \approx \theta$. For parts whose boundaries consist of line and arc segments, there are six types of distance constraints, with which we can model all contact and clearance specifications: 1. vertex-line; 2. edge-line; 3. arc-line; 4. arc-arc; 5. conditional edge-edge, and; 6. conditional arc-arc. The assembly in Figure 1 has all the six types of constraints. For each constraint type we write in parenthesis the number of degrees of freedom it constrains.

1. *vertex-line constraint* (1): this constraint is used to describe distance and angle relationships between two linear features. For example, in Figure 1, the flush relationship between the top of parts $P_6$ and $P_4$ is described with a vertex-line constraint between the left vertex of part $P_6$ and the line supporting the upper edge of $P_4$.

2. *edge-line constraint* (2): this constraint is used to describe a distance relationship between two edges. For example, in Figure 1, the contact relationship between the left edge of $P_6$ and the right edge of $P_4$ is described with an edge-line constraint. The edge-line constraints are expressed as two vertex-line constrains, one for each vertex of the edge.

3. *arc-line constraint* (1): this constraint is used to describe distance or contact relationships between an arc and an edge, such as the contact between parts $P_4$ and $P_5$ in Figure 1. An arc-line constraint entails a linear equation defining the distance of the arc center to the line supporting the edge as the required distance plus the arc radius. In our tolerancing model [Ostrovsky-Berman, 2004], circular arc segments are specified by the two endpoint vertices $v_1$, $v_2$, and either the radius $r$ or the arc angle $\alpha$, all of which are functions of the variational parameters.

4. *arc-arc constraint* (1): this constraint is used to describe contacts between two arcs, which are common in mechanisms with rotating parts, such as the contact between parts $P_3$ and $P_4$ in Figure 1.

5. *conditional edge-edge constraint* (1): this constraint is used to specify contacts between nominally parallel edges. In the nominal case, two parallel edges make contact in a line segment; in toleranced assemblies, the contact is usually a point. For example, consider parts $P_1$ and $P_5$ in Figure 1. The design intent is to make

contact between both pairs of edges: first with the horizontal edges (which are wider and therefore provide more stable contacts), then with the vertical edges. The former edges are termed the primary mating edges, and the latter two are termed the secondary mating edge and the conditional edge (the conditional edge is typically the shorter edge). The secondary contact is generally between the secondary mating edge and a vertex of the conditional edge, but which vertex makes contact depends on the instance of the parts. For example, if the vertical edge of $P_1$ leans to the right and $P_5$ is nominal, then the upper vertex of $P_5$ will be in contact; else, when the edge leans to the left, the lower vertex will be in contact.

6.  *conditional arc-arc constraint* (1): this constraint is used to specify contact between arcs of the same radius. The nominal contact between arcs of the same radius, as in parts $P_2$ and $P_3$ in Figure 1, is a circular arc, but when the geometries vary, there are three possible solutions: contact between the interiors of the circular arcs, and contact between an arc and either the first or second endpoint of the other arc, termed the conditional arc. Note that conditional edge-edge and arc-arc constraints specify that contact should occur between the two features, but do not specify which vertex is in contact.

The six types of relative position constraints can be used to capture the design intent and model all contact and clearance specifications. However, the designer must ensure that the constraints do not yield unsolvable equations. We propose the following modeling guidelines to prevent this situation:

1.  Features not participating in the constraints may overlap, even in the nominal solution. The designer should identify them and prevent the overlap by including them in the constraints, as is the case in Fig. 1 for the constraint between $P_4$ and $P_5$.
2.  Lines participating in three constraints must not all be parallel, as this results in dependent equations. For example, in Figure 1, $P_5$ cannot simultaneously contact both the left and right edges of the cavity of $P_1$.
3.  Arc-arc constraints may result in equations whose solutions are imaginary. To avoid this, the constraints must be feasible under small variations of the parts.

### 2.2.2 Computation of the sensitivity matrices

We now present a four-step algorithm to compute the vertex sensitivity matrices. The steps are: 1. model the relations between the two parts; 2. construct the corresponding system of equations; 3. compute the transformation relating the parts and its partial derivatives according to the variational parameters, and; 4. apply the transformation on the vertices of the free part to obtain the sensitivity matrix of each vertex.

In step 1, the relations are determined by three constraints of the six types described in Section 2.2.1. In step 2, the equations are constructed according to Section 2.2.1. For efficiency, we precompute and store the coefficients and their partial derivatives according to the variational parameters of parts $A$ and $B$. There is no need to evaluate the partial derivatives of the coefficient functions, because they can be obtained from the original vertex nominal values and partial derivatives, which were given as input to the algorithm.

                    Y. Ostrovsky-Berman and L. Joskowicz

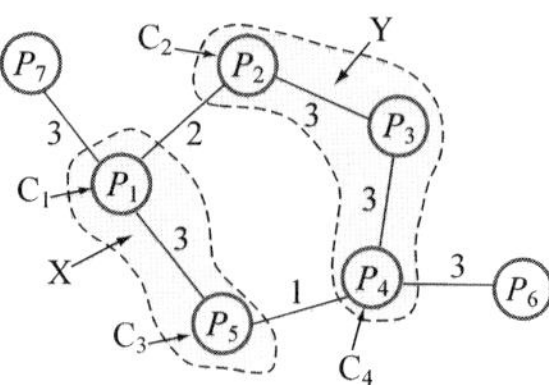

*Fig. 3. The assembly graph of the example in Fig. 1. The distances between part features that are connected by an edge are zero. Parts $C_1$, $C_3$ and $C_2$, $C_4$ of rigid bodies X and Y, respectively, contain features constraining the relative position constraints of the two bodies.*

In step 3, we solve the system of abstract equations constructed in step 2 by substituting the coefficients into general solution templates that we derived for the following three types of systems: 1. three linear equations; 2. two linear equations plus one quadratic equation; 3. two quadratic equations plus one linear equation. The system of equations resulting from $0 \leq z \leq 2$ arc-arc (quadratic) constraints and 3-$z$ linear constraints has $2^z$ solutions. However, only one of them corresponds to the nominal positions of the parts. The correct solution is identified by comparing the transformed vertices with the nominal vertex positions. For the partial derivatives of each of the template solutions, we derived corresponding templates, consisting of the coefficients and their partial derivatives. Since these were computed in step 2, the nominal solution $T = (t_x, t_Y, \theta)$ and its derivatives $\partial T/\partial p_j = (\partial t_x/\partial p_j, \partial t_y/\partial p_j, \partial \theta/\partial p_j)$ are computed with a constant number of elementary arithmetic operations.

In step 4, we use the transformation derivatives to compute the sensitivity matrices of the vertices of $B$. Each vertex $u \in B$ undergoes the transformation $T$ in order to satisfy the relations.

When one of the constraints is conditional, the transformation $T$, which is correct for all instances of the assembly, is computed as follows. First, we solve the system of equations once for each of the cases (two or three solutions), and denote the resulting transformations $T_i = (t_{ix}, t_{iy}, \theta_i)$. An infinitesimal change in a single parameter $p_j$ either results in one of $T_i$ being the correct solution, or leaves all solutions correct. In the latter case, $\partial T/\partial p_j = \partial T_1/\partial p_j = \partial T_2/\partial p_j = \partial T_3/\partial p_j$. In the former case, we determine which of the solutions is correct (checking distance relations) for an infinitesimal increase and decrease of $p_j$. We then compute the left-hand and right-hand derivatives: $\partial T^+/\partial p_j = \partial T_i^+/\partial p_j$, and $\partial T^-/\partial p_j = \partial T_i^-/\partial p_j$, where $T_i^+$ ($T_i^-$) is the correct solution for an increase (decrease) in $p_j$, and $\partial T^+/\partial p_j$ ($\partial T^-/\partial p_j$) is the right-hand (left-hand) derivative of a $T_i$.

### 2.3. Relative positions of parts in an assembly

We now describe how to model the relative position of parts in the entire assembly. Previous work by Latombe et al. [Latombe et al., 1997] introduces the *relation graph* to describe the relative position constraints between nominal parts with two degrees of freedom each. We extend this graph to include cycles and support parts with general tolerances and three degrees of freedom, and call it the assembly graph.

---

**Input:** Assembly graph, toleranced parts models
1.  Find a path in the assembly graph between $P_i$ and $P_j$.
2.  Iterate on the path edges $e = (P_k, P_l)$ in order:
    **If** weight($e$) = 3 **then** compute transformation $T_{kl}$ positioning $P_l$ relative to $P_k$ (Section 2.2).
    **Else if** weight($e$) < 3 (cycle edge) **then**
    - i.   Find rigid bodies $X$ and $Y$ from graph cycle ($X$ contains $P_k$).
    - ii.  Identify parts with constrained features $C_1, C_2, C_3, C_4$ (as in Figure 3).
    - iii. Compute transformations positioning parts in $X$ relative to $P_i$.
    - iv.  Compute transformations positioning parts in $Y$ relative to $C_2$.
    - v.   Compute transformation $T_{XY}$ positioning $Y$ relative to $X$ according to constraints in $C_1, C_2, C_3, C_4$.
    - vi.  Continue path from the exit edge (if it exists)
3.  **For** each $u \in P_j$
    - a.  **For** each variational parameter $p_k$ in parts from $P_i$ to $P_j$
        - i.   Find the two transformations that depend on $p_k$.
        - ii.  Apply each transformation on $u$.
        - iii. Sum the derivatives of the previous step for the $k^{\text{th}}$ column of $S_u$.

**Output:** Sensitivity matrices of part vertices

*Table 1. Algorithm for computing the sensitivity matrix of $P_j$ relative to $P_i$*

Graph nodes correspond to parts and undirected edges correspond to constraints between parts. Edge weights are 1, 2, or 3, and indicate the number of degrees of freedom constrained between the two parts. The edge data structure holds additional information about each constraint, such as the feature names of parts $A$ and $B$, the value or parametric expression of the distance between these features, and the type of constraint. Figure 3 shows the assembly graph of the assembly in Figure 1.

The assembly specification is well-constrained if it is both complete and non-redundant. The specification is complete if for every assembly instance the relative position of all pairs can be determined from the constraints. It is non-redundant if the removal of any constraint results in incompleteness. Our framework only supports well-constrained assemblies. A necessary and sufficient condition for a well-constrained assembly of $N$ parts is that the sum of edge weights is $3(N - 1)$, and that for each cycle in the graph with $N_c$ nodes, the sum of weights is $3(N_c - 1)$ and there is exactly one edge of weight 2 and one edge of weight 1 (a cycle with three edges of weight 2 results in a non-linear system of six equations with no general solution). The above conditions are a special case of the Grübler equation for planar mechanisms [Erdman, 1997]. Well-constrained assembly graphs have two important properties:

1.  When two parts are connected by a chain of edges of weight 3, their relative position is determined link by link, where each link is solved as in Section 2.2. Such a chain of parts can be regarded as a single rigid part, because any rigid transformation on the parts as a group preserves the relation constraints.

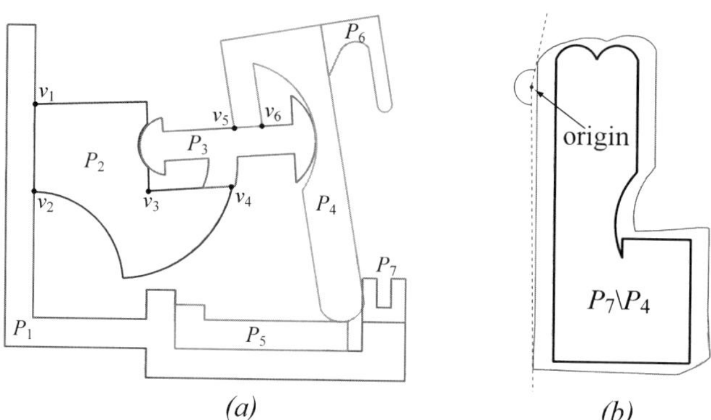



*Fig. 4. (a) Infeasible instance of the mechanism when vertices $v_1, v_2, v_3, v_4$ of $P_3$ and $v_5, v_6$ of $P_4$ vary in 1mm from their nominal positions. (b) The configuration space obstacle $P_7 \setminus P_4$ (thick curve), its tolerance envelope (thin curve), and the cone of blocked translation directions.*

2.  A cycle of $N_c$ parts has exactly $N_c$-2 edges of weight 3, one edge of weight 2, and one edge of weight 1. The last two edges divide the cycle into two non-intersecting sets of parts $X$ and $Y$ connected by the edges. The relative position between parts connected by edges of weight 1 or 2 cannot be determined because it is under-constrained, but the relative positions between the rigid bodies corresponding to $X$ and $Y$ is well constrained. Figure 3 shows the sets $X$, $Y$ for the assembly in Figure 1.

Table 1 shows the algorithm to compute sensitivity matrices of part $P_j$ relative to $P_i$. The algorithm computes the relative position transformations between pairs of parts on the path from $P_i$ to $P_j$, and from them computes the sensitivity matrices. Its complexity is $O(n_j q r_{ij})$, where $n_j$ is the complexity of $P_j$, $q$ is the maximal number of parameters affecting a single vertex, and $r_{ij}$ is the number of parts in the path from $P_i$ to $P_j$. Note that this result is a generalization of the result of [Cazals, 1997] for parts with two degrees of freedom, and since in their model the vertices are linear functions of the variation parameters, the approximation is in fact exact.

## 3.   APPLICATIONS AND EXAMPLE

The sensitivity matrix of a vertex in a toleranced assembly describes the effect of the parameter variations on the position of the vertex, relative to the chosen source part, or the datum. When used as input to the algorithm in [Ostrovsky-Berman, 2004], the resulting tolerance envelope bounds the area occupied by the part under all possible assembly variations. One very useful property of the sensitivity matrices is their additivity – it is possible to combine matrices of vertices with shared parameter dependency to obtain the correct combined sensitivity. Without respecting parameter dependencies, the stack-up analysis of feature tolerance zones is overly conservative, as it ignores parameters whose effect on the variability of two features cancel each other out. The additivity is especially important in the computation of configuration spaces (C-space) of toleranced assemblies [Sacks, 1998].

In the C-space approach for assembly analysis [Halperin et al., 2000, Inui et al., 1996, Latombe, 1991, Sacks, 1998], the space describing the degrees of freedom of a part or a group of parts is partitioned into free space and blocked space. For motion planning with limited translations, the C-space of a part $P_i$ is two dimensional, and the obstacle made by part $P_j$ is computed using the Minkowski difference of sets: $P_j \setminus P_i = \{v_j - v_i \mid v_i \in P_i, v_j \in P_j\}$. The outer boundary of the obstacle is obtained by first computing the boundary features (vertices and line and arc segments in our model) pairwise Minkowsky difference and then computing the outer cell of the resulting arrangement of curves. This outer bounds the C-space obstacle of the parts.

When the parts are toleranced, we can first compute each part's tolerance envelope, and then compute the configuration space obstacle using the envelopes. However, this analysis is overly conservative because it ignores parameter dependencies. The correct method is to construct the *C-space envelopes* for pairs of parts as follows. First, compute the pairwise feature Minkowski difference as before. The vertices of segments bounding the Minkowski difference have explicit representations as functions of the vertices of $P_i$ and $P_j$. Thus it is possible to compute their nominal positions and their partial derivatives. Next, we compute the outer tolerance envelopes of the pairwise features, and obtain an arrangement of elementary features, whose outer cell is the C-space envelope of the obstacle. When C-space envelopes of obstacles are used in assembly analysis methods instead of the nominal obstacles, the analysis accounts for parameter dependency in the worst case variation of the parts in the assembly.

We have implemented the relative position computation and assembly graph data structure using MATLAB, and ran it on several examples of assemblies. To illustrate, we describe the results we obtained on the assembly in Figure 1. The input consists of $N=7$ parts, $m=58$ variational parameters. The maximum part complexity is $n=14$, maximal path length $r=7$ (from $P_6$ to $P_7$ and vice versa, including cycle parts), and maximal number of local parameters $q=3$. The CPU time to compute the sensitivity matrix of all the parts relative to $P1$ was 1.37 seconds on a Pentium IV 2.4GHz with 512 MB RAM. Figure 4(a) shows an instance of the mechanism when only six of the variational parameters are allowed to vary within $\pm 1mm$ tolerance intervals (about 1% from the average feature length in the assembly). Even though the nominal horizontal clearance between $P_4$ and $P_7$ is of $20mm$ and four of the vertices translate vertically, the instance represents an infeasible assembly because the parts overlap.

Even with smaller tolerance intervals, tolerancing significantly affects the assembly. Figure 4(b) shows part $P_7$ as the configurations space obstacle of $P_4$, without tolerances, and when each variational parameter has a $\pm 0.3mm$ tolerance interval. The obstacle and the origin determine the directions in which $P_4$ is free to move without colliding with $P_7$ (shown in Figure 4(b) as part of the unit circle). Thus, unlike the nominal assembly, there are instances of the assembly in which of $P_7$ blocks the directions separating $P_4$ from $P_3$, and therefore $P_7$ must be removed first in the disassembly sequence.

## 4. CONCLUSION

We have presented a framework for worst case analysis of toleranced planar assemblies. The framework is more general than existing ones in terms of the geometry of parts (line and arc segment boundaries) and the tolerancing model used. The sensitivity matrices of each part vertex that are computed by our algorithm can used to compute the tolerance envelopes bounding the areas occupied by the parts under all possible assembly instances. The envelopes provide a characterization of geometric uncertainty that is more accurate than those produced by Monte Carlo methods and is useful in assembly planning and mechanism design. Directions for future work include modeling mechanisms of interest to industry, extension to three-dimensional parts, starting with polyhedra, and optimal part placement with respect to objective functions [Li, 2001, Sodhi, 1994] instead of pre-determined contacts.

## REFERENCES

[**Cazals, 1997**] Cazals, F. and Latombe, J.-C., Effect of tolerancing on the relative positions of parts in an assembly, *IEEE Int. Conf. on Rob. and Automation*, 1997.

[**Erdman, 1997**] Erdman, A.G. and Sandor, G.N., Mechanism Design: Analysis and Synthesis, Volume 1. Third edition, PrenticeHall 1997.

[**Halperin et al., 2000**] Halperin, D., Latombe, J.-C. and Wilson, R.H., *A general framework for assembly planning: The motion space approach.*, *Algorithmica*, Vol. 26, pp 577-601, 2000.

[**Inui et al., 1996**] Inui, M., Miura, M. and Kimura, F., Positioning conditions of parts with tolerances in an assembly, *IEEE Int. Conf. on Robotics and Automation*, 1996.

[**Latombe, 1991**] Latombe, J.-C., *Robot motion planning*, Kluwer Ac. Publishers, 1991.

[**Latombe et al., 1997**] Latombe, J.-C., Wilson, R.H., Cazals, F., Assembly sequencing with toleranced parts, *Computer-aided Design*, Vol. 29(2), pp 159-174, 1997.

[**Li, 2001**] Li, B. and Roy, U., Relative positioning of toleranced polyhedral parts in an assembly, *IIE Transactions*, Vol. 33(4), pp 323-336, 2001.

[**Ostrovsky-Berman, 2004**] Ostrovsky-Berman, Y. and Joskowicz, L., Tolerance envelopes of planar mechanical parts, *9th ACM Symposium on Solid Modeling and Applications,* pp 135-143, 2004.

[**Sacks, 1998**] Sacks, E. and Joskowicz L., Parametric tolerance analysis of part contacts in general planar assemblies, *CAD*, Vol. 30(9), pp 707-714, 1998.

[**Sodhi, 1994**] Sodhi, R. and Turner, J.U., Relative positioning of variational part models for design analysis, *CAD*, Vol. 26(5), pp 366-378, 1994.

[**Turner, 1990**] Turner, J.U., Relative positioning of parts in assemblies using mathematical programming, *CAD*, Vol. 22(7), pp 394-400, 1990.

# Geometrical Variations Management in a Multi-Disciplinary Environment with the Jacobian-Torsor Model


**A. Desrochers**
*Université de Sherbrooke,*
*2500, boul. de l'Université*
*J1K 2R1, Québec, Canada*
*Alain.Desrochers@USherbrooke.ca*



**Abstract:** In recent years, several computational tools for tolerance specification and analysis have emerged to help the designer in the specification of geometrical product requirements. This paper proposes to broaden the scope of such a tool, called the Jacobian-Torsor model, so as to embrace a wider range of variations and uncertainties. In that perspective, the paper presents an original typology of geometrical variations and uncertainties. These variations are classified according to their own properties and features. Among the features that must be dealt with regarding variations and uncertainties are the following: cumulative versus independent and required versus random or predictable. In turn, the variations or uncertainties must themselves be identified through either, standards, measurements or models. Finally, in terms of simulation, these variations and uncertainties can be addressed through deterministic or stochastic approaches, or both, with the Jacobian-Torsor model.
**Keywords**: Jacobian-torsor, uncertainties, tolerance, multi-disciplinary, PLM.


## 1. INTRODUCTION

For some years now, research efforts have been devoted to representing tolerances and assessing their effects on clearances within a given mechanism. However, tolerances are simply the expression of uncertainties at the design stages of the product development process. They provide bounds on dimensional requirements against which the final product must be checked.

The purpose of this paper is to extend the same principles to other types of uncertainties, and subsequent phases of the product lifecycle. To this end, a typology of uncertainties at different stages of product maturity will be proposed. Details will also be given as to how these can be integrated in the Jacobian-torsor model for analysis purposes.

The Jacobian-torsor model is presented in details in [Desrochers et al., 2003]. However, its main features are presented in one following sub-section, to help understand how it could be adapted to suit the need of the proposed typology in a




*J.K. Davidson (ed.), Models for Computer Aided Tolerancing in Design and Manufacturing, 75–84.*
© 2007 *Springer.*

Product Lifecycle Management (PLM) perspective. Before that however, a brief survey of the literature on the subject will be exposed.

### 1.1. Litterature review

Little work as been done on the relation between uncertainties, multi-disciplinary environment and Product Lifecyle Management (PLM). Many work and commercially available software are focusing on the design phase and the representation of their uncertainties as tolerances on parts and clearances on assemblies. For that purpose, they propose deterministic as well as stochastic approaches to address analysis (computing clearances from tolerances) and synthesis (tolerance allocation for a given clearance) problems [Hong et al., 2002].

The other phases of the product lifecycle have also been separately investigated. For instance, many researchers have been addressing the issue of uncertainty modelling in manufacturing [Legoff et al., 2001]. The domains of metrology and assembly are also rich in works that address the topic of uncertainties and variations. At the operation phase of the product lifecycle spectrum, researchers have been modelling the effects of loads on assemblies [Merkley, 1998][Samper & al., 1998].

The topic of geometrical variations management has started being studied in recent years. Linares [Linares et al., 2003], for instance has examined the effects of specification uncertainties on tolerance zone using the concept of Statistical Confidence Boundaries. Marguet [Marguet et al., 2001], has exposed the relation from requirements to key characteristics and to specifications. However, these works remain closely tied to the design phase of the product lifecycle.

Hence, there has not been much efforts to propose a coherent information stream and model linking design, production and operation uncertainties and variations. This paper will attempt to do just so in a PLM and multi-disciplinary perspective. This would be the main originality of the proposed work.

### 1.2. The Jacobian-torsor model

The Jacobian-torsor model is basically a matrix equation, relating Functional Requirements (FR) or clearances at the assembly level to Functional Elements (FE) or tolerances at the part level. This is done using a simple matrix product with a Jacobian (J) expressing the geometrical relation of the FE with respect to the FR (equation (1)).

$$[FR] = [J][FEs] \tag{1}$$

In mathematical terms, FE and FR are represented as torsors, with their respective components being bounded by intervals, according to the size of the corresponding uncertainty zone (tolerance or clearance). This is shown in equation (2), where the terms of equation (1) are expressed with the interval formulation.

Additionally, Functional Elements can relate surfaces belonging either to the same part or to two distinct parts establishing a contact or a play (fit or gap). In the first case, the FE is said to represent an *internal pair*, whereas in the second instance, it expresses a *kinematic pair*. In both cases, however, the mathematical representation is based on interval bounded torsor components.

$$
\begin{bmatrix} [\underline{u},\overline{u}] \\ [\underline{v},\overline{v}] \\ [\underline{w},\overline{w}] \\ [\underline{\alpha},\overline{\alpha}] \\ [\underline{\beta},\overline{\beta}] \\ [\underline{\delta},\overline{\delta}] \end{bmatrix}_{FR} = \left[ \left[ J_1 J_2 J_3 J_4 J_5 J_6 \right]_{FE1} \cdots \left[ J_1 J_2 J_3 J_4 J_5 J_6 \right]_{FEN} \right] \cdot \begin{bmatrix} \begin{bmatrix} [\underline{u},\overline{u}] \\ [\underline{v},\overline{v}] \\ [\underline{w},\overline{w}] \\ [\underline{\alpha},\overline{\alpha}] \\ [\underline{\beta},\overline{\beta}] \\ [\underline{\delta},\overline{\delta}] \end{bmatrix}_{FE1} \\ \cdots \\ \begin{bmatrix} [\underline{u},\overline{u}] \\ [\underline{v},\overline{v}] \\ [\underline{w},\overline{w}] \\ [\underline{\alpha},\overline{\alpha}] \\ [\underline{\beta},\overline{\beta}] \\ [\underline{\delta},\overline{\delta}] \end{bmatrix}_{FEN} \end{bmatrix} \tag{2}
$$

Where:

$[FR] = \begin{bmatrix} [\underline{u},\overline{u}] \\ [\underline{v},\overline{v}] \\ [\underline{w},\overline{w}] \\ [\underline{\alpha},\overline{\alpha}] \\ [\underline{\beta},\overline{\beta}] \\ [\underline{\delta},\overline{\delta}] \end{bmatrix}_{FR}$, $[FEi] = \begin{bmatrix} [\underline{u},\overline{u}] \\ [\underline{v},\overline{v}] \\ [\underline{w},\overline{w}] \\ [\underline{\alpha},\overline{\alpha}] \\ [\underline{\beta},\overline{\beta}] \\ [\underline{\delta},\overline{\delta}] \end{bmatrix}_{FEi}$ : Small displacements torsors associated to some functional requirement (play, gap, clearance) represented as a [FR] vector or some Functional Element uncertainties (tolerance, kinematic link, ….) also represented as [FE] vectors ;

$\left[ J_1 J_2 J_3 J_4 J_5 J_6 \right]_{FEi}$ : Jacobian matrix expressing a geometrical relation between a [FR] vector and some corresponding [FE] vector;

N : Number of torsors in a kinematic chain;

$\underline{u},\underline{v},\underline{w},\underline{\alpha},\underline{\beta},\underline{\delta}$ : Lower limit of $u,v,w,\alpha,\beta,\delta$ ;

$\overline{u},\overline{v},\overline{w},\overline{\alpha},\overline{\beta},\overline{\delta}$ : Upper limit of $u,v,w,\alpha,\beta,\delta$ .

The details on the construction of the Jacobian will not be presented in this paper and the reader will be referred to [Desrochers et al., 2003] for that purpose.

In the following section, the proposed typology will be now presented and its classes and features explained.

## 2. UNCERTAINTY TYPOLOGY

The proposed classification (figure 1) has a tree like structure, suitable with object–oriented programming. It is also comprehensive in many ways. Indeed, it accounts for the specific nature of each type of uncertainty or geometrical variation. As it will be shown, these can in turn be represented with the Jacobian-torsor model by FR or FE vectors, allowing the computation of clearances and gap in assemblies.

The following sub sections present the details and features of this typology structure.

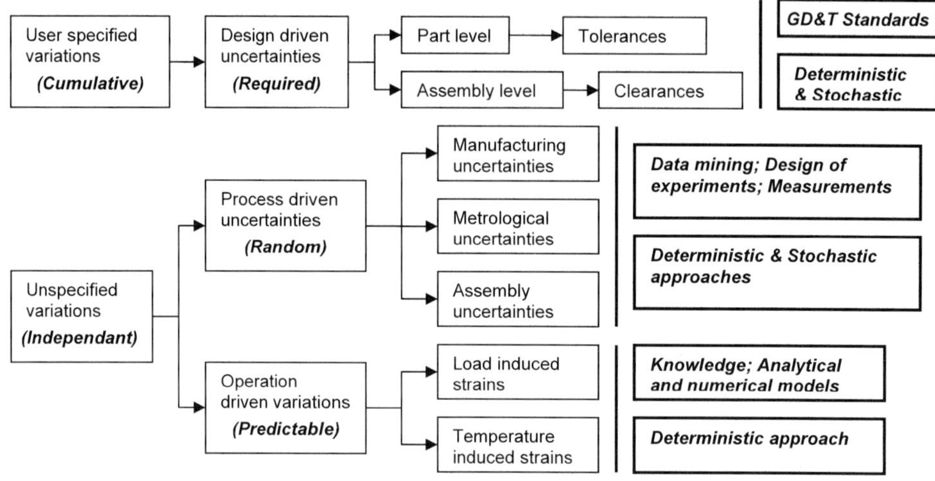


*Figure 1; Uncertainty typology structure.*

## 2.1. Specified vs. unspecified geometrical variations

The first level of the typology distinguishes between user specified and unspecified geometrical variations. The user specified class, as its name plainly implies, contains uncertainties in the form of tolerances and clearances which are typically specified on a drawing by the designer and which become ultimately part of the product requirement. Specified variations are generally cumulative. Indeed, according to the standards, form tolerances are included in orientation tolerance zones, themselves being contained within location tolerance zones. The envelope principle and the Maximum Material Condition also support the same idea; additional specifications are not likely to change the overall size of the uncertainty zone but rather to add additional layers within it.

Unspecified variations obviously capture all other uncertainties and variations that do not fall in the first category outlined above. As opposed to the specified geometrical variations, they are generally independent. In other words, every unspecified variation will participate to a global uncertainty zone which will quite simply be the sum of all contributors for a given surface.

## 2.2. Design driven uncertainties

They are the only member of the user specified class of uncertainties, as described in the preceding section. They are labelled as *"Required"* in figure 1 because they represent the designer's expectations in terms of part and assembly precision. For this reason, they can be considered as *requirements* against which the part must be checked.

Practically speaking, these requirements are defined at two levels; assembly and part level. At the assembly level, clearances and fit are specified. In the Jacobian-torsor model, fits are modelled as kinematic pairs, i.e. six component torsors expressing the amount of play or interference between two cylindrical features on different parts. The expression of fits requirements by the designer follows a standard classification relating the amount of clearance or interference to predefined classes such as clearance fits, transition fits and interference fits.

In the case of clearances, they are also labelled as Functional Requirements and are the result of the analysis process. Again, they are expressed as six components torsors but appear on the left side member of the matrix equation in the Jacobian-torsor model (equation 2). They are generally specified on the assembly drawings between two planar faces on different parts and indicate the minimum and maximum values of the admissible play.

Conversely, at the part level, the designer specifies tolerances. These are naturally converted into six component torsors expressing the corresponding tolerance zones as internal pairs and labelled as "Functional Elements" in the Jacobian-torsor model. The expression of tolerance requirements is naturally governed by standards such as those of the ASME or ISO.

Overall, design driven uncertainties define, in the virtual world, the characteristics of the ideal product, i.e. one that would meet the expectations of the designer at the geometrical level. They can be used in deterministic or stochastic computations with the Jacobian-torsor model.

### 2.3. Process driven uncertainties

Process driven uncertainties are part of the unspecified variations. They are unspecified in the sense that they are simply not part of the designer's product requirements. They originate from the "real world" as they are being generated by the various processes used in the production and verification of the final physical product.

Process uncertainties are considered as random (as opposed to "required" in the preceding category) because they cannot be predicted and depend on the process itself. Their values are rather obtained through data mining in the specification data sheets provided by the manufacturers of the process equipments or they can also be determined through quality control measurements on the products at the outcome.

These uncertainties include manufacturing uncertainties on part, assembly uncertainties related to fastening techniques and metrological uncertainties. In short, every step of the production cycle introduces some uncertainties that add up and increase the dimensional spectrum of the final product.

More specifically, manufacturing uncertainties can be modelled as six components torsors in the form of functional elements (FE) originating from internal pairs (on the same part) in the Jacobian-torsor model. Assembly uncertainties can also be expressed as torsor components but in the form of kinematic pairs in functional elements (i.e. between two parts). Finally, metrological uncertainties will generate internal pair FE torsors, as measurements are generally taken on isolated parts.

As for design driven variations, process driven uncertainties can lead to deterministic or stochastic analysis. However, owing to the random nature of these uncertainties, they are particularly well suited for statistical approaches.

### 2.4.  Operation driven variations

Operation driven variations reflect mainly load and temperature induced geometrical variations on parts and assemblies. The term "variation" is used here rather than "uncertainties", because load and temperature strains are predictable as they can be computed using material strength equations or finite element models. For the same reason they are naturally better suited for deterministic analysis.

In the case of load strains, these variations or deviations can be translated in terms of torsor components and intervals, according to corresponding six degrees of freedom load values. However, the effects of operation driven variations may also be realistically assimilated to dimensional shifts; displacing the corresponding process uncertainty zones in one direction or another depending on load values.

Finally, operation variations address the issue of flexible parts as they encompass load induced strains.

Temperature variations are slightly different, as their effects are non directional. Indeed, temperature changes act more as scale factors, increasing or decreasing feature sizes proportionally in all directions. Nonetheless, they can too be represented as torsors, the difference being that the six degrees of freedom interval values or dimensional shift components will not be independent of each other.

## 3.  VARIATION MANAGEMENT IN A PLM PERSPECTIVE

The preceding classification can naturally be ordered sequentially, according to the various phases of the product lifecycle. This is illustrated in figure 2, where the interactions between the various phases, in terms of variation and uncertainty management, are shown. Most interesting is the fact that these can be combined and compared therefore leading to the validation of both product design and processes.

From a computational point of view, the preceding section as shown that all the uncertainty sources can be represented by six components torsors, making them compatible for computations with the Jacobian-torsor model. Such computations will be performed in the scope of the validation tasks, as presented in the next sub-sections.

### 3.1.  Process validation

Regarding process validation, the procedure involves the identification of Geometric Dimensioning & Tolerancing (GD & T) specifications on the various parts. These specifications must obviously reflect the clearance requirements at the assembly level. From there, at the modelling stage, tolerance zones for parts and sub-assemblies (if required) are defined. Meanwhile, on the production side, manufacturing and metrology uncertainties must be measured or estimated and combined to generate part uncertainty zones. If appropriate, fastening uncertainties may also be taken into account. In this instance however, it becomes necessary to distribute the global uncertainty zone at the assembly interface among the external surfaces of the sub-asssembly. Fastening uncertainties must be accounted for, when the corresponding processes introduce additional variations that could not be modelled at the part level. From there, a comparison can be made between the tolerance zones on parts or sub-assemblies with their corresponding production phase uncertainty zones. Finally, if the latter is included in the former, than the process can be validated.

### 3.2.  Product design validation

For product design validation, the comparison is made between the design phase and the operation phase. At the design stage, the designer is expected to specify clearance values that must be met by the mechanism in operation, i.e. taking into account load and temperature conditions. Their effects on part geometry must in turn be computed or simulated so that deflection or strain values can be obtained for both loads and temperature. Corresponding total part deviations can then be used with the Jacobian-torsor model to predict clearances in operation. Product design validation is then completed by comparing the predicted clearance in operation with the corresponding design requirement.

The link between product and process validation is at the identification stage of the design phase (fat arrow in figure 2). Indeed, this is where clearance values at assembly (for process validation) and clearance values in operation (for product validation) are defined. Obviously, these two parameters are not independent. The next section addresses specifically the issue by proposing a "PLM synthesis procedure" enabling the designer to estimate corresponding assembly clearances from functional requirements of the product in operation.

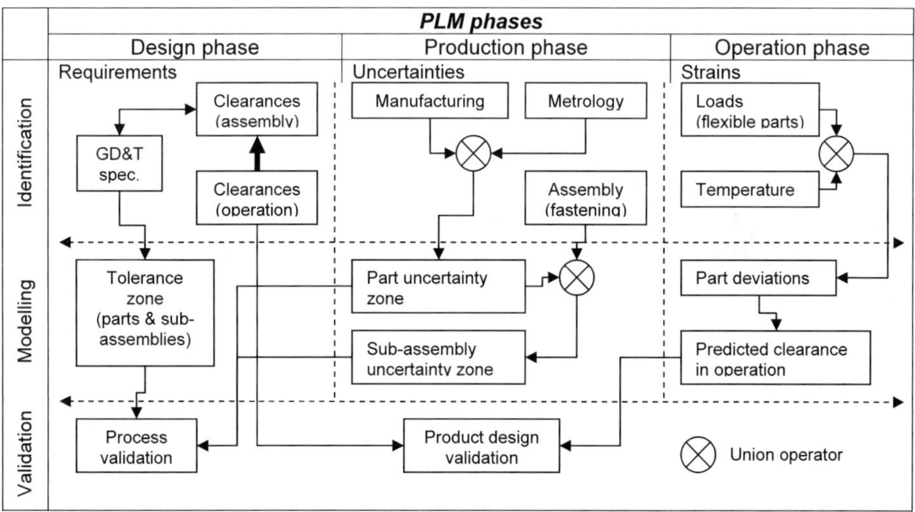


*Figure 2; Product and processes validation in a PLM perspective.*

## 4.  PLM TOLERANCE SYNTHESIS PROCEDURE

As mentioned in the preceding section, one last important link in the proposed architecture is the one which establishes the relation between design requirements in operation and design requirements at the production phase.  The proposed approach to address this issue is shown in figure 3 as a "PLM tolerance synthesis procedure".

### 4.1.  Description of the proposed methodology

The method uses Functional Requirements (clearances) in operation as a starting point. With this information, an uncertainty allocation process (synthesis) is performed, yielding a set of part uncertainties *at operation conditions*.  At the same time, computations are performed, using analytical or finite element models, to estimate the strains on each part in operation.  Dimension values are also tuned, so that the corresponding chains of dimensions yield the nominal clearance, as specified by the requirements *at operation conditions*.

The goal of the following step in the procedure is to convert geometrical data from operation conditions to assembly conditions.  For that purpose, the values of the strains are being utilized to compensate the corresponding dimensions on the parts so as to reflect their original values *at assembly*.  For instance, if loads and temperature increase the length of a part, then the corresponding strain should be subtracted from the nominal length of the part *in operation*, therefore providing the original dimension *at assembly*. Part uncertainties need not to be compensated since such compensations would be equivalent to second order variations (% elongation applied on size of uncertainty zone).  It is worth noting that torsor based uncertainty models also use first order linearization for their computations.

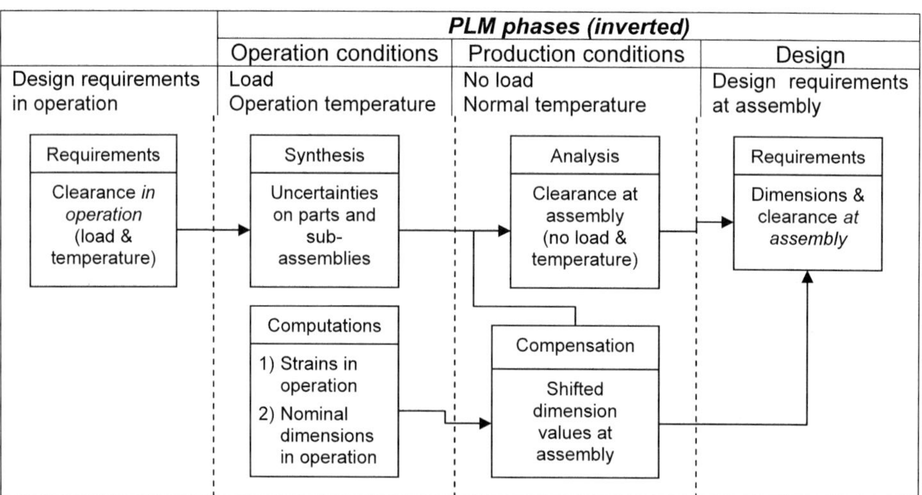

*Figure 3; Product requirements from operation to assembly conditions.*

From there, compensated dimensions and uncertainties are fed to an analysis model, such as the Jacobian-torsor, therefore providing a corresponding clearance value *at assembly*. The uncertainties can then be converted into part specifications or tolerances so as to formalize the design intent and the requirements for production.

Lastly, it should be noted that the whole procedure remains compatible and coherent with the Jacobian-torsor model, since part uncertainties may be expressed in terms of small displacement torsor components, just like tolerances.

## 5. CONCLUSION

The attempt of this paper was to broaden the scope of current tolerancing research from two distinct but complementary points of view. The first was by adopting a multidisciplinary approach, embracing all types of uncertainties and variations in a concurrent engineering perspective. For that purpose, a novel typology of variations and uncertainties was proposed, along with its relation to the Jacobian-torsor model.

The second was to rearrange the proposed typology sequentially, according to the Product Lifecycle Management (PLM) timeline. Embedded in the proposed architecture was also a workflow for the validation of both product design and processes. This included a specific methodology to translate product requirements in operation into design requirements for production, therefore closing the loop between design and manufacturing. Again, the procedures were presented along with explanations on their compatibility with the torsor representation.

However, it should be noted that the work presented is still in its preliminary form as an actual implementation with the Jacobian-torsor model has not yet been completed. Nevertheless, the author remains confident that the actual Jacobian-torsor prototype can be adapted to suit the needs of the presented work in a relatively short period of time. This assumption is based on the fact that all uncertainties and variations presented in this paper can be modelled as torsors, either in the form of Functional Elements or Functional Requirements. From there, the Jacobian-torsor model can handle deterministic or stochastic approaches, in analysis and even synthesis mode. The biggest remaining challenge then becomes the construction of a proper user interface, reflecting concurrent engineering practices in a PLM perspective.

In the longer term, this work may introduce the general idea of associating uncertainty data to the nominal geometry, across disciplines and throughout the product lifecycle, therefore leading to more realistic product models.

## ACKNOWLEDGMENTS

The author would like to express special thanks to Walid Ghie, post doctoral fellow, for his help in gathering some of the bibliographical references used in this paper.

REFERENCES

**[Desrochers et al., 2003]** Desrochers, A.; Ghie, W.; Laperrière, L.; "Application of a Unified Jacobian-Torsor Model for Tolerance Analysis"; *Special Issue on Computing Technologies to Support Geometric Dimensioning & Tolerancing (GD & T), Journal of Computing and Information Science in Engineering*, vol 3, no1, March 2003, pp. 2-14

**[Hong et al., 2002]** Hong, Y.S.; Chang, T.C.; "A comprehensive review of tolerancing research"; *International Journal of Production Research*, vol 40, no11, pp. 2425-2459

**[Legoff et al., 2001]** Legoff, O.; Villeneuve, F.; "Three-dimensional geometrical tolerancing: quantification of machining defects"; In: *Proceedings of the 7th International Seminar on Computer Aided Tolerancing*, pp. 201-212; Cachan, France, 2001

**[Linares et al., 2003]** Linares, J.M.; Bachmann, J.; Sprauel, J.M.; Bourdet, P.; "Propagation of specification uncertainties in tolerancing"; In: *Proceedings of the 8th International Seminar on Computer Aided Tolerancing*, pp. 301-310; Charlotte, USA, 2003

**[Marguet et al., 2001]** Marguet, B.; Mathieu, L.; "Method for Geometric Variation Management from Key Characteristics to Specification"; In: *Proceedings of the 7th International Seminar on Computer Aided Tolerancing*, pp. 121-130; Cachan, France, 2001

**[Merkley, 1998]** Merkley, K.; "Tolerance Analysis of Compliant Assemblies"; *Ph.D. thesis, Brigham Young Unversity*; 1998

**[Samper et al., 1998]** Samper, S.; Giordano, M.; "Taking into account elastic displacements in 3D tolerancing"; *Journal of Materials Processing Technology*, 78 (1998), pp. 156-162

# Tolerance Analysis and Synthesis by Means of Deviation Domains, Axi-Symmetric Cases

**M. Giordano, S. Samper, J. P. Petit**
*LMECA (Laboratoire de Mécanique Appliquée)*
*Ecole Supérieure d'Ingénieurs d'Annecy,*
*Université de Savoie, BP 806, 74016 Annecy cedex 16, France*
*Max. Giordano@esia.univ-savoie.fr*

**Abstract:** The small displacement torsors are generally used for the representation of the geometrical deviations. The standardised tolerances can then be translated by a set of inequalities between the components of a deviation torsor. In the case of cylindrical tolerance zones, and surfaces of revolution, the inequalities are quadratic and the stack up tolerances are more difficult to calculate. But the axi-symmetric case makes it possible to reduce the space to three dimensions at the maximum instead of six in the general case. Topological operations like the Minkowski sum to carry out the domains for stack up analyse of tolerances are then reduced to operations on polyhedrons. Moreover, we propose a way of taking into account the size tolerances. The first presented application relates to metro-logic inspection for a specification with maximum material condition on both the toleranced surface and the datum. The second example makes it possible to determine the deviation between two surfaces belonging to two different parts after mating them by two contact features.

**Keywords**: tolerance analysis, tolerance synthesis, axi-symmetry, domains.

## 1.  INTRODUCTION

The small displacement torsor is used for the modelisation of the geometrical deviation of mechanical parts [Bourdet et al., 1996], [Desrochers et al., 2003].
The components of a small displacement torsor can also be seen as differential parameters for orientation and location. First, it is necessary to precise what is the deviation torsor.

### 1.1. Deviation torsor

A position or orientation tolerance can be translated by the limitations on the components of the deviation torsor. The precise definition of the deviation torsor can be achieved on the following form: a theoretical element is associated to the real toleranced feature (surface or axis). A frame is attached to this theoretical feature. We call it



*J.K. Davidson (ed.), Models for Computer Aided Tolerancing in Design and Manufacturing, 85–94.*
© 2007 *Springer.*

toleranced frame. A datum frame is built from one or more other features. The displacement from the datum frame to the toleranced frame can be seen as the sum of a displacement that define the position and orientation of the toleranced feature compared to the datum frame, and an other displacement that characterise the deviation. The deviation torsor is the first order approximation of this deviation displacement.

The datum frame can be built from only one elementary feature like a plane or the axis of a cylinder, but it is also possible to build a datum frame from different features: common datum or ordered datum system. More generally, the different possibilities allowed by the standard can be modelised by the deviation torsor model.

The datum frame and the frame attached to the toleranced feature can have some degrees of freedom. The components of the deviation torsor corresponding to these degrees of freedom cannot be measured and are indefinite values. They can also be considered as zero because there is no deviation for this degrees of freedom. In the same way a size tolerance between two parallel planes in the nominal configuration can be modelised thank to a deviation torsor.

The deviation torsor can be expressed by two vectors for a given point O. The vectorial expression of the torsor is noted $\mathbf{E} = (\delta\theta, \delta\mathbf{O})$ where $\delta\theta$ is the vector for the infinitesimal rotation and $\delta\mathbf{O}$ for the linear displacement of the point O.

For a given frame (i) constituted by a point $O_i$ and a vectorial base $\mathbf{x}_i$, $\mathbf{y}_i$, $\mathbf{z}_i$, the torsor can be written under a scalar form $E_i=(rx_i, ry_i, rz_i, tx_i, ty_i, tz_i)$ where the three first terms are the components of the rotation vector and the others are the components of the displacement of the point $O_i$.

The great advantage of the torsors for the displacement modelisation is due to the fact that  the set of torsors is a vectorial space (dimension 6). The composition of small displacements is obtained by a sum of torsors. This operation is commutative and simple to compute, while the composition of two displacements in the general case (great displacements) is a non commutative operation. For the operations of tolerance transfer or stack up tolerances of assembled parts, the deviations have to be composed. The sum of torsors allows to modelise these operations.

### 1.2. Deviation and clearance domain

A dimensional or geometrical tolerance compatible with the ISO or ANSI standard make it possible to limit the deviation torsor. Considering that any point of the theoretical feature must be inside the tolerance zone, we obtain inequalities on the components of the deviation torsor.

In the configuration space of dimension 6, a point corresponds to a particular value of the torsor. Therefore, the set of the deviation torsor components according to the tolerance is a domain in this configuration space. The boundary of this domain is a hyper-surface. The equation of these hyper-surfaces on an implicit form are obtained by replacing the inequalities by equalities.

The use of the configuration space to modelise the tolerances has been used by different authors although they are often introduced with different forms.

Desrochers use the screw parameters or torsor components of small displacements. The position and orientation of a feature is describe by these parameters [Desrochers et al., 2003].

Sacks and Jokowicz, [Sacks and Joskowicz, 1998] describe a tolerance analysis method that computes tolerance zones of parts in configuration space. But the method is limited to planar kinematical systems.

Mujezinovic and Davidson use Tolerance maps, a convex volume of points corresponding to different possible positions of toleranced feature. The study is limited to plane faces [Mujezinovic et al., 2004] and to line or axis [Davidson and Shah, 2002]. The areal coordinate are introduce to describe any configuration of the feature in its tolerance zone as a linear combination of particular configurations. But in fact areal coordinate for position and orientation are the components of small displacement torsor.

The proposed model allows to translate the tolerance zone into a domain in the configuration space those mathematic representation is a set of inequalities on differential parameters. This model enables stack up analysis in an assembly in the worst case. It is the only model in my knowledge that permits to do this on a systematic method even if the computation are complex in the general case [Petit, 2004].

The clearance in a joint between two parts, can also be modelised by a small displacement torsor called clearance torsor. A frame is attached to each part of the joint so that the frames coincide for a particular configuration. For example for a cylindrical joint the particular configuration is obtained when the two axes are in coincidence. The clearance torsor  represent the relative displacement of the frames permitted by the clearance in the joint. The clearance domain is also defined in the configuration space. The contact conditions give the inequalities associated to this domain.

The torsor components depends on the frame chosen to express the torsor on a scalar form. We call it the projection frame. It must not be confused with the datum frame used to define the torsor itself. Therefore, the domain depends also of the projection frame. In particular cases, the domain has some characteristics of symmetry, so that the domain is invariant for particular changes of the projection frame. The case of a symmetry of revolution for the geometry and the tolerances is investigated in detail in this paper. But  previously, the general case of  change the projection frame is considered.

### 1.3. Change the projection frame for a torsor

The frame (0) is constituted by the point $O_0$ and a vectorial base $x_0$, $y_0$, $z_0$. The scalar representation of a torsor E in this frame is noted $E_0 = (R_0, T_0)$, where $R_0 = (rx_0, ry_0, rz_0)^T$ is the colon matrix of the co-ordinate of the small rotation vector in the vectorial base (0).

The frame (1) is constituted by the point $O_1$ and a vectorial base $x_1$, $y_1$, $z_1$. The transfer matrix from the base (0) to (1) is noted $P_{01}$, and $U_{01}$ is the colon matrix of the co-ordinates of the point $O_1$ in the frame (0).

We note $E_1 = (R_1, T_1)$ the components of the same torsor E but defined in the projection base (1). Then the following matrix expressions give the new components of the torsor:

$$R_1 = P_{10} R_0 \qquad \text{and} \qquad T_1 = P_{10} T_0 + \hat{U}_{10} R_0 \tag{1}$$

With $P_{10} = P_{01}^{-1} = P_{01}^{T}$ and $\hat{U}_{10}$ is the anti-symmetric matrices of the cross-product, associated to the colon matrix: $U_{10} = - P_{10}U_{01}$.

$$\text{If} \quad U_{01} = ( x, y, z ) \qquad \text{then} \qquad \hat{U}_{10} = \begin{bmatrix} 0 & -z & y \\ z & 0 & -x \\ -y & x & 0 \end{bmatrix}$$

## 2.   AXI-SYMMETRIC CASES

A toleranced system is considered as axi-symmetric when:
- all the functional surfaces taken into account in the tolerances are surfaces of revolution around the same axis **z**,
- the tolerance zones corresponding to the given tolerances are volumes of revolution around the same axis **z**.

In consequence, the datum frames are centred on the axis of revolution, and the components of the deviation torsors for the rotation around the axis are always zero. Figure (1a) shows some examples of axi-symmetric geometric tolerances.

The deviation torsors have the following form: $E=(rx, ry, 0, tx, ty, tz)$. Different parts can be assembled. The contact surfaces are surfaces of revolution. The clearance torsor is in the form : $J=( rx, ry, wz, tx, ty, tz)$. The component wz is the rotation around the axis and is not limited because it is a degree of freedom of the mechanism. The other components are limited by the contact conditions.

When the projection frame is changed, by a rotation of an angle $\theta$, around the z axis, and a translation z along the same axis, the matrices that define the change of frame are:

$$U_{01} = \quad (0, 0, z) \quad \text{et} \qquad P_{01} = \begin{bmatrix} Cos\theta & -Sin\theta & 0 \\ Sin\theta & Cos\theta & 0 \\ 0 & 0 & 1 \end{bmatrix}$$

Then the components of a torsor in the frame (1) can be expressed from the components in the frame (0).

$$rx_1 = Cos\theta\ rx_0 + Sin\theta\ ry_0 \qquad tx_1 = Cos\theta\ tx_0 + Sin\theta\ ty_0 + z\ ry_1$$
$$ry_1 = -Sin\theta\ rx_0 + Cos\theta\ ry_0 \qquad ty_1 = -Sin\theta\ tx_0 + Cos\theta\ ty_0 - z\ rx_1 \tag{2}$$
$$tz_1 = tz_0$$

### 2.1 Case of a cylindrical tolerance zone around the axis of revolution Oz

We find this case for example for a coaxiality of a cylinder or of any surface of revolution when the datum is an axis built from one or more surfaces of revolution around the same axis.

The deviation torsor is in the form $E=(rx,ry,0,tx,ty,0)$ in a frame O, **x**, **y**, **z**. The displacement vector of the points $P_1$ et $P_2$ at the extremity of the axis:

$$\delta P_1 = (tx + ry\ h/2)\ \mathbf{x} + (ty - rx\ h/2)\ \mathbf{y}$$

$\boldsymbol{\delta P_2} = (tx\text{-}ry\ h/2)\ \mathbf{x} + (ty + rx\ h/2)\ \mathbf{y}$

For the model, the interpretation of the coaxiality tolerance leads these two points to be inside the circles $C_1$ and $C_2$ respectively, at the extremity of the cylindrical tolerance zone. The diameter of this cylinder is t and is length is h. (Fig.1b) So the tolerance is translated by the inequalities:

$$(2\ tx\ /h + ry\ )^2 + (2\ ty/h - rx\ )^2 < (t/h)^2$$
$$\text{and}\quad (2\ tx\ /h - ry\ )^2 + (2\ ty/h + rx\ )^2 < (t/h)^2 \tag{3}$$

## 2.2 Case of toleranced plane perpendicular to the axis in the nominal configuration

The tolerance zone is also a cylinder but the value t of the tolerance is the thickness of the cylinder while the diameter D is the one of the circle that delimits the face. It can be an orientation tolerance, the datum is then an axis, or a position tolerance. In that case the datum defines an axis and a point on this axis. Let P be a point on the plane face, belonging to the circle (fig. 1c). The polar co-ordinates of this point are (D/2, $\theta$). The projection frame is centred on the $\mathbf{z}$ axis. The co-ordinates of the deviation torsor in this frame is (rx, ry, 0, tx, ty, tz).

The displacement of the point P is $\boldsymbol{\delta P} = (0, 0, tz+rx\ \mathrm{Sin}\theta - ry\ \mathrm{Cos}\theta\ )$. This point must be inside the tolerance zone so that :

$$(rx\ \mathrm{Sin}\theta - ry\ \mathrm{Cos}\theta) < t/D - 2tz/D \qquad \text{and} \qquad (-rx\ \mathrm{Sin}\theta + ry\ \mathrm{Cos}\theta) < t/D + 2tz/D$$

These inequalities must be satisfied for all the possible values of $\theta$.

For any value of tz, each inequality is satisfied for all the points of a half space limited by a straight line at the constant distance from the origin. The intersection of all these half spaces is limited by the envelope of all the straight lines. Therefore, it is a circle. Its diameter is equal to (t/D – 2tz/D) if tz is positive and (t/D + 2tz/D) if tz is negative. In all the cases, tz is in the interval [-t/2, t/2].

The deviation domain is then defined by the following quadratic inequalities:

$$(ry)^2 + (rx)^2 < (t/D - 2tz/D)^2 \quad \text{and} \quad (ry)^2 + (rx)^2 < (t/D + 2tz/D)^2 \tag{4}$$

For an orientation tolerance, only the angular components are limited:

$$(ry)^2 + (rx)^2 < (t/D)^2 \tag{5}$$

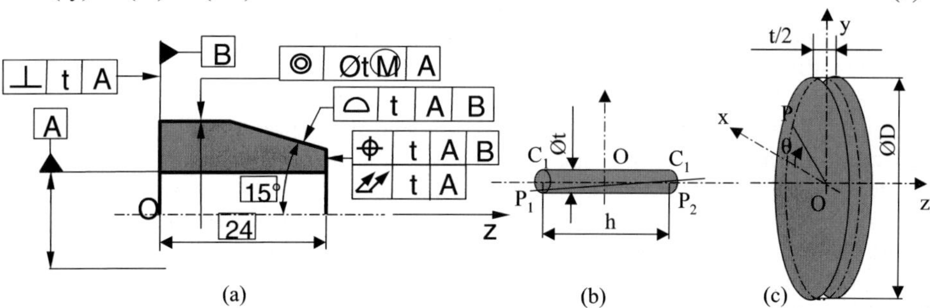


*Figure 1; Axi-symmetric cases and two examples of tolerance zones*

The general case of the different inequalities can be written on the form:

$$(a\,tx + ry)^2 + (a\,ty - rx)^2 < (t/b - ctz)^2 \qquad (6)$$

The three previous cases correspond to the following values of the coefficient a, b and D:

1.) $a = 2/h$ ;     $b = h$ ; $c = 0$     and     $a = -\,2/h$ ;     $b = h$ ; $c = 0$

2.) $a = 0$ ;     $b = D$ ; $c = 2/D$     and     $a = 0$ ;     $b = D$ ; $c = -2/D$

3.) $a = 0$ ;     $b = D$ ; $c = 0$

### 2.3 Axi-symetric domain properties

The domain remains unchanged if the projection frame rotates around the z axis and translate on the z axis.

Let us consider the general inequality:

$$(a\,tx_1 + ry_1)^2 + (a\,ty_1 - rx_1)^2 < (t/b - ctz_1)^2$$

Then, using the equations (2) and after development and simplifications, the inequality becomes:

$$(a\,tx_0 + ry_0)^2 + (a\,ty_0 - rx_0)^2 < (t/b - ctz_0)^2$$

This property proves that the deviation domain remains without change. The deviation domains have a symmetry of revolution in their own configuration space. But it is not the usual symmetry of revolution for figures in the 3D space. The deviation domains defined by a set of inequalities like in (6) can be represented in a five dimension space because they have only five components or less. The intersection of these domains by an hyper-plane defined by the equations tx=constant, ry=constant and tz=constant is a domain in the 2D space. It can be represented in a plane by the domain inside two parallel lines, symmetrical around the origin. The cartesian inequalities for this domain are:

$$a\,ty - rx < f(tx, ry, tz) \qquad \text{and} \qquad a\,ty - rx > -\,f(tx, ry, tz) \qquad (7)$$

The slope between the "ty" axis and these lines is the coefficient a.

The cross section of an axi-symmetric domain defined by one inequality (6) with an hyper-plane, where tx, ty and tz are constant, is the disk inside the circle of radius $(t/b - ctz_0)$. If the domain is defined by several inequalities, the cross section of the domain is the intersection of all the disks built from each inequality. The cross section with an hyper-plane defined by constant rx, ry and tz is also a disk if "a" is different from zero. These properties characterise the axi-symmetric domains.

When different parts are assembled, the geometrical deviations stack up. The functional requirements concern generally features not belonging to the same part. In the "worst case" approach, the deviation domain that characterise the functional requirement must be inside the Minkowski sum of the different domains of the stack up. See the references [Teissandier et al., 1999], [Giordano et al., 2001], [Petit, 2004] about the applications of Minkowski sum in tolerancing. It can be proved that the Minkowski sum of two axi-symmetric domains is also an axi-symmetric domain. So, it is sufficient to compute the Minkowski sum only in a three dimensional space. Three components are considered: rx, ty, tz or tx, ry, tz. The quadratic inequality (6) is then replaced by the two linear inequalities (7).

## 3.   APPLICATION TO THE INSPECTION OF SPECIFICATIONS

The example concerns a coaxiality tolerance for the axis of a cylinder. The datum is the axis of an other cylinder. The maximal material modifier is used both for the toleranced feature and the datum (fig.2). The proposed method is quite different from those presented in the literature [Davidson and Shah, 2003].

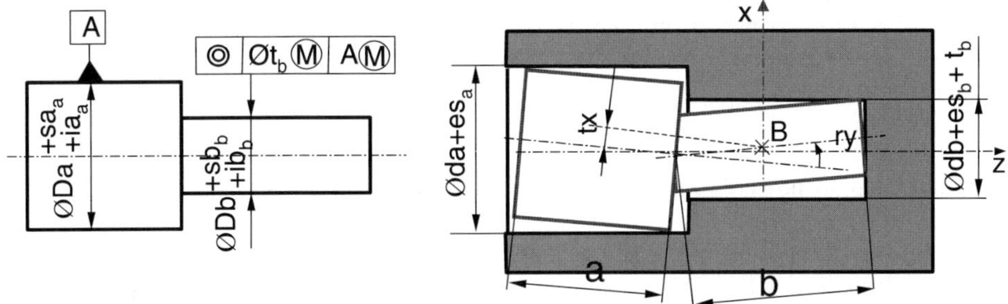


*Figure 2; coaxiality tolerance and interpretation*

The problem is to determine the relations between the deviation parameters that can be measured (components of the deviation torsors) so that the part is in accordance with the tolerance.

It is assumed that after the measurement, the following parameters are known :

- the diameters of the associated cylinders : (Da + ea) and (Db + eb),

- the deviation components for the rotation and the position of the cylinder B with respect to the cylinder A : rx,ry, tx and ty.

We suppose that the dimensional tolerances are defined :

ia < ea < sa, and ib < eb < sb,

and the coaxiality tolerance specified value is tb.

The part is in conformity with the specifications if the tolerances of size are checked and if the part can be assembled in its theoretical gauge. For the gauge, the dimension of the diameter of the cylinder B is the maximum value added to the geometric tolerance Db + sb + tb, while the diameter of A is the maximum value Da + sa. We consider the mechanism constituted by the part assembled in the gauge. The subscript (0) is for the gauge and (1) for the part.

Since the two cylindric joints A and B and the two parts constitute a single loop mechanism, the clearance and deviation torsors are linked by the relation:

$$J_{0A1} + E_{A1B} + J_{1B0} + E_{B0A} = 0$$

Where $J_{0A1}$ and $J_{1B0}$ are the clearance torsors for the two joints and $E_{A1B}$ and $E_{B0A}$ are the deviation torsors of the parts. The geometry of the gauge is perfect, therefore $E_{B0A} = 0$.

The equivalence between the geometric tolerance and the assembly is translated into the condition between the domains: $[E_{A1B}] = [J_{0B1}] + [J_{1A0}]$ . The sum of Minkowski of the domain. The domain is built in the plane rx, ty (fig.3).

The domain $[J_{1A0}]$ is determined at the point B. The parameters that characterize the domain $[E_{A1B}]$ allows to write the inequalities between the deviation parameters.

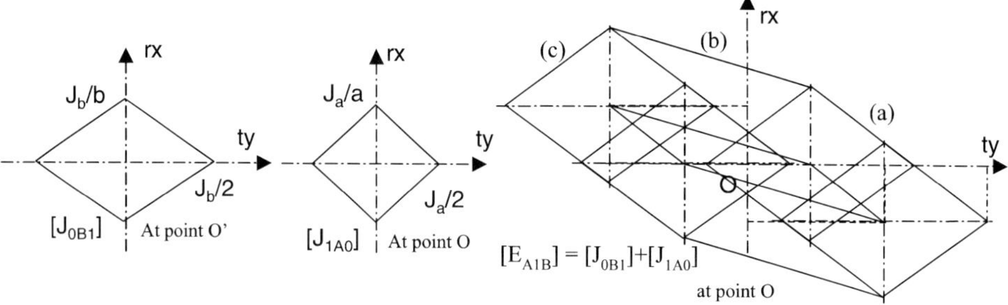

*Figure 3 ; Deviation domain and their Minkowski sum*

## 4.   APPLICATION TO TOLERANCE SYNTHESIS

In the second example the problem is to compare the deviation resulting from the assembly of two parts, with a requirement between two feature belonging to the different parts. The parts are toleranced in a qualitative form but the values of the tolerance zones are unknown. The qualitative tolerances are defined in figure 4. They are justified by technological choices for the joint between the two parts. The head 2 is in contact with the barrel 1, on the plane A maintained by tie rods and is centred in the "short" cylinder C fitted with a clearance in the cylinder B.

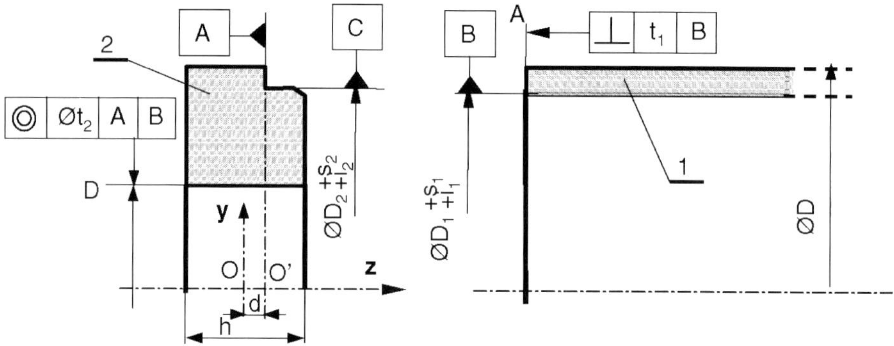

*Figure 4; Qualitative tolerances of axi-symmetric parts for tolerance synthesis*

We notice :
-    B: a frame attached to the cylinder B,
-    (1): a frame attached to the plane A but centred on the B axis of the part (1),
-    (2): a datum ordered frame attached to the part (2), so that the plane A is the primary datum and cylinder C is the secondary datum.
-    D: the datum attached to the cylinder D.

The deviation torsors are linked by the relation:         $E_{BD} = E_{B1} + E_{12} + E_{2D}$

Each torsor represent the small displacements to go from a frame to the other. $E_{B2}$ is the deviation of the plane with regard to the cylinder. The domain $[E_{B1}]$ is built from the perpendicularity tolerance. It is characterised by the parameter $t_1$. The domain is defined at the point O' in the plane ty, rx. If the torsors are expressed at the point O, the domain represented by a segment is changed into an other segment.

$E_{21}$ is the deviation torsor of the frame (2) with regard to the frame (1). This deviation is due to the clearance between the cylinder C and B. We notice J the difference between the two diameters of the cylinders.

Since the planes A of the two parts are in coincidence, the relative displacement is a translation in the plane. The domain $[E_{12}]$ is independent on the point chosen to define the torsor. $[E_{2D}]$ is the deviation domain for the D toleranced feature. The datum frame is in fact the frame (2). It is defined at the point O. The three domains are represented in the figure 5.

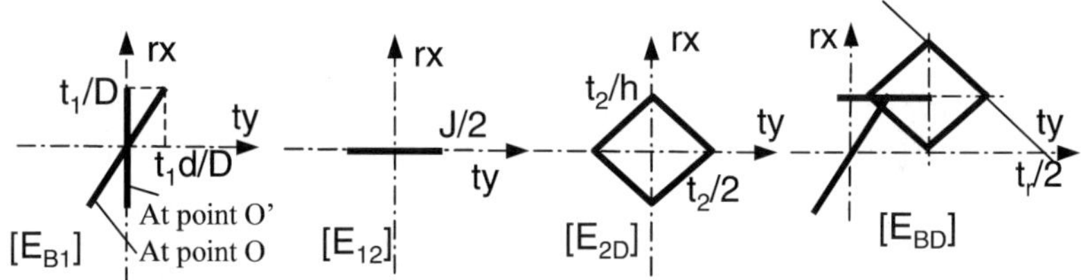


*Figure 5; Deviation domains and stack up tolerances*

The three deviation torsors have any value in their domain. When the parts are assembled the torsor sum $E_{BD}$ will belong to the domain obtained by the Minkowski sum of the three previous domains defined at the same point O.

The resulting domain does not correspond to a tolerance zone. The stack up tolerance of geometric standardised tolerances is not in general case a standardised tolerance. Nevertheless, if the functional requirement between the feature B and D is given using a coaxiality tolerance, it is possible to determine the relation between the value $t_r$ of this tolerance and the values of the intermediate tolerances. The relation is easy to determine thanks to the domain operations in the plane rx, ty (fig.5) :

$t_r = t_1(h+2d)/D + J + t_2$

Different solutions are possible. Others criteria must be given for the repartition of the three values J, $t_2$ and $t_1$. Since J depends of the actual size, the worst case is when J is maximum, so the clearance has a unfavourable effect on the requirement. The least material condition can then be used.

## 5.  CONCLUSION

The method based on deviation and clearance domain is very efficient for the tolerance analysis. Tolerance synthesis needs to compute Minkowski sum on the domains. In the general case the operation is complex but in the case of axi-symmetric systems, the problem can be solved with simple linear operations. From qualitative tolerances, it is possible to determine analytic relations that permits to determine the tolerances that ensure given requirements. This synthesis tolerancing approach can be applied to different practical systems.


## REFERENCES

**[Bourdet et al., 1996]** Bourdet P., Mathieu L., Lartigue C., Ballu A., « The concept of small displacement torsor in metrology » Advanced mathematical tool in metrology, Advances in mathematics for applied sciences, World Scientific, Vol 40, 1996.

**[Desrochers et al., 2003]** Desrochers A., Béron V. and Laperrière L., « Revisiting Screw Parameter Formulation for Accurate Modeling of Planar Tolerance Zones », the 8th international CIRP Seminar on Computer Aided Tolerancing, Charlotte, North Carolina, USA, 28-29 April 2003, pp.239-248.

**[Sacks and Joskowicz, 1998]** L., Sacks E., Joskowicz L., "Parametric kinematic tolerance analysis of general planar systems. Computer-aided design 1998, vol 30(9), pp. 707-714.

**[Mujezinovic et al., 2004]** Mujezinovic A., Davidson J.K., Shah J.J., « A New Mathematical Model for Geometric Tolerances as Applied to Polygonal Faces », Transaction of the ASME, J. of Mech. Design, Vol. 126, May 2004, pp. 504-518.

**[Davidson and Shah, 2002]** Davidson J.K, and Shah J.J, « Geometric tolerances : a new application for line geometry and screws » ImechE Journal of Mechanical Engineering Sciences, Vol.216, Part C, 2002, pp.95-104.

**[Petit, 2004]** Petit J.P., « Spécification géométrique des produits : méthode d'analyse de tolérances. Application en conception assistée par ordinateur. » P.h.d. theses, Université de Savoie, France, 17 déc. 2004.

**[Teissandier et al., 1999]** Teissandier D., Delos V., Couetard Y., "Operations on polytopes: application to tolerance analysis", 6th CIRP inter. Seminar on computer-aided tolerancing, Univ. of Twente, Enschede, The Netherlands, 22-24 march 1999.

**[Giordano et al., 2001]** Giordano M., Kataya B., Pairel E., « Tolerance analysis and synthesis by means of clearance and deviation spaces », Geometric product Specifcation and verification, selected conf. papers of the 7th CIRP seminar on Computer Aided Tolerancing, April 2001, Kluwer Academic Pub., pp.145-154.

**[Davidson and Shah, 2003]** Davidson J.K. and Shah J.J., « Using Tolerance-Maps to represent material condition on both a feature and a datum », the 8th international CIRP Seminar on Computer Aided Tolerancing, Charlotte, North Carolina, USA, 28-29 April 2003, pp.92-101.

# Re-Design of Mechanical Assemblies using the Unified Jacobian-Torsor Model for Tolerance Analysis


**W. Ghie*, L. Laperrière*, A. Desrochers****

*Université du Québec à Trois-Rivières, Department of Mechanical Engineering
Ghie@uqtr.ca ; Luc_Laperriere@uqtr.ca
**Université de Sherbrooke, Department of Mechanical Engineering
Alain.Desrochers@USherbrooke.ca



**Abstract:** this paper describes how the unified Jacobian-Torsor model can be used for the redesign of assembly tolerances. Through a numerical and graphical approach, the designer is guided into choosing his tolerances. After having identified a functional requirement (FR) and the functional elements (FEs) of the dimensional chain, it becomes possible to compute the percentage contribution of each FE to the FR. A percentage contribution chart tells the designer how each dimension has contributed to the total FR. Once these contributions have been identified, we proceed to modify the most critical tolerance zones to respect the requirements imposed by the designer. The results are evaluated by comparing the predicted variation with the corresponding specified design requirements. Then, the contributions can be used to help decide which tolerance to tighten or loosen. This study has been developed for Worst-Case approach. Finally, an example application is presented to illustrate our methodology.

**Keywords**: Tolerance, Torsor, Jacobian, Analysis, Design, Statistical and Synthesis.


## 1. INTRODUCTION

Design procedures mainly include two phases: functional design and manufacturing design. Tolerance design directly influences the functionality of assemblies and its cost. Tolerance synthesis is a procedure that distributes assembly tolerances between components. Tolerance synthesis is an essential step in all design phases to assure quality conformity and economic manufacturing. A tighter tolerance is normally preferred for product functionality; however, manufacturing costs usually increases due to the requirement of more rigorous operations. On the contrary, a loose tolerance is less costly; however, it may cause inferior quality. Therefore, it is necessary to make trade-offs between tolerance and cost. Traditionally, designers allocate tolerance based on their experience and information contained in design handbooks or standards. This approach does not guarantee functionality or assemblability; nor does it minimize cost. The focus of this paper is to guarantee product functionality. Then the tolerance specification is an important part of mechanical design. Design tolerances strongly influence the functional performance and manufacturing cost of a mechanical product.




*J.K. Davidson (ed.), Models for Computer Aided Tolerancing in Design and Manufacturing*, 95–104.
© 2007 *Springer*.

Computer-aided tolerance analysis programs allow a designer to verify the relations among all design tolerances to produce a consistent and feasible design. To produce the designed component within the specified tolerance, design teams often ask the question: *where in the assembly should I focus my intention to reduce or increase the tolerance? How to deal with parts with multiple characteristics?* Their decision should affect the contributions of the various characteristics. This paper presents a novel approach for determining the critical tolerances given the inherent uncertainty at the early design phases.


## 2.   LITERATURE REVIEW

Many different approaches for Computer Aided Tolerancing (CAT) have been developed over the years, for example variational classes [Requicha 1984,1983], vectorial tolerancing [Wirtz 1998][Weber et al.1999] or linear programming approaches [Pradeep et al. 2005][Angus 2001][WEI 1997] just to name a few. Of concern in this paper are the Jacobian and torsor approaches [Ghie 2004]. [Desrochers et al. 03]. The former is based on small displacements modeling of points using transformation matrices of open kinematic chains in robotics. The latter models the extreme limits of 3-D tolerance zones resulting from a feature's small displacements using a torsor representation with constraints.  Both of them fall under the category of kinematic approaches to tolerancing.  The kinematic approach is also used by many other researchers and is also at the base of at least one successful commercial implementation of a CAT system [Hong et al. 2002].  Thus the kinematic CAT model is built on top of existing CAD nominal information and its matrix formulation makes it compatible for performing the required computations natively in the CAD system. Barraja M [Bararaja et al. 2005] presents a method for allocating tolerances to the dimensions of the kinematic couplings. The objective is to reduce the manufacturing cost without exceeding limits on the variation of the coupled position and orientation. This method uses parametric models of the contacting surfaces and a solution for the resting position of the coupled bodies. The tolerances of the coupled bodies are related to manufacturing costs via tolerance-cost relations for common processes. Chase [Chase et al. 1995] implements a Direct Linearization Method (DLM) that relates to the tolerance analysis The linearization has to do with a simplification in the way the functional relationships between the deviations of a part's features and the deviations of some assembly's points of interests are expressed. The resulting linear equations are simulated by:

$$\Delta U_i = \sum_{j=1}^{n} \left| S_{ij} \right| tol_j \le T_{asm} \tag{1}$$

Where:

tolj;   :  Is the tolerance of the jth manufactured dimension;
Tasm  :  Is the design specification for the ith assembly variable;
Sij     :  Is the tolerance sensitivity matrix of the assembly:
ΔUi   :  Is the sum of the participating tolerances; i,j,n are the assembly variables

Tolerance synthesis problems have attracted attention for many years. Instead of using traditional methods, recently researchers have developed tolerance synthesis approaches for optimal tolerance design. Tolerance synthesis problems are usually formulated as non-linear programming models [Chen 2001].


## 3.   PROPOSED RE-DESIGN APPROACH


### 3.1 Prior work

In previous papers we have presented a tool for deterministic tolerance analysis [Ghie 2004], [Desrochers et al. 2003] which uses an interval arithmetic formulation:

$$
\begin{bmatrix} [\underline{u},\bar{u}] \\ [\underline{v},\bar{v}] \\ [\underline{w},\bar{w}] \\ [\underline{\alpha},\bar{\alpha}] \\ [\underline{\beta},\bar{\beta}] \\ [\underline{\delta},\bar{\delta}] \end{bmatrix}_{FR} = \left[ \left[ J_1 J_2 J_3 J_4 J_5 J_6 \right]_{FE1} \cdots \left[ J_1 J_2 J_3 J_4 J_5 J_6 \right]_{FEN} \right] \cdot \begin{bmatrix} \begin{bmatrix} [\underline{u},\bar{u}] \\ [\underline{v},\bar{v}] \\ [\underline{w},\bar{w}] \\ [\underline{\alpha},\bar{\alpha}] \\ [\underline{\beta},\bar{\beta}] \\ [\underline{\delta},\bar{\delta}] \end{bmatrix}_{FE1} \\ \cdots\cdots \\ \begin{bmatrix} [\underline{u},\bar{u}] \\ [\underline{v},\bar{v}] \\ [\underline{w},\bar{w}] \\ [\underline{\alpha},\bar{\alpha}] \\ [\underline{\beta},\bar{\beta}] \\ [\underline{\delta},\bar{\delta}] \end{bmatrix}_{FEN} \end{bmatrix}
\tag{2}
$$

Where:

$[FR] = \begin{bmatrix} [\underline{u},\bar{u}] \\ [\underline{v},\bar{v}] \\ [\underline{w},\bar{w}] \\ [\underline{\alpha},\bar{\alpha}] \\ [\underline{\beta},\bar{\beta}] \\ [\underline{\delta},\bar{\delta}] \end{bmatrix}_{FR}$ , $[FEi] = \begin{bmatrix} [\underline{u},\bar{u}] \\ [\underline{v},\bar{v}] \\ [\underline{w},\bar{w}] \\ [\underline{\alpha},\bar{\alpha}] \\ [\underline{\beta},\bar{\beta}] \\ [\underline{\delta},\bar{\delta}] \end{bmatrix}_{FEi}$ : Small displacements torsors associated to some functional requirement (Play, gap, clearance) represented as a [FR] vector or some Functional Element uncertainties (tolerance, kinematic link, ….) also represented as [FE] vectors ; with N representing the number of torsors in a kinematic chain;

$\left[ J_1 J_2 J_3 J_4 J_5 J_6 \right]_{FEi}$ : Jacobian matrix expressing a geometrical relation between a [FR] vector and some corresponding [FE] vector;

$\left( \underline{u}, \underline{v}, \underline{w}, \underline{\alpha}, \underline{\beta}, \underline{\delta} \right)$ and $\left( \bar{u}, \bar{v}, \bar{w}, \bar{\alpha}, \bar{\beta}, \bar{\delta} \right)$ : Lower and Upper limits of $u, v, w, \alpha, \beta, \delta$ .


As can be seen, dispersions around a functional condition are represented by the column matrix [FR] in which the six small displacements are bounded by interval values. Similarly, the various functional elements encountered in the tolerancing chain are represented by corresponding column matrix [FEs] where intervals are again used to represent variations on each element. Naturally, the terms in this expression remain the same as those with "conventional" Jacobian modeling [Ghie 2004].

### 3.2 Percent Contribution Charts

Equation (2) gives the accumulation of the tolerance contributions from each $C_{FEi}$ to the total FR. Then, for a better understanding, we rewrite the mathematical formulation which was introduced in equation (3) :

$$\begin{bmatrix} [\underline{u},\bar{u}] \\ [\underline{v},\bar{v}] \\ [\underline{w},\bar{w}] \\ [\underline{\alpha},\bar{\alpha}] \\ [\underline{\beta},\bar{\beta}] \\ [\underline{\delta},\bar{\delta}] \end{bmatrix}_{FR} = C_{FE1} + \cdots\cdots + C_{FEN} \tag{3a}$$

Using the law from equation (2), each element in the previous equation then becomes:

$$C_{FEi} = \begin{bmatrix} [\underline{C}_{iu},\bar{C}_{iu}] \\ [\underline{C}_{iv},\bar{C}_{iv}] \\ [\underline{C}_{iw},\bar{C}_{iw}] \\ [\underline{C}_{i\alpha},\bar{C}_{i\alpha}] \\ [\underline{C}_{i\beta},\bar{C}_{i\beta}] \\ [\underline{C}_{i\delta},\bar{C}_{i\delta}] \end{bmatrix}_{FEi} = \begin{bmatrix} J_1 J_2 J_3 J_4 J_5 J_6 \end{bmatrix}_{FEi} \begin{bmatrix} [\underline{u},\bar{u}] \\ [\underline{v},\bar{v}] \\ [\underline{w},\bar{w}] \\ [\underline{\alpha},\bar{\alpha}] \\ [\underline{\beta},\bar{\beta}] \\ [\underline{\delta},\bar{\delta}] \end{bmatrix}_{FEi} \tag{3b}$$

Where

$C_{FEi}$      :   The contribution of the FE i to the FR;

$[\underline{C}_{ik},\bar{C}_{ik}]$   :   Represent the interval contribution in a direction k, with k= $u,v,w,\alpha,\beta,\delta$ ;

i           :   Represent the reference of FE i in the dimensional chain with i=1,2,3,...N;

Using the contribution of each FE, it is easy to obtain a chart histogram of contribution with percentage values. The percentage contribution chart tells the designer how each dimension contributes to the total variation (FR). The contribution includes the effect of both dimensional and geometric tolerances. The Worst Case model assumes all the component dimensions occur at their worst limits simultaneously. It is used by the designers to insure that all the tolerances will meet the specified assembly limit. However, as the number of parts in the assembly increases, the component tolerances must be greatly reduced in order to meet the assembly limit, requiring higher production costs. Then, the contribution is obtained by equation 4:

$$\% C_{FEik} = \frac{\left| \bar{C}_{FEik} - \underline{C}_{FEik} \right|}{\sum_{j=1}^{N} \left| \bar{C}_{FEjk} - \underline{C}_{FEjk} \right|} \tag{4}$$

Where

$\% C_{FEik}$ : Represent the percentage contribution of FE i in the direction k to the FR. With, i=1, 2,..., N and k= u,v,w,$\alpha$, $\beta$ or $\delta$;

After having identified a FR and all FEs of the dimensional chain, it becomes possible to compute the percentage contribution of each FE to the FR. This it is possible using the equations above (equation 4). Hence, we can obtain the percentage contribution chart; a histogram of the contribution of each tolerance specified in the chain. Using the contribution chart, the designer can identify the most critical tolerances, i.e, those which contributed most to the total FR. Once these contributions have been identified, we proceed to modify the most critical tolerance so as to respect the FR imposed on the designer. The results are then evaluated by comparing the predicted variation with the corresponding specified design requirements. Depending if the computed variation is greater or less than the specified assembly tolerance, the contributions can further be used to help decide which tolerance to tighten or loosen.

## 4.   STEPS FOR THE RE-DESIGN OF ASSEMBLY

The goal of tolerance redesign is to verify and adjust the tolerances assigned at the design stage so as to meet a given FR target at the assembly level (gap or clearance). More specifically, the tolerance redesign method consists of the flowing steps:

1. Establish a critical FR and specify the design limits;
2. Identify the component dimensions which contribute to the FR;
3. Specify tolerances for each dimension in the dimensional chain around the FR which has been specified in the first step. Then, impose initial values (extremes limits) for each tolerance in the chain;
4. In this step, apply the unified model (equation 2) firstly to obtain the variation of the FR (calculated FR) and secondly to obtain the contribution percentage (equation 4) of each FE to the FR;
5. The role of this step is to compare the result of FR values obtained in step 4 with the design specifications imposed in the first step. If the assigned FR sought by the designer is not met, the tolerances of critical FE need to be reassigned using tolerances allocation techniques. Calculated FR values can then be adjusted to meet the designer's intent by tuning the values (equation 4) of the corresponding critical tolerances (largest or smallest contributors). So, if the calculated FR is smaller than the imposed FR, then, by augmenting the smallest contributor tolerance obtained in step 4, we increase the value of the calculated FR. However, if the calculated FR is greater than the imposed FR, then we need to decrease the calculated FR. This can be done through a reduction of the biggest contributor tolerance. Of course it then becomes important to check the values of the manufacturing tolerances on the vendor-supplied parts because these can obviously not be modified. The procedure is repeated until the calculated FR falls within specification.

The steps illustrated above describe a comprehensive method for creating and validating assembly tolerances. The procedure can be assimilated to an iterative synthesis method targeted at helping the designer select the right values for the tolerances of each element in a mechanical assembly.

## 5. NUMERICAL EXAMPLE

The *centering pin mechanism* in figure 1 will be used to demonstrate the use of the tool. In this figure, we have labelled some key tolerances from ta to tg. The mechanism features three parts with two functional conditions shown in figure 2. Table I presents some initial, manually generated values for the variables ta to tg. These initial values are all expressed as intervals in equation 5. In this example, the designer is being imposed a first FR1, related to the vertical centering of the pin, with a precision of ±1mm.

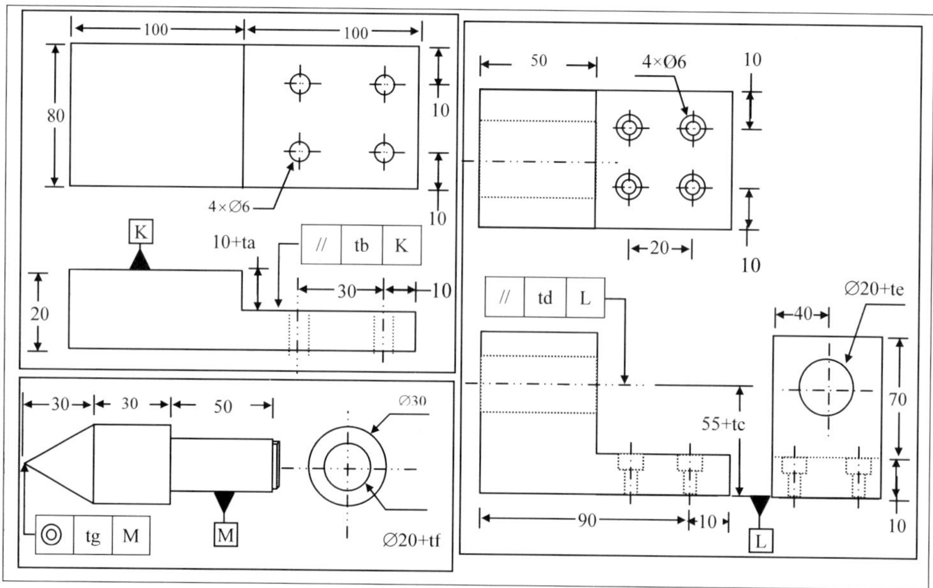

*Figure 1: A three parts centering pin mechanism*

Prior to applying the principles set forth in the unified model proposed in [Desrochers et al. 03] one needs a chain of dimensions. This chain can be identified using a connection graph. In figure 2, contact surfaces are first identified between the parts, before a connection graph can be constructed to establish the dimension chain around the FR1 condition on the mechanism. Then we have to distribute the FR1 condition between the different tolerance zones identified in the corresponding dimensional chain. More specifically, the kinematic chain obtained contains three internal pairs (FE0, FE1), (FE2, FE3), (FE4, FE5) as well as one kinematic pair (FE1, FE2). Note that there are two functional requirements: FR1, which applies between (FE0, FE5) and FR2 between (FE3, FE4). We also see in the tolerance chain, that FR1 and FR2 are dependant.

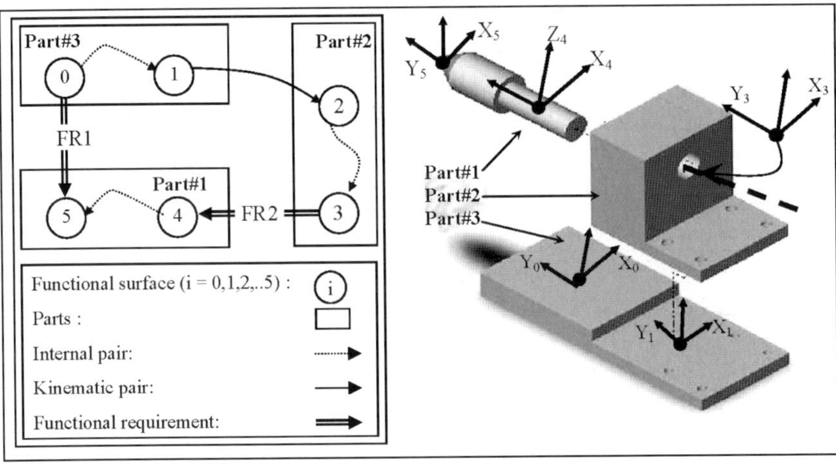


*Figure 2: Kinematic chain identification*

In this example, we assume that the reference frames are in the middle of the tolerance or contact uncertainty zone and that they are associated to the second element in the pairs defined above. The kinematic torsor for (FE1, FE2) is considered null because the contact between the two planes is assumed to be perfect and form tolerances are not specified in this instance. We must therefore determine the remaining four Small Displacement Torsor Intervals: T1/0 for FE1 relative to FE0, T3/2 relating FE3 to FE2, T5/4 relating FE5 to FE4, as well as T4/3 for the inclusion of FR2 in the analysis. From this, using the initial tolerance values from table I (first section), the final expression of the Jacobian-Torsor formulation with intervals is then represented by equation 5.

$$
\begin{bmatrix} [u,\bar{u}] \\ [v,\bar{v}] \\ [w,\bar{w}] \\ [\alpha,\bar{\alpha}] \\ [\beta,\bar{\beta}] \\ [\delta,\bar{\delta}] \end{bmatrix}_{FR} =
\begin{bmatrix} 1 & 0 & 0 & 0 & -50 & 100 \\ 0 & 1 & 0 & 50 & 0 & 0 \\ 0 & 0 & 1 & -100 & 0 & 0 \\ 0 & 0 & 0 & 1 & 0 & 0 \\ 0 & 0 & 0 & 0 & 1 & 0 \\ 0 & 0 & 0 & 0 & 0 & 1 \end{bmatrix}_{FE1}
\begin{bmatrix} 1 & 0 & 0 & 0 & 0 & 82.50 \\ 0 & 1 & 0 & 0 & 0 & 0 \\ 0 & 0 & 1 & -82.50 & 0 & 0 \\ 0 & 0 & 0 & 1 & 0 & 0 \\ 0 & 0 & 0 & 0 & 1 & 0 \\ 0 & 0 & 0 & 0 & 0 & 1 \end{bmatrix}_{FE2}
\begin{bmatrix} 1 & 0 & 0 & 0 & 0 & 80 \\ 0 & 1 & 0 & 0 & 0 & 0 \\ 0 & 0 & 1 & -80 & 0 & 0 \\ 0 & 0 & 0 & 1 & 0 & 0 \\ 0 & 0 & 0 & 0 & 1 & 0 \\ 0 & 0 & 0 & 0 & 0 & 1 \end{bmatrix}_{FE3}
\begin{bmatrix} 1 & 0 & 0 & 0 & 0 & 0 \\ 0 & 1 & 0 & 0 & 0 & 0 \\ 0 & 0 & 1 & 0 & 0 & 0 \\ 0 & 0 & 0 & 1 & 0 & 0 \\ 0 & 0 & 0 & 0 & 1 & 0 \\ 0 & 0 & 0 & 0 & 0 & 1 \end{bmatrix}_{FE4}
\cdot
\begin{bmatrix}
\begin{bmatrix} [0,0] \\ [0,0] \\ [-0.2,+0.2] \\ [-0.0013,+0.0013] \\ [-0.0025,+0.0025] \\ [0,0] \end{bmatrix}_{FE1} \\
\begin{bmatrix} [-0.1,+0.1] \\ [0,0] \\ [-0.1,+0.1] \\ [-0.002,+0.002] \\ [0,0] \\ [-0.002,+0.002] \end{bmatrix}_{FE2} \\
\begin{bmatrix} [-0.0265,+0.0265] \\ [0,0] \\ [-0.0265,+0.0265] \\ [-0.0011,+0.0011] \\ [0,0] \\ [-0.0011,+0.0011] \end{bmatrix}_{FE3} \\
\begin{bmatrix} [-0.05,+0.05] \\ [0,0] \\ [-0.05,+0.05] \\ [-0.003\bar{3},+0.003\bar{3}] \\ [0,0] \\ [-0.003\bar{3},+0.003\bar{3}] \end{bmatrix}_{FE4}
\end{bmatrix}
\tag{5}
$$

| #Try | FE | Name | Values | % FR | Contribution Chart |
|---|---|---|---|---|---|
| **Sec. 1 : Initial guess** | 1 | ta | -0.2000/0.2000 | 34.26 | 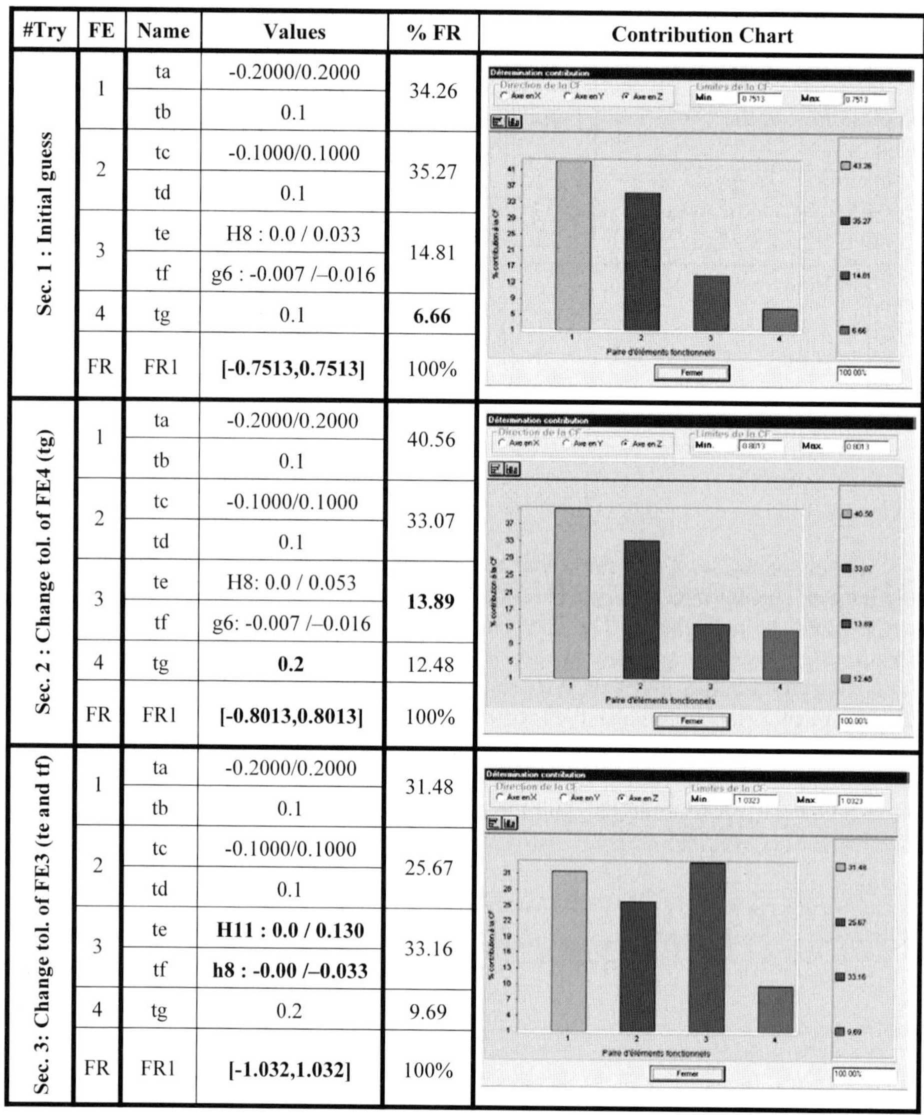 |
| | | tb | 0.1 | | |
| | 2 | tc | -0.1000/0.1000 | 35.27 | |
| | | td | 0.1 | | |
| | 3 | te | H8 : 0.0 / 0.033 | 14.81 | |
| | | tf | g6 : -0.007 /−0.016 | | |
| | 4 | tg | 0.1 | **6.66** | |
| | FR | FR1 | **[-0.7513,0.7513]** | 100% | |
| **Sec. 2 : Change tol. of FE4 (tg)** | 1 | ta | -0.2000/0.2000 | 40.56 | |
| | | tb | 0.1 | | |
| | 2 | tc | -0.1000/0.1000 | 33.07 | |
| | | td | 0.1 | | |
| | 3 | te | H8: 0.0 / 0.053 | **13.89** | |
| | | tf | g6: -0.007 /−0.016 | | |
| | 4 | tg | **0.2** | 12.48 | |
| | FR | FR1 | **[-0.8013,0.8013]** | 100% | |
| **Sec. 3: Change tol. of FE3 (te and tf)** | 1 | ta | -0.2000/0.2000 | 31.48 | |
| | | tb | 0.1 | | |
| | 2 | tc | -0.1000/0.1000 | 25.67 | |
| | | td | 0.1 | | |
| | 3 | te | **H11 : 0.0 / 0.130** | 33.16 | |
| | | tf | **h8 : -0.00 /−0.033** | | |
| | 4 | tg | 0.2 | 9.69 | |
| | FR | FR1 | **[-1.032,1.032]** | 100% | |

*Table I: Tolerances of the three parts centering pin mechanism*

With the initial tolerances taken from the first section of table I, we obtain a corresponding contribution chart, again shown in table I. In this example, we used the equation (4) to obtain the contributors tolerances. According to the fourth column of table I (in section 1), the FR1 calculated is ±0.7513. This value is much more stringent than the designer specification of ±1mm. Therefore we need to change the tolerances in the dimensional chain to respect that. In this case, efforts are being focused at the lesser value (6.66%) in the contribution chart, so as to make its contribution more significant. In the first section of table I, the smaller contributors correspond to FE4. Consequently, its tolerance will be increased in an objective to decrease the cost of the assembly. Therefore the tolerance tg of 0.1 will be replaced by 0.2 and the step 4 will be executed to obtain a new value of ±0.8013 for FR. This value is still significantly below the ±1mm objective targeted for FR by the designer. As a result, the contribution chart is consulted again to adjust another tolerance in the dimensional chain. The critical contributor is now FE3 (contribution of 13.89%) so the tolerance te and tf are reassigned values (see section 3 in table I, H11 and h8) and the corresponding FR finally increases to ±1.032.

After three iterations, the FR computed (±1.032) by the unified model is in line with the designer's specification of ±1mm. The corresponding tolerance values shown in the last section (section 3) of table I are then accepted accordingly.


## 6.   CONCLUSION

A unified Jacobian-Torsor model with intervals has been used to allocate tolerances in a dimensional chain. Through a numerical (percentage contribution) and graphical (chart, histogram) approach, the proposed methodology guides the designer into choosing the proper tolerances. The percentage contribution chart tells the designer how each tolerance has contributed to the total FR. Once these contributions have been identified, the designer proceeds to modify the most critical tolerance zones so as to respect the Functional Requirement imposed. Finally, this general synthesis method can be used in both a deterministic (WC) and easily adapted to statistical context. This is the subject of current developments. In statistical approach, contribution's functional element (equation (4)) is replaced by statistical contribution using the theory of Root-Sum-Squares in statistical. In the example that was presented, changes in the various intervals were manually specified and the model was run for each new set of values. The changes in the various tolerance values were based on the relative contribution of each Functional Element. Cost and the optimal tolerance values for the assembly were not taken into consideration. This is the subject of current developments. The objective is to have a model that can perform the search for optimal values automatically using some cost tolerance function proposed in the literature and the contribution percentage of each FE to the total FR in the assembly.

## REFERENCES

**[Angus 2001]** Angus, J.; "Combined parameter and tolerance design optimization with quality and cost"; In: *International Journal of Production Research*, v39, n5, pp. 923-952; 2001.

**[Bararaja and al. 2005]** Barraja, M; Vallance, R.; "Tolerancing kinematic couplings"; In: *Precision Engineering*; v29, pp. 101-112,2005.

**[Chen 2001]** Chen, M.-C.; "Tolerance synthesis by neural learning and non-linear programming"; In: *International Journal of Production Economics*; v70; pp.55–65; 2001.

**[Chase et al., 1995]** Chase, K.W.; Greenwood, W.H.; "Design issues in mechanical tolerance analysis"; In: *Manufacturing Review*, ASME;v1;n1;pp. 50 –59;1995.

**[Desrochers et al., 2003]** Desrochers, A.; Ghie, W.; Laperrière, L.; "Application of a unified Jacobian –Torsor model for tolerance analysis"; In: *Journal of Computing and Information Science in Engineering*; v3;n1; pp. 2-14; 2003.

**[Ghie 2004]** Ghie W.; "Modèle unifié Jacobien-Torseur pour le tolérancement assisté par ordinateur "; In: *Thèse de doctorat*; Université à Sherbrooke Canada; 2004.

**[Hong et al. 2002]** Hong, Y. S.; Chang, T. C.; "A comprehensive review of tolerancing research"; In: *International Journal of Production Research*, v40, n11, pp.2425-2459, 2002.

**[Jinsong et al. 1998]** Jinsong, G.; Kenneth, W; Chase, K.W; Spencer, P.M.; "Generalized 3-D tolerance analysis of mechanical assemblies with small kinematic adjustments"; In: *IIE Transactions*; v30, pp.367-377; 1998.

**[Pradeep et al. 2005]** Pradeep K.S.; Satish, C. J.; Pramod, K.J.; "Advanced optimal tolerance design of mechanical assemblies with interrelated dimension chains and process precision limits"; In: *Computers in Industry*; in press; 2005.

**[Requicha 1984]** Requicha, A. A. G.; "Representation of tolerances in solid modeling: issues and alternatives approaches"; In: *Solid modelling by computer*; pp. 3-19; 1984.

**[Requicha 1983]** Requicha, A. A. G.; "Toward a Theory of Geometrical Tolerancing", In: *International Journal Robotics* Research; v2,n4, pp. 45-60; 1983.

**[SPO 73]** Spotts, M.F.; "Allocation of Tolerances to Minimize Cost of Assembly"; In: *Journal of Engineering for Industry*; ASME, v95, pp. 762-764; 1973.

**[Weber et al.1999]** Weber, C.; Britten, W; Thome, O.; "Conversion of geometrical tolerances into vectorial tolerance representations a major step towards computer aided tolerancing"; In: *International Design Conference*, Dubrovnik, May 19-22;1998.

**[Wirtz 1998]** Wirtz, A.;"Vectorial tolerancing"; In: *International Conference on CAD/CAM and AMT*; CIRP; Israël, Jérusalem; decembre11-14; 1998.

**[WEI 1997]** WEI, C-C; "*Allocation tolerances to minimize cost of nonconforming assembly*"; In: Assembly Automation, n4, pp303-306; 1997.

# Complex Mechanical Structure Tolerancing by Means of Hyper-graphs

*M. Giordano, E. Pairel, P. Hernandez*
*LMECA (Laboratoire de Mécanique Appliquée)*
*Ecole Supérieure d'Ingénieurs d'Annecy,*
*Université de Savoie, BP 806, 74016 Annecy cedex16, France*
*Max. Giordano@esia.univ-savoie.fr*


**Abstract:** The topological structure of a mechanism is often represented using graphs. In the field of tolerance, the graph constitutes a support for the analysis or the qualitative synthesis of dimensional and geometric tolerances. Unfortunately, the number of elementary features and relations is important even for usual mechanical systems, and they are represented by very complex graphs. In this paper, we propose to represent the mechanisms by hyper-graphs whose vertices themselves are graphs. One creates a set of different graphs, one in the other, like "nested dolls" from the low level of an elementary graph, to the highest level and the most detailed. This structure of hyper-graph is used as a support with a method of tolerances synthesis. This approach allows to deal with a complex tolerancing problem like a succession of more simple problems of tolerance synthesis. An example will illustrate this approach for the qualitative aspects. The graph allows to represent the topological structure of the mechanism and to consider the functional data such as the type of assembly and the functional requirements. Although already used by various authors, the graphs are today generally considered only as tutorial to guide the choice of the specifications. Our intention is to give to the graphs a role much more significant and even fundamental for the qualitative and quantitative analysis and synthesis of the tolerances. To our knowledge, the concept of hyper-graph is a very new approach in this field.

**Keywords**: tolerancing, graph, hyper-graphs.


## 1.   INTRODUCTION

The topological structure of a mechanism is often represented by graphs. This method is current for multi-body dynamic analysis [Samin and Fisette, 1997]. The graph enable to classify mechanisms in different types and to contribute to their synthesis for applications for example for planar kinematic chains in robotics [Mitrouchev, 2000]. In this case, generally, the vertices of the graph are the joints and the edges link the joints of the same part. Graphs are also used for the representation of the assembly sequences



*J.K. Davidson (ed.), Models for Computer Aided Tolerancing in Design and Manufacturing,* 105–114.
© 2007 Springer.

[Xinwen Niu et al. 2003]. Each vertex is a part or a subset of parts, and each edge is an assembly operation.

In the field of tolerancing, the graph constitutes a support for the analysis or the qualitative synthesis of dimensional and geometric tolerances. Generally, the vertices of the graph represent the geometric features to be toleranced while the edges represent contacts or functional requirements or the geometrical or topological relations between the features. Ballu and Mathieu uses a such representation [Ballu and Mathieu, 1999]. Kandikjan, Shah and Davidson add others edges in the graph, to represent specific geometric relations between the features. They aim to check the validity of given geometric tolerances, in a qualitative aspect [Kandikjan et al., 2001]. But in our point of view, the graph have only to represent the topological structure of the mechanism but not its geometry, otherwise, the graph is more complex than the drawing and then is not useful.

In all the cases, unfortunately, the number of elementary features and relations is important even for simple mechanical systems, and they are represented by very complex graphs. If the graph is too complex, it will not be really useful for the representation of the mechanical structure of for tolerances synthesis.

The graph is for the 3D model of tolerances what the chain of dimensions is for the uni-dimensional model. Indeed, the chains of dimensions constitute a particular case where the solids and the joints form a simple loop. The traditional method of the chains of dimensions consists in plotting a chain of vectors associated with each functional dimension. In the general 3D case there are often several contacts on the same part and the simple loops are rare. The graphs defined here allows to represent the topology of these connections.

Work on the concept of Technologically and Topologically Related Surfaces (TTRS) showed the importance of the topology and the functionality of contact faces in a mechanism for tolerance analysis and synthesis [Clément et al., 1991], [Salomon et al., 1996]. The authors show that the surfaces must be associated according to the functional topology and requirements. The basic TTRS are the functional features. A relation of equivalence anable to classify them according to seven classes. An operation of reclassification of two related TTRS creates a new TTRS belonging to one of the seven classes. A binary tree is thus built whose leaves are the functional features. But no method allows to build this tree automatically nor to deduce the standardized specifications from it.

Although based on the concept of associated feature, the here presented notions are different on many points. Functional features will be vertices of a graph which is not necessarily a tree. From this graph a tolerance scheme will correspond to standardized specifications. For example, one will build a hierarchical related features in order to consider a datum reference system. One will be able to associate features at the same level for the tolerance of a group of features or for several features in a common zone or to take a group as datum.

In this communication, we propose first to represent the geometric tolerance scheme of a part, thanks to a "tolerancing graph". Then the mechanical structure, the contacts between the features, the functional requirements and the tolerances will be represented by an hyper-graph. In an hyper-graph, the vertices are graphs themselves and one edge can be a subset of several edges. Thus, one creates a set of graphs, each one in the other, like "nested dolls" from the lowest level of an elementary graph until the highest level and most detailed. At each level, the vertices are graphs of the higher level and each arc is a set of arcs of the higher level. This method has already been used for tolerancing qualitative approach by Giordano and Hernandez [Giordano et al., 2003].

The rules which allows to build the graphs starting from the functional data and the geometrical properties of the surfaces are not described here. Only is defined the way of describing the topology of the mechanism, the functionalities and the associations of surfaces and finally the tolerances.

The contact graphs and the tolerancing graphs are first defined, then, the hyper-graphs are described from an example. Finally, the qualitative tolerance synthesis will be considered, through the same example.


## 2.   CONTACT GRAPHS AND TOLERANCING GRAPHS

For the graph of the mechanism, the classic method is used: the vertices represent the parts and the edges are the joints between two parts. It can be a multi-graph when there are several joints between the two same parts.

### 2.1 Contact graph

The contact graph of a mechanism is a graph where the vertices are the functional features of the part. The contact faces that are implied in the joints are generally considered as functional features but also the features concerned by functional requirements. Two types of edges can be distinguished: the contact between two features of different parts, linked together by a joint, and the edges between the features of the same part.

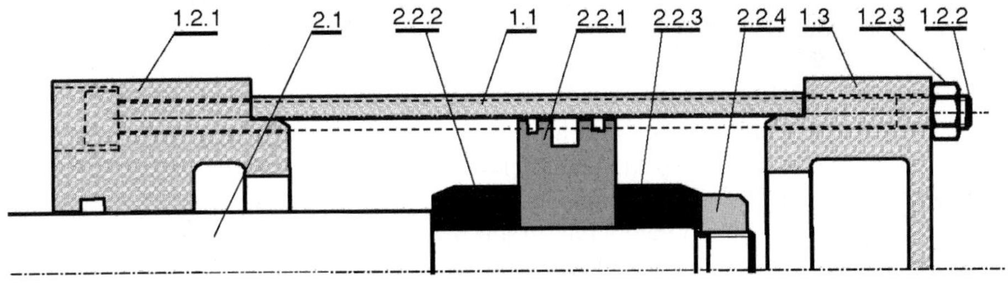


*Fig.1; hydraulic cylinder*

The figure 2 is the contact graph for an hydraulic cylinder represented in fig.1. It is a very simple mechanism with only a translation motion between two subsets. Ten parts are taken into account, (the seals are not represented),  but there are 36 different features represented by small circles (the vertices of the graph), and 17 joints between the features, represented by segments (the edges of the graph). The parts are represented by the big circles.

- The four edges, plotted in dotted lines, represent functional requirements:
- The rod piston (2.1) must have a linear motion with regard to the head (1.2.1) and to the cap (1.3) assembled on a support (functional requirements FR1 and FR2).

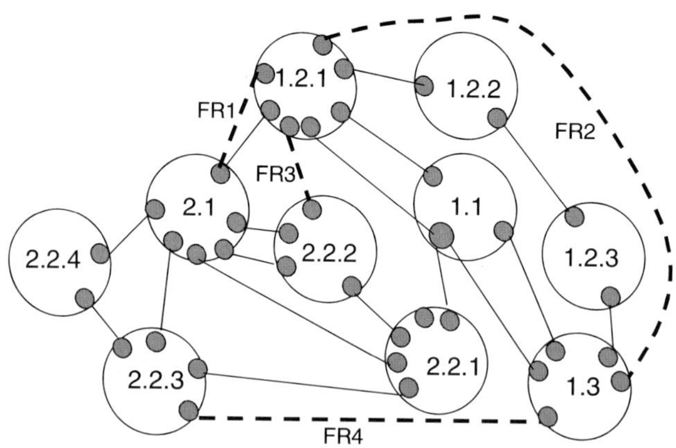


*Fig.2; Contact graph of the hydraulic cylinder fig.1*

The cushioning bushings (2.2.2) and (2.2.3) must be able to pass into the holes of the head (1.2.1) and the cap (1.3) with a clearance lying between two defined values, a minimum and a maximum value.

Generally, a circle is drawn to precise what are the different features attached to the same part. Yet, the edges between the features are not clearly identified. It will be interesting to use the general properties of a graph, so it is necessary to define a formal graph. Thus, the edges between the features have to be defined. For example the graph can allow to find the different paths between two vertices

### 2.2 Tolerancing graph

The tolerancing graph for each part translate the geometrical tolerances, orientation and position, on a qualitative aspect. The toleranced features and the datum are defined on the tolerancing graph. From a standardised tolerance, it is always possible to build the tolerancing graph. The tolerancing graph of the complete mechanism is the concatenation of the contact graph and the tolerancing graphs for each part.

In a tolerancing graph, the vertices are toleranced feature or geometrical or dimensional tolerances. The edges are particular links between two tolerances, or between a feature and a tolerance. These links precise the semantic of the tolerances. The edges are oriented. Two types of edges are considered: the first ones that precise the toleranced feature, oriented from the tolerance to the feature, and the second ones define the datum, oriented from the datum to the tolerance. The figure 3 shows an elementary case of a geometric tolerance with a simple datum feature.

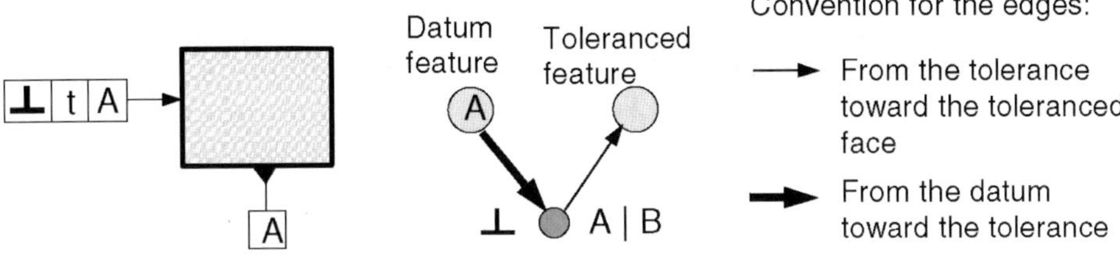



*Fig.3; Simple tolerancing graph*

Formally, a cartesian ortho-normal frame is associated to each vertex. A frame is attached to each toleranced feature of the part. An arbitrary rule can be chosen to define the origin and the three axes of the frame, according to the type of feature (figure 4).

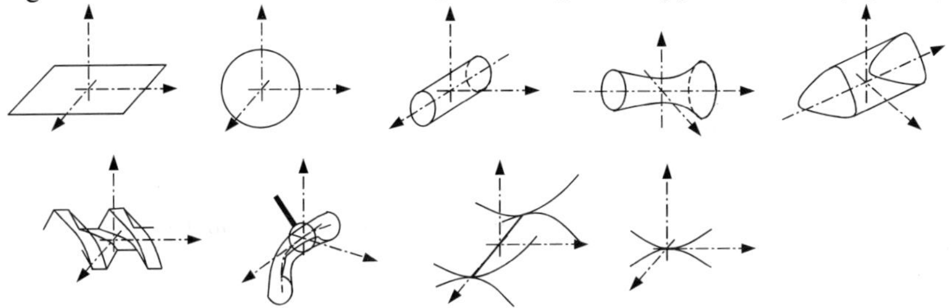

*Fig.4; A frame attached to a feature*

This frame is built from the toleranced feature and from the datum by a conventional method. If the datum is a datum system built from different features, it is also represented by a vertex. For example, (fig.5), it can be considered that the frame attached to the tolerance of perpendicularity is the datum system where A is the primary datum and B the secondary datum. The frame attached to the vertex A has the three degrees of freedom of the plane, while the datum system attached to the tolerance of perpendicularity has only one degree of freedom: the translation parallel to the two planes A and B. More generally, a cartesian frame with particular degrees of freedom is attached to each vertex.

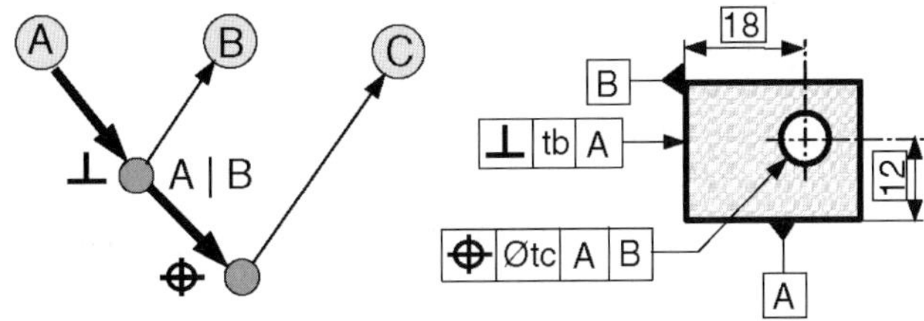



*Fig.5; A toleranced part and the corresponding tolerancing graph*

A new feature can be associated to a tolerance. This feature is built from the toleranced feature and the datum. In the example shown fig.6, for a size tolerance between two parallel planes, there is no datum. The two planes are toleranced between each others. It is the symmetric plane that is associated to this size tolerance. This symmetric plane is the datum (A) for the geometric tolerance of two others planes. For this symmetry tolerance, the toleranced feature is built from the two parallel planes features. More exactly from the two actual features parallel in the nominal position.

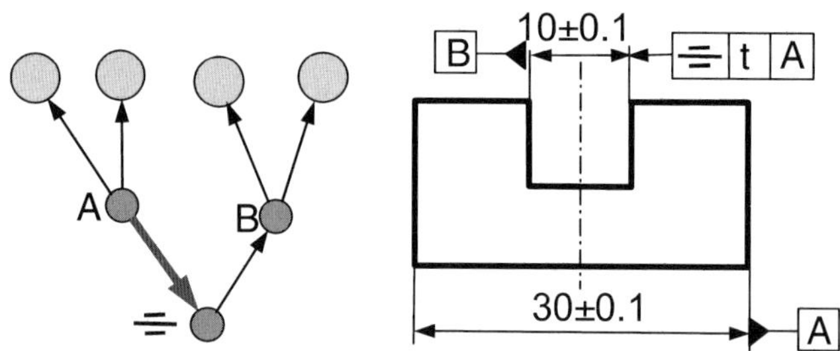

*Fig.6; A tolerance can define a new toleranced feature*

The method can be generalised to all types of tolerances including the case where the toleranced feature or the datum system is a group of features.
From any toleranced part, it is also possible to build a tolerancing graph, but reciprocally from a tolerancing graph, different solutions in conformity with the standard representation of the tolerances can be found. For some graphs, it is not possible to find standardised tolerances. It is then necessary to modify the graph to make possible the translation. For example, the standard do not allow to tolerance any group of feature but only some particular cases like, for example, a pattern of holes or a group of parallel planes. In this case the solution consists in defining a tolerance for each feature of the group.
The tolerancing graph can be used for example to check the consistency of a given tolerancing scheme, and to check if each functional surface is toleranced.
In a first step, tolerancing graph can be designed from functional requirement. Then different solutions for standardised geometric tolerances can be proposed  for the same tolerancing graph.


## 3.   TOLERANCING HYPER-GRAPHS

From the definition of a contact graph for a mechanism and of a tolerancing graph for a part, it is easy to define the tolerancing graph for a mechanism. We call a complete tolerancing graph of a mechanism, a graph where all the contact features are represented. Each feature of the graph must be toleranced. And they must be toleranced each one, with respect to a datum built from other features.

But even for simple mechanisms where the number of parts is small, the tolerancing graph is complex. For the example of the hydraulic cylinder, in the kinematic graph there are only two vertices representing the piston and the cylinder and one edge representing the cylindrical joint between the two parts. Functional requirement are not represented here. Between these two types of graphs, from the simplest, the kinematic graph, to the more complex, the complete tolerancing graph, it is possible to consider different levels of complexity, with a graph for each level.

These various levels form a tree of graphs. At the root, on the low level, the graph consists of only one vertex. The kinematic graph is at the level just above, and only the principal parts in relative motions and the joints between these parts appear. At the highest level, the vertices of the graph are frames attached to functional features and the edges represent links between functional features making it possible to structure the tolerances from the qualitative point of view.

*Figure 7; graph of the hydraulic cylinder, level 1.*

It is easy to build the graph from a level to the upper level just above. The vertices belonging to the same part form a sub-graph. This sub-graph is replaced by a vertex of the upper level. One obtains a multi-graph in the general case, but the edges between two same vertices are then replaced by a simple edge to make a simple graph. For example the graph figure 8 can be easily transformed into the graph figure 7.

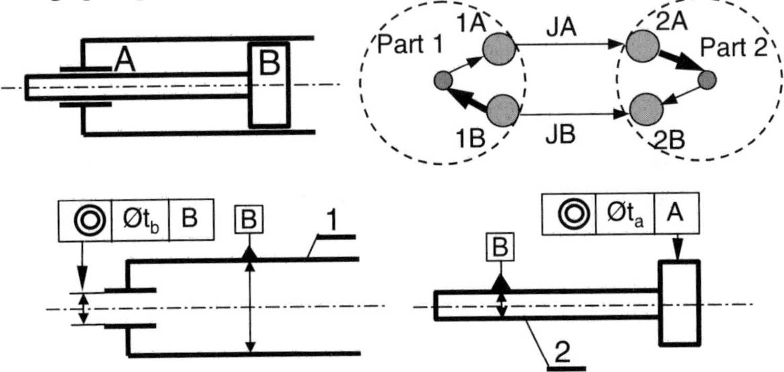


*Figure 8; graph of the hydraulic cylinder, level 2.*

On the same way, the graph in the fig.9 is the vertex P2 of the graph fig.8 and fig.7. For each level it is possible to built the graph of the lower level by a simplification that consists in regrouping vertices or edges. However, the reciprocity is not so easy. The vertices representing the feature are directly defined from the technological choices, but for the tolerances it is necessary to imagine a strategy that generates them from these technological choices.

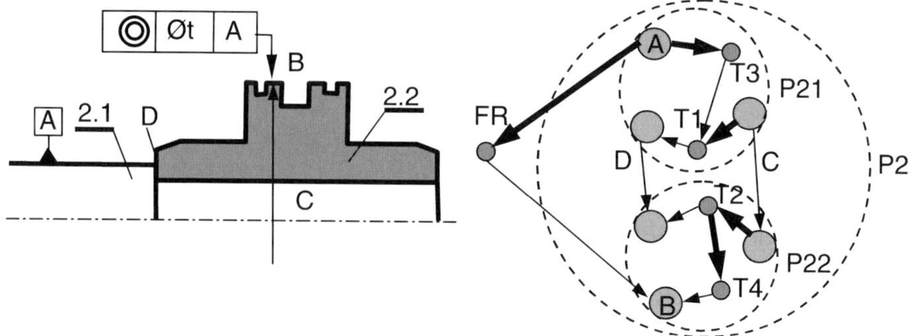

*Figure 9; theoretical sub-graph for the piston and the rod.*

## 4.    APPLICATION TO TOLERANCE SYNTHESIS

This structure of hyper-graph is used as a support for a tolerance synthesis method. Each joint between the parts and the functional requirements are represented by edges of the graph. They will become functional requirements for the higher level, up to the top level where the conditions will be tolerances, written by using the standardised syntax. This Cartesian approach makes it possible to deal with a complex problem like a succession of simpler problems. In the example fig.9, the tolerances T1 and T2 are due to the joint between parts P21 and P22 by the features C and D. The theoretical method to link the feature A to the group C and D consists in adding the tolerances T3 and T4 for the part P22. It is difficult to find a standardised tolerance corresponding to the sub-graph P21, so it is transformed into the sub-graph and the associated tolerances shown fig.10.

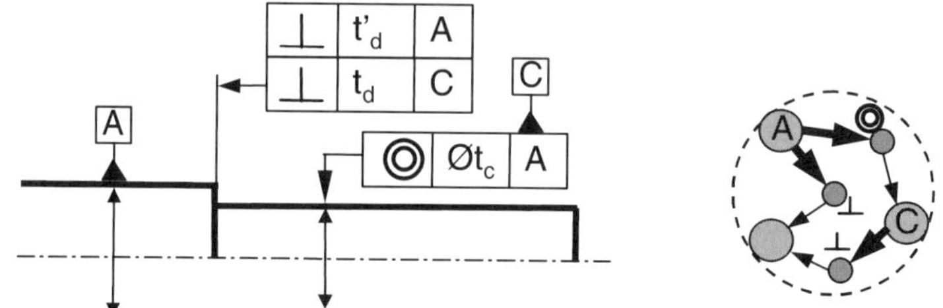

*Figure 10; modified graph and standardised tolerances.*

The method makes possible a qualitative synthesis of the tolerances, but also is used as a support with a quantitative synthesis. Since a Cartesian frame is attached to each vertex of the graph, a displacement can be associated to each edge. The torsor of small displacements allows to determine the displacement between any two vertices of the graph if the displacement associated to each edge is known. Reciprocally, if a functional requirement is defined like a relation between two vertices of the graph, the qualitative problem consists in searching the relations associated to the edges which contribute for this requirement.


## 5.  CONCLUSION

The tolerancing hyper-graph representation is a very efficient tool to analyse different levels of complexity in the same data structure. The graphs can be an essential support for the analysis and the synthesis of tolerances.  It has been shown that they can be built in a progressive way starting from the functional requierements and from the selections of the designer. By defining the joints between the parts of the mechanism and the role of each functional feature, one builds a set of graphs from the simplest level to the most complete level containing the tolerance scheme. From these graphs, and by referring to a standard (ISO or ASME Y14.5) it is possible to define a  tolerance scheme, thus to define the tolerances on the qualitative aspect.  The graph will be used for the computation of the stack up tolerances, and to evaluate the functional and assembly requirements.
Consequently the hyper-graph representation is a support to define a method of tolerance synthesis. However, two important questions of the method must be considered in a more general case: How to build automatically, the tolerance graph from the contact topology at each transfer from a level to the higher level? How to generate standardised tolerances from the tolerancing graph of each part?  This communication gave partial answers to these questions through a simple example.
However, the tolerance hyper-graph seems a very powerful tool for tolerance synthesis of complex mechanisms.


## REFERENCES

[**Samin and Fisette, 1997**] Samin J.C., Fisette P., "The benefits of symbolic generation in modelling, identifying and simulating multi-body systems", *European Journal Mech. Eng.* Vol 42, N°3, pp.137-143.

[**Mitrouchev, 2000**] Mitrouchev P.; "Symbolic structural synthesis and a description method for planar kinematic chains in robotics"; *Eur. Journal Mech. A/Solids, Elsevier,* Vol 20, (2001), pp. 777-794.

[**Kandikjan et al., 2001**] Kandikjan T, Shah J.J, Davidson J.K.; "A mechanism for validating dimensioning and tolerancing schemes in CAD systems"; *Computer-Aided Design, Elsevier, Vol 33 (2001)*, pp. 721-737.

[**Xinwen Niu et al. 2003**] Xinwen Niu, Han Ding, Youlun Xiong; "A hierarchical approach to generating precedence graphs for assembly planning"; *International journal of Machine Tools & Manufacture Elsevier,* Vol. 43 (2003) pp 1473-1486.

**[Giordano et al., 2003]** Max Giordano, Pascal Hernandez, Serge Samper, Eric Pairel "Expression of geometric functional requirements dedicated to concurrent engineering"; acts of the CIRP Design Seminar, Grenoble France, May 2003.

**[Ballu and Mathieu, 1999]** Ballu A. and Mathieu L., Choice of functional specifications using graphs within the framework of education, Proceedings of the $6^{th}$ CIRP Int. seminar on Computer-Aided Tolerancing, Uni, Twente, Ensched, The Netherlands, March 1999, pp.197-206.

**[Clément et al., 1991]** A. Clément, A. Desrochers, and A. Rivière, Theory and practice of 3D tolerancing for assembly", Proceeding of the CIRP Seminar on Computer Aided Tolerancing, Penn State University, U.S.A., May 1991.

**[Salomon et al., 1996]** O.W. Salomons, H.J. Jonge Poerink, F.J. Haalboom, F. van Slooten, F.J.A.M. van Houten, H.J.J. Kals, "A computer aided tolerancing tool I : Tolerance specification", Computers in Industry, 31, pp. 161-174, Elsevier 1996.

**[Salomon et al., 1996]** O.W. Salomons, H.J. Jonge Poerink, F.J. Haalboom, F. van Slooten, F.J.A.M. van Houten, H.J.J. Kals, "A computer aided tolerancing tool II : Tolerance analysis", Computers in Industry 31 pp. 175-186, Elsevier 1996.

# An Efficient Solution to the Discrete Least-Cost Tolerance Allocation Problem with General Loss Functions


*J. Lööf*, T. Hermansson**, R. Söderberg****

**Chalmers University of Technology, Product and Production Development, SE-412 96 Göteborg Sweden,* johan.loof@chalmers.se

***Fraunhofer-Chalmers Research Centre, Chalmers Science Park, SE-412 88 Göteborg Sweden,* tomas.hermansson@fcc.chalmers.se

****Chalmers University of Technology, Product and Production Development, SE-412 96 Göteborg Sweden,* rikard.soderberg@me.chalmers.se



**Abstract:** The tolerance allocation problem consists of choosing tolerances on dimensions of a complex assembly so that they combine into an 'optimal state' while retaining certain requirements. This optimal state often coincides with the minimum manufacturing cost of the product. Sometimes it is balanced with an artificial cost that the deviation from target induces on the quality of the product.

This paper analyses and suggests a solution to the discrete allocation problem. It also extends the problem to include treating general loss functions. General loss in this paper means an arbitrary polynomial function of a certain degree. We also briefly review the current work that has been made on solving the tolerance allocation problem.

**Keywords:** Tolerance allocation, quality loss, discrete optimization.


## 1. INTRODUCTION

As competition between manufacturing industries progresses, it becomes increasingly more important that the products be manufactured at a low price and be of generally high quality. Otherwise, customer confidence cannot be maintained. The variation of individual parts accumulates when they are assembled, and affects the variation of critical measures on the product, such as the gap between a door and a fender on a car. These relationships can be analyzed by CAT (Computer Aided Tolerancing) software. Adjusting individual part tolerances can reduce the variation of critical measures. However, this often results in a higher manufacturing cost. Tolerance allocation consists of choosing the tolerances in an optimal way (i.e. such that the assembled variation is controlled, at the same time minimizing the manufacturing cost in a quality aspect). The

115


*J.K. Davidson (ed.), Models for Computer Aided Tolerancing in Design and Manufacturing,* 115–124.
© 2007 *Springer.*

                    J. Lööf, T. Hermansson and R. Söderberg

general tolerance allocation problem allows the tolerances to be chosen within a tolerance band. The discrete tolerance allocation problem, on the other hand, chooses from a finite set of available tolerances.

In this paper, we aim to concentrate on the discrete tolerance allocation problem. That problem can be interpreted as a semi-assignment optimization problem with budget constraints, and is further explained in Section 3. This can be justified by the fact that it is in principle unrealistic that manufacturers can adjust the tolerances perfectly according to the optimal (continuous) solution. More often, the manufacturer can only choose between a finite number of machines and/or part manufacturers. This suggests that a discrete set of tolerances is more suitable. However, this is computationally a much harder problem than in the general case.

## 2.  REVIEW OF TOLERANCE ANALYSIS AND ALLOCATION METHODS

This section reviews earlier and related work on the subject. Many authors have studied tolerance allocation (also called tolerance synthesis) over a long time period. We first supply a short introduction to tolerance analysis.

### 2.1. Tolerance analysis

Let a product be defined by $n$ dimensions represented by stochastic variables $x_1, ..., x_n$, where each $x_i$ is related to a statistical distribution with expected value $\mu_i$ and standard deviation $\sigma_i$. Let $D_i$ be the nominal value for $x_i$ and $t_i$ its tolerance, such that $x_i \in (D_i - t_i, D_i + t_i)$ must hold. Furthermore, let $y$ be a critical measure on the product. The relation between y and $x_1, ..., x_n$ can be described by an assembly function $f$:

$$y = f(x_1, ..., x_n).  \tag{1}$$

$f$ is often approximated by a linearization about the expectance values $(\mu_1, ..., \mu_n)$:

$$y = f(\mu_1, ..., \mu_n) + \sum_{i=1}^{n} a_i(x_i - \mu_i),  \tag{2}$$

where $a_i = \dfrac{\partial f}{\partial x_i}(\mu_i)$ are called the sensitivity coefficients. The assembled tolerance $\tau$ for $y$ can now be estimated (statistically) by

$$\tau^2 = C_{p_y}^2 \sum_{i=1}^{n} \left( \frac{a_i t_i}{C_{p_i}} \right)^2,$$

where $C_{p_y}, C_{p_i}$ are the process capability indices.

### 2.2. Tolerance allocation

In an early design phase, the nominal values are assigned to each dimension of the product. From design requirements, an upper bound on the variation of each critical measure is known. The task is then to specify tolerances such that (a) these bounds are not violated and (b) a certain property is optimized. This property does not necessarily have to coincide with the manufacturing cost in any way; it could be a pure geometric characteristic that is desired to be minimized/maximized. Another application could be to allocate tolerances such that they each contribute 'as equally as possible' to the critical measure variation. Hermansson and Lööf [Hermansson and Lööf, 2005] modeled and implemented this particular instance as a GUI (Graphical User Interface) coupled with commercial CAT software [RD&T, 2002].

The most common application of tolerance allocation is nevertheless to minimize the manufacturing cost. This can be modeled as

$$\min \quad \sum_i C_i(t_i)$$
$$\tau_j \leq V_j \tag{3}$$
$$l_i \leq t_i \leq u_i,$$

where a cost function $C_i$ has to be available for each tolerance $i$. Note that we have not yet in this model restricted us to considering only discrete tolerance values; rather, we allow tolerances to be allocated within a whole tolerance band $(l_i, u_i)$. $V_j$ is the upper bound on the variation on measure j that must not be exceeded.

Many different cost functions have been suggested to use together with the continuous tolerance optimization problem (3). Table 1 lists a selection of cost functions that have occurred in literature.

| Cost Model | Name | Author |
|---|---|---|
| $A - B \cdot t$ | Linear | Edel & Auer |
| $A + \dfrac{B}{t}$ | Reciprocal | Chase & Greenwood Parkinson |
| $A + \dfrac{B}{t^2}$ | Reciprocal squared | Spotts |
| $A \cdot e^{-B \cdot t}$ | Exponential | Speckhart |

*Table 1; A Selection of Cost Models [Chase and Parkinson, 1991]*

The coefficient $A$ represents a fixed cost, such as machine set-up, material, etc. If one of these functions describes the true manufacturing cost correctly, solving the optimization problem (3) will indeed result in a minimized production cost for the manufacturer.

### 2.3. Quality loss

Taking into account the quality loss the assembled variation brings to the customer, it is favorable to add a penalty function $y \mapsto L(y)$ to the objective function. The penalty function punishes a deviation from target (T) for each measure.

Taguchi used a quadratic penalty function (4) [Taguchi *et al.,* 1989].

$$L(y) = z(y - T)^2 \tag{4}$$

Choi used the quadratic loss function (4) for the tolerance allocation problem [Choi *et al.,* 2000]. Söderberg refined this approach further by introducing the monotonic loss function [Söderberg, 1995]. The asymmetric quadratic loss function (5) punishes the deviation from the target differently depending on the direction.

$$L(y) = \begin{array}{ll} z_1(y - T)^2 & y < T \\ z_2(y - T)^2 & y \geq T \end{array} \tag{5}$$

Since a loss function on this form has the (stochastic) variable y as input, whereas our optimization problem (3) is stated in the tolerances $t_i$, the expectance of L constitutes a good measure on the quality loss in terms of (3):

$$E[L(y)] = \int_{-\infty}^{\infty} L(y)F(y)dy, \tag{6}$$

where F(y) is the probability density function. In this paper, we will not restrict ourselves to a specific loss function. We will accept general loss functions in the sense that they can be described by an arbitrary polynomial of a certain degree. It is worth noting, however, that an implicit expression of the expected quality loss of (4) is trivial to derive:

$$\begin{aligned} E[L(y)] &= E\big[z(y - T)^2\big] = zE\big[(y - \mu)^2 + (\mu - T)^2 + 2(y - \mu)(\mu - T)\big] \\ &= z\sigma^2 + z(\mu - T)^2 + 2z(\mu - T)(E[y] - \mu) = z\sigma^2 + z(\mu - T)^2. \end{aligned} \tag{7}$$

Otherwise, these calculations can be carried out with an appropriate quadrature rule, a method for approximating integral calculations [Heath, 1997].

Solution methods used on the continuous allocation problem are, among others, SQP (Sequential Quadratic Programming), simulated annealing, genetic algorithms, and iterations on Lagrangian multipliers [Hong and Chang, 2002].

So far in this paper, tolerances may be chosen from within an enclosed region. However, it is more likely that the manufacturer may only choose between a finite number of machines and/or manufacturers. The optimal solution is also closely related to which cost function one aims to use (See Table 1). This suggests that considering a discrete set of tolerances could be more suitable to a real-life situation. This does not make the problem any easier, despite what one might expect. Before we introduce the discrete tolerance allocation problem, we need a brief review of the field of discrete optimization and some common solution methods.

## 3.  DISCRETE OPTIMIZATION

A discrete optimization problem is to

$$\text{minimize } f(x) \text{ over all } x \in S ,$$

or short

$$\min f(x)$$
$$x \in S, \tag{8}$$

where $f : X \supset S \mapsto R$ is an objective function and $S$ is a set of objects of some discrete structure in the underlying space $X$. $S \subset X$ is called the *feasible set*, and an object $x \in S$ is called a *feasible solution*. An *optimal solution* $x^* \in S$ is a feasible solution such that

$$f(x^*) \le f(x) \text{ for all } x \in S .$$

An optimization problem can either have an optimal solution or the problem is infeasible, that is if $S$ is empty or if the problem is unbounded. If $X$ is the set of all integers, (8) is referred to as an *integer program* (IP). This text will mainly focus on IPs. If, in addition, both $f$ is linear and $S$ is linearly constrained, they define together an *integer linear program* (ILP).

### 3.1. Solution methods

Since IPs are NP-hard [Do and Ko, 2000], there exist no general computationally effective algorithms for solving them. Solution methods can be categorized into *optimal methods* and *heuristics*.

### 3.1.1. Heuristics

Heuristics (or approximate algorithms) are designed to quickly find good (but not necessarily optimal) solutions to specific types of problems. This class of algorithms includes *simulated annealing* and *genetic algorithms*. The latter has been widely adapted to the tolerance allocation problem ([Iannuzzi and Sandgren, 1994], [Chen and Fischer, 2000], and [Prabhaharan *et. al.,* 2004]). *Dual heuristics* has been shown to be very useful: they can provide good lower bounds on the optimal value in a branch & bound algorithm.

### Lagrangian and LP relaxation

Consider the problem

$$f^* = \min f(x)$$
$$c(x) \ge 0 \tag{9}$$
$$x \in X.$$

We define the Lagrangian function $L : X \times R_+^m \mapsto R$ as follows:

$$L(x,\mu) := f(x) + \sum_{i=1}^m \mu_i^T c_i(x) = f(x) + \mu^T c(x).$$

The *Lagrangian dual problem* to (9) is then to

$$q \;= \min q(\mu) \atop \mu \in R_+^m, \tag{10}$$

where the dual function $q(\mu) := \max_{x \in X} L(x,\mu)$. The concavity property always holds for (10), which often (but not always) makes this problem easy to solve. There are then methods to transform this dual solution to an approximate primal feasible solution.

Weak duality states $q^* \leq f^*$, and in general IP's strong duality ($q^* = f^*$) cannot be expected. The advantage of Lagrange relaxation is that one can relax certain constraints and keep others in the underlying space and always receive a lower bound on the optimal value to the original (primal) problem. Each configuration will yield a different duality gap. A special case of Lagrangian relaxation is *LP relaxation* of an ILP, where the integrality condition is removed and the resulting LP is solved. Again, since the feasible set is expanded, this solution bounds the optimal value of the ILP from below. This weak duality property suggests that a dual heuristic is promising to use in an *optimal method*.

### 3.1.2. Optimal methods

Optimal methods guarantee finding the optimal solution. An initial approach to trying to find an optimal solution to (8) would be to carry out a *complete enumeration* of the feasible set $S$ in an intelligent way and pick the optimal feasible solution. *Branch & bound algorithms* consist of two main components:

- A successful partitioning of the feasible set into mutually disjoint subsets $S_i$ (branching)
- An algorithm is available to calculate (lower) bounds $z_i$ of the objective function on these subsets $S_i$ (bounding).

Now, if the lower bound $z_i$ on the subset $S_i$ is higher than the optimal feasible solution found so far, we can completely rule out the optimal solution being in $S_i$. In the worst case, however, we still have to enumerate all the solutions.

We present the discrete tolerance allocation problem formulation in the next section and propose an optimal method for how to solve it in an efficient way.

## 4.  THE DISCRETE TOLERANCE ALLOCATION PROBLEM

### 4.1. Problem formulation

We now present our formulation of the discrete tolerance allocation problem. For each dimension $i$, denote the available tolerance choices $t_{ij}$ together with associated costs $c_{ij}$. The assignment of a tolerance $t_{ij}$ to a dimension $i$ is represented by the binary variable $x_{ij}$. Furthermore, let $a_{ik}$ be the sensitivity coefficients with respect to the critical measure $k$. We also assume that there is a function $L_k$ for each measure $k$ that somehow measures the quality loss a deviation of $k$ from target induces (Sec. 2.3). The problem becomes to

$$
\begin{aligned}
& \min \sum_{i,j} c_{ij} x_{ij} + \sum_k E[L_k] && & (a) \\
& \sum_{i,j} (a_{ik} t_{ij})^2 x_{ij} \le V_k^2 && k = 1,...,m & (b) \\
& \sum_j x_{ij} = 1 && i = 1,...,n & (c) \\
& x_{ij} \in \{0,1\} && i = 1,...,n, \ j = 1,...,n_i & (d)
\end{aligned}
\qquad (11)
$$

($11b$) is *a budget constraint*. It insures that the limit on the variation of each of the measures is not violated. ($11c$) is called an *assignment constraint*. It guarantees that exactly one tolerance is assigned to each dimension. Note that restricting the loss functions to polynomials together with (11d) preserves (11) as an ILP.

Ostwald and Huang solved the problem without loss function using an additive algorithm by Balas [Ostwald and Huang, 1977]. Lee and Woo developed an algorithm for ensuring the optimal selection among tolerances by exploiting the special structure of the constraints [Lee and Woo, 1989].

### 4.2. Problem solution method

We hereby propose an efficient optimal method to the tolerance allocation problem stated in equation (11). It is sometimes called Dakin's method. The method follows the optimal method framework presented in section 3.2, and consists of three main steps: (a) start with a relaxed LP, (b) branch on the binary variables by adding constraints, and (c) successively retrieve lower bounds by solving a series of 'warm-started' dual LPs until an optimal integer solution is found:

```
0.  Set z* := +inf, the best solution found so far
1.  Solve the relaxed linear program P yielding a solution X
2.  Pick a variable xi in X that is not 0 or 1
3.  Partition P into two sub-problems P1 and P2 by adding a constraint (xi =
        0 for P1 or xi = 1 for P2) for P respectively
4.  For each sub-problem Ps (s = 1,2),
        a.  Calculate a lower bound zs to the optimal solution by solving the
            dual to Ps
        b.  If Xs is pure binary and zs < z*, record Xs as the best solution
            found so far and set z* := zs
        c.  else if zs < z*, repeat from 2. with P := Ps and X := Xs
5.  return X*
```

The algorithm terminates when an optimal solution is encountered. The Dual Simplex Method [Nash and Sofer, 1996] solves the dual LPs in each step. Since adding a constraint in the primal LP corresponds to adding a variable in the dual LP, the solution to the previous dual LP is still a basic feasible solution. Hence, the initialization step of Simplex, namely finding a basic solution, is already done (a 'warm start'). This results in heavily reduced computational time. Branching in a more sophisticated way can further reduce the time. One example of that is considering the sub-problem with greatest bound $z$ first.

The proposed algorithm has been implemented in C++ as a module coupled with commercial CAT software [RD&T, 2002].

### 4.3. Test results and discussion

The module has been tested on a real car model (Figure 1). This test model consists of eight measures and 21 dimensions with five defined tolerance-cost couples each. This implies that we have 105 binary variables in our optimization problem. Data about the model has been collected from the CAT software RD&T [RD&T, 2002].

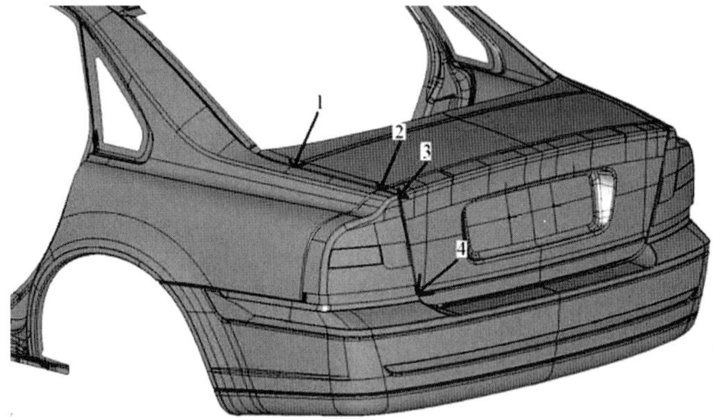


*Figure 1: Test model.*

The test has been divided into two sections: (i) an analysis of how the proposed algorithm fulfils the constraints on the measures, and (ii) a comparison between the suggested algorithm and exhaustive search in terms of computational time and optimal value, of which the latter of course should coincide.

Data from the first test is illustrated in Table 2. Eight measures are considered, four in the flush direction and four in the gap direction. Each measure has a number associated with it that corresponds to the arrows in Figure 1. For example, Flush_1 is the size of the flush between the body and the bootlid by arrow 1 in Figure 1. $\tau\_D$ is the accumulated measure variations/tolerances. V is the upper bound that must not be exceeded. The optimal cost in this very case is 25.1.

| Measures | V | $\tau\_D$ |
|---|---|---|
| Flush_1 | 2 | 1.9859 |
| Flush_2 | 2 | 1.8265 |
| Flush_3 | 2 | 1.5965 |
| Flush_4 | 2 | 1.7875 |
| Gap_1 | 3 | 2.9106 |
| Gap_2 | 3 | 2.8932 |
| Gap_3 | 3 | 2.7987 |
| Gap_4 | 3 | 2.7871 |

*Table 2; Measure variation in optimal solution*

| # mea. | z_D | z_E | Time_D | Time_E |
|---|---|---|---|---|
| 8 | 25.1 | 25.1 | 0.67 | 79.9 |
| 4 | 24.5 | 24.5 | 0.11 | 37.7 |
| 3 | 24.2 | 24.2 | 0.17 | 37.2 |
| 2 | 23.9 | 23.9 | 0.25 | 26.2 |
| 1 | 23.6 | 23.6 | 0.12 | 20.4 |

*Table 3; A Comparison*

The solution satisfies the constraints on the critical measures. In addition, it chooses tolerances in a way such that the accumulated measure is close to the defined limits. This is expected, since larger tolerances are associated with lower costs. Table 3 shows the optimal solution and the calculation times for the proposed algorithm (z_D). It also shows the exhaustive search (z_E) on a series of tests containing different numbers of measures.

The optimal solutions are the same for the two algorithms as expected. The big difference is the time it takes to evaluate the solution. For example, the proposed algorithm for eight measures is about 120 times faster than exhaustive search.

## 5.   CONCLUSION

The tolerance allocation of a mechanical assembly has great impact on production cost, functionality and quality. This paper includes a review of earlier work on the subject of tolerance allocation. An algorithm is proposed that solves the discrete tolerance allocation problem in an efficient way and *guarantees* that an optimal solution is found (as opposed to for instance genetic algorithms and simulated annealing). A module has been implemented in C++ together with the algorithm. The implementation is such that the tolerances can be allocated in a way that fulfils defined critical measures and minimizes the manufacturing cost. It is also possible to add loss functions that imply a high quality product. Finally, the allocation module is very effective. It solves a model that consists of eight measures and 21 dimensions with five defined tolerance-cost couples about 120 times faster than exhaustive search.

REFERENCES

**[Chase and Parkinson, 1991]** Chase, K.; Parkinson, A. R.; "A Survey of Research in the Application of Tolerance Analysis to the Design of Mechanical Assemblies"; ADCATS Report No. 91-1 April 1991.

**[Choi et al., 2000]** Choi, H-G. R.; Park, M.-H.; Salisbury, E.; "Optimal Tolerance Allocation With Loss Functions"; Journal of Manufacturing Science and Engineering; ASME; vol. 122; pp. 529-535; August 2000.

**[Chen and Fischer, 2000]** Chen, T-C.; Fischer, G.W.; "A GA-Based Search Method for the Tolerance Allocation Problem"; Artificial Intelligence in Engineering; Vol. 14; No. 2; pp.133-141; April 2000.

**[Do and Ko, 2000]** Do, D.-Z.; Ko, K.-I.; "Theory of Computational Complexity"; New York; Wiley; 2000; ISBN 0-471-34506-7.

**[Heath, 1997]** Heath, M., T.; "Scientific Computing: An Introduction Survey"; pp. 245-249; McGraw-Hill 1997; ISBN 0-07-115336-5.

**[Hermansson and Lööf, 2005]** Hermansson, T.; Lööf, J.; "Allokering av symmetriska toleranser i komplexa sammansättningar"; Master Thesis; Göteborgs Universitet; 2005.

**[Hong and Chang, 2002]** Hong, Y. S.; Chang, T-C.; "A Comprehensive Review of Tolerancing Research"; International Journal of Production Research; 2002; Vol. 40; No. 11; pp. 2425-2459.

**[Iannuzzi and Sandgren, 1994]** Iannuzzi, M.; Sandgren, E.; "Optimal Tolerancing: The Link Between Design and Manufacturing Productivity"; Design Theory and Methodology; ASME DTM ´94, pp. 29-42.

**[Lee and Woo, 1989]** Lee, W-J.; Woo T. C.; "Optimum Selection of Discrete Tolerances"; Journal of Mechanisms, Transmission, and Automation in Design; ASME; June 1989; Vol. 111; pp. 243-251.

**[Nash and Sofer, 1996]** Nash, S. G.; Sofer, A.; "Linear and Nonlinear Programming"; pp. 144-180; McGraw-Hill 1996; ISBN 0-07-114537-0.

**[Ostwald and Huang, 1977]** Ostwald, P. F.; Huang, J.; "A Method for Optimal Tolerance Selection"; Journal of Engineering for Industry; ASME, Vol.99, Aug. 1977, pp. 558-565.

**[Prabhaharan et. al., 2004]** Prabhaharan, G.; Asokan, P.; Ramesh, P.; Rajendran, S.; "Genetic-Algorithm-Based Optimal Tolerance Allocation Using a Least-Cost Model"; The International Journal of Advanced Manufacturing Technology"; November 2004; Vol. 24; No. 9-10; pp. 647-660.

**[RD&T, 2002]** RD&T Software Manual, RD&T Technology AB, Mölndal, Sweden, 2002.

**[Söderberg, 1995]** Söderberg, R.; "On Functional Tolerances in Machine Design"; Machine and Vehicle Design, Chalmers University of Technology, Göteborg, Sweden, 1995; ISBN 91-7197-129-7.

**[Taguchi et. al., 1989]** Taguchi, G.; Elsayed, E. A.; Hsiang T. C.; "Quality Engineering in Production Systems"; McGraw-Hill; 1989; ISBN 0-07-062830-0.

# Monitoring Coordinate Measuring Machines by User-Defined Calibrated Parts

*Prof. Dr.-Ing. Dr.h.c.mult. A. Weckenmann, Dr.-Ing. S. Beetz, Dipl.-Ing. J. Lorz*
*University Erlangen-Nuremberg, Chair Quality Management and Manufacturing,*
*Metrology, Naegelsbachstraße 25, D-91052 Erlangen*
*weckenmann@qfm.uni-erlangen.de*


**Abstract:** Coordinate measuring machines (CMM) are essential for quality assurance and production control in modern manufacturing. Due to the necessity of assuring traceability during the use of CMM interim checks with calibrated objects have to be carried out periodically. For this purpose usually special artifacts like standardized ball plates, hole plates, ball bars or step gages are measured. Measuring calibrated series parts would be more advantageous. Applying the substitution method acc. ISO 15530-3 [ISO 15530-3, 2000] such parts can be used. It is less cost intensive and less time consuming than measuring expensive special standardized objects in special programmed measurement routines. Moreover, measurement results can be directly compared with the calibration values; thus direct information on systematic measurement deviations and uncertainty of the features to be measured are available. The paper deals with a procedure for monitoring horizontal-arm CMMs with calibrated sheet metal series parts.
**Keywords**: measurement uncertainty, CMM, monitoring accuracy.


## 1. INTRODUCTION

In manufacturing the continuously increasing demands on geometrical accuracy led to the necessity of testing conformance more frequently. Consequently the measurement process was shifted from the measuring room to the shop floor and the number of measuring instruments was increased. Besides the requirement that the uncertainty of measurement results must be adequate to the tolerance, traceability must be ensured for all measurement results.

This requirement is defined in ISO 9001 [ISO 9001, 2000]: "The organization shall establish processes to ensure that monitoring and measurement can be carried out in a manner that is consistent with the monitoring and measurement requirements.

Where necessary to ensure valid results, measuring equipment shall

    a) be carried out or verified at specified intervals, or prior to use, against measurement standards traceable to international or national measurement standards exist, the basis used for calibration or verification shall recorded;

    b) be adjusted or re-adjusted as necessary;

    c) ......."

125

*J.K. Davidson (ed.), Models for Computer Aided Tolerancing in Design and Manufacturing*, 125–134.

In order to fulfill these requirements, also for the production of sheet metal parts, it is necessary to monitor the CMMs where they are applied. Monitoring the accuracy of CMMs requires the application of calibrated artifacts, normally ball plates, hole plates and step gages as defined in given standards (e.g. ISO 10360) or guidelines (e.g. VDI/VDE 2617). CMMs with a large measuring volume, such as typical horizontal arm CMMs, have to be checked with ball bars. For all artifacts special expensive fixtures, CNC routines, evaluation programs and training courses for the operators are necessary. Both time consuming positioning and fixing routines for each measurement and the purchase of standardized artifacts is quite expensive. Furthermore, the measurement of artifacts interrupts the regular operation on the CMM.

In contrast to these disadvantages of the regular procedure of acceptance tests and re-verification tests the application of the substitution method offers an economical procedure to monitor CMMs in practice, but calibrated series parts are required. These parts can be selected from the mostly measured series workpieces. For the calibration routine the same measurement strategy is used as for the series measurements, thus no additional CNC program has to be prepared. But the main advantage is the fact, that the comparison between measurement and calibration results directly indicates systematic errors of the CMM measurement process. The recording of the deviations together with the relevant measurement conditions in a database points out the influences. Another benefit is that the experimental procedure to estimate the feature specific measurement uncertainty for measurements on series parts can be applied according to the rules of the ISO-Guide to the Expression of Uncertainty in Measurement [GUM, 1995].

The only problem is, to make available normal workpieces traceably calibrated with the specification of measurement uncertainty (acc. to GUM) for each feature and characteristic to be considered.

By the application of calibrated sheet metal parts in this particular case, two more impacts must be considered. First of all the fixture system has an effect on the calibrated values. Therefore the calibration routine must include the complete fixture and the fixing process (influence of the operator). Furthermore, the monitoring of large CMMs with small sheet metal parts reduces the informational value. A solution for this problem is presented.


## 2.  SUBSTITUTION METHOD

The substitution method, where the CMM is used as a comparator, is a well-established principle in metrology. The principle bases on the replacement measurement of the workpiece and an identical or similar calibrated object. Such object, also named as working standard, can be a calibrated workpiece or a gage. The comparison of gained measurement values from the working standard and the calibrated values indicates the systematic deviations of the measurement device, which are subsequently used to correct the results of corresponding measurements at the series parts.

### 2.1. Substitution measurement process

A sequence of a substitution measurement comprises the handling (e.g. fixing) of the workpiece and of the working standard, as well as the repeated measurement of the parts. Beside of the geometrical affinity between standard and workpiece the measurement conditions shall be equal during the measurements, e.g. used material, temperatures and applied measurement strategy. The definition of the features to be measured at the working standard and the orientation of the standard within the measuring volume of the CMM are arbitrary. The feature specific measurement uncertainty is experimentally determined and includes all influences, which impact the measurement process on series workpieces. The GUM [GUM, 1995] outlines the necessary steps for the correct calculation of measurement uncertainties. The following methods are in accordance to GUM:

- Uncertainty budget [ISO 14253-2, 1999]
- Simulation [VDI 2617-7, 2006; ISO/CD 15530-4, 2006]
- Uncertainty estimation with calibrated workpieces [ISO 15530, 2000]

### 2.2. Consequences of measurement uncertainty for the inspection of the workpiece

The uncertainty zone designates an interval about the indicated value in which the unknown "true value" is located with a certain statistical probability. If a value is measured that is less distant than the magnitude of uncertainty from the specification limits, no reliable information can be obtained, whether the "true value" is inside or outside the specification limits. A reliable statement can only be given, if the indicated value is located inside the conformance zone or in the non-conformance zone (figure 1).

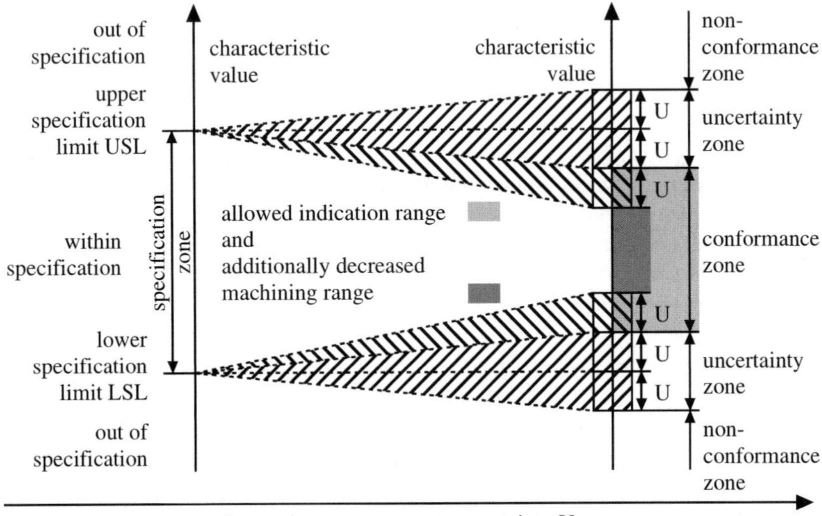


*Figure 1; Consequences of uncertainty on the assessment of measurement results*

In addition to the evaluation of measurement results, it is crucial that the uncertainty is always the drawback of the party responsible for the evidence. A supplier, who wants to prove that a workpiece is compliant with the specifications, has to show that the measured characteristics are inside the conformance zone. Otherwise, a customer, who wants to reject a workpiece, has to prove that at least one characteristic is in the non-conformance zone.

The standardized basis for this decision is defined in [ISO 14253-2, 1998] "Decision rules for proving conformance or non-conformance with specifications".

## 3. DETERMINING MEASUREMENT UNCERTAINTY WITH CALIBRATED WORKPIECES

To verify the capability of CMMs the feature specific measurement uncertainty has to be known and in an acceptable ratio to the tolerance of the workpiece. In coordinate metrology the main contributors to the uncertainty are CMM, workpiece, environment and operator including his/her measurement strategy [Weckenmann, 2001].

The experimental method described in [ISO 15530-3, 2000] is characterized by the summarization of all influences on the uncertainty and their superposition during the measurement. It is not necessary to know the separate influences. They are combined in the result of the experiment. Only systematic experiments allow identifying the actual impact of the variable influences. Indeed, this requires a high effort.

To detect the influences of material and manufacturing process the ISO standard foresees that more then one calibrated part has to be measured. The determined roughness and form deviation provide information about the dispersion of the manufacturing process and allow calculating its influence. These characteristics can be gathered more economically from un-calibrated series workpieces rather than a high number of calibrated workpieces.

### 3.1. Influences on the measurement uncertainty

Based on the measured characteristics of the standard measurement process the measurement uncertainty budget for the series measurement process can be determined. Only the experience gained form the measured characteristics allow a statement about the quality of the measurement result, even there is only an incomplete knowledge of the actual measurement.

The standard uncertainties, which are due to the individual influence factors, are determined according to the method defined in the GUM. In addition, possible correlations between several factors have to be analyzed on measurements. The achieved results are structured and saved in a database, thus they are easily accessible for further automatic application in different measurement circumstances. However, the uncertainties $u_{kal}$ (from calibration of the reference workpiece), $u_{prz}$ (from the measurement process determined by repeated measurements) and $u_{ws}$ (from the differences between master part and series workpiece) contribute to the feature specific measurement uncertainty.

### 3.2. Calculation of the measurement uncertainty

The standard uncertainty $u_{kal}$ can be derived from the extended standard uncertainty $U_{kal}$ and the extension factor k, which are indicated in the calibration certificate:

$$u_{kal} = \frac{U_{kal}}{k} \tag{1}$$

The standard uncertainty of the measurement process on the CMM $u_{prz}$ determined by repeated measurements for each scenario, in each monitoring field (figure 3) and for the series of observations ($p_1$, $p_2$....$p_k$) can be determined by

$$u_{prz} = \frac{s(p_k)}{\sqrt{n}} \tag{2}$$

$$s(p_k) = \sqrt{\frac{1}{n-1}\sum(p_k - \bar{p})^2} \quad \text{with} \quad \bar{p} = \frac{1}{n}\sum p_k \tag{3}$$

The standard uncertainty $u_{ws}$ caused by the dispersion of the influences roughness $u_{RZ}$, form deviation $u_F$, temperature $u_T$ and elasticity $u_E$ between series workpiece and calibrated workpiece can be calculated by

$$u_{ws} = \sqrt{u_{RZ}^2 + u_F^2 + u_T^2 + u_E^2} \tag{4}$$

with the assumption that the influences do not correlate. For rigid workpieces the influence $u_E$ can be disregarded. For expansion of the series of observations ($p_1$, $p_2$....$p_k$) with un-calibrated workpieces the factors $u_F$ and $u_{Rz}$ can be disregarded, too.

In coordinate metrology measurement results are also related to the reference temperature of 20 °C according to ISO 1. If the temperature of the workpiece differs from the reference temperature the workpiece expands and causes geometrical deviations. Current CMMs are capable to record the workpiece temperature and to compensate the thermally caused expansion of the workpiece, if the expansion coefficient $\alpha_{WS}$ is known. Beside the deviation of the workpiece temperature ($\Delta T_{WS}$) and the uncertainty of the temperature measurement ($u(\Delta T)$) the measurement uncertainty is caused by the uncertainty of the expansion coefficient ($u(\alpha)$). The contribution $u_T$, due to the uncertain knowledge of the temperature and the coefficient of linear thermal expansion, can be calculated by

$$u_T = \sqrt{L_{WS}^2 \cdot \alpha_{WS}^2 \cdot u^2(\Delta T_{WS}) + L_{WS}^2 \cdot \Delta T_{WS}^2 \cdot u^2(\alpha_{WS})} \tag{5}$$

The extended uncertainty $U_y$ of each measured feature is calculated from these standard uncertainties as

$$U_y = k_p \cdot \sqrt{u_{kal}^2 + u_{prz}^2 + u_{ws}^2} + |b| \tag{6}$$

Mostly a significant systematic deviation $b$ between the indicated values $p_i$ of the CMM and the calibrated value $x_{kal}$ of the working standard can be observed. If this systematic deviation $b$ cannot be corrected, it must be added to the combined standard uncertainty.

$$b = \bar{p} - x_{kal} \tag{7}$$

This extended uncertainty $U_y$ is calculated by the multiplication of the associated combined standard uncertainties with an extension factor. The extension factor is selected concerning to the distribution of the values (probability density distribution), which is assigned to the result $y$ that the interval

$$I_y = [\, y - U_y \, ; y + U_y \,]$$ 
(8)

covers a probability of $P \geq 95$ %. The extended uncertainty is defined as half of the interval.

### 3.3. Characteristic for a successful interim check

Periodical interim checks build the basis for the decision, if the CMM is capable to measure the workpieces to be inspected. For an interim check reduced performance verification may be applied to demonstrate the capability that the CMM conforms to specified requirements. The extent of the verification as described for the acceptance test and re-verification tests [ISO 10360-2, 2001] may be reduced by: number of measurements, location and orientation being performed. Also the specifications to be checked can be defined by the operator.

The aim of the interim check is to prove if the measurement instrument has not changed its characteristics and thus its influences on the measurement uncertainty. Since all other components of the extended uncertainty $U_y$ are influenced by other factors then the CMM, only the deviations $b$ between calibrated value and measurement results have to be checked. These deviations have to be smaller than the extended measurement uncertainty $U_y$, which is determined during verification or re-verification tests. If the feature specific deviations $b$ are smaller, it can be assumed that the influence of the CMM has not changed, otherwise the CMM has to be re-verified.

### 3.4. Knowledge base for evaluating the impact of influences and generating a long time revision

The feature specific deviations are stored in a SQL database together with information about significant influences as environmental conditions, selected measurement strategy and operator. By means of data processing the scenarios, containing deviations and related influences, a knowledge base is created. This knowledge base will subsequently define the pre-conditions for the evaluation of the influences' impact. Automated evaluation routines and a management console were set up on the basis of the acquired knowledge. The management console enables the access to data in the knowledge base regarding the long-run behavior of the CMM, in order to identify changes of the CMM accuracy in the time response. The inspections of the long-run behavior can be transparently documented in quality control charts. Thereby several influences, e.g. measurement strategy, evaluation criteria and environmental conditions, are explicitly pointed out, thus become transparent and can be subsequently assessed in regard to their correlations. This is the basis to reduce the measurement uncertainty by decreasing the several influences' effect systematically.

## 4.   MONITORING OF CMMS WITH CAR PARTS

In order to verify the above mentioned procedure a horizontal-arm CMM for measuring car sheet metal parts in series production of a German automotive company has been monitored. As representative part a shaft collet has been selected. The working standard was calibrated at the reference CMM of the measurement centre of the Chair Quality Management and Manufacturing Metrology (QFM).

### 4.1. Calibration of a shaft collet

The measurement centre of QFM is accredited by PTB[1] as DKD[2]-calibration laboratory (DKD-K-36501) for DKD-calibration of standard geometrical features on prismatic workpieces according to ISO 17025. The measuring room (Class 1 acc. guideline VDI/VDE 2627) has almost perfect environmental conditions, e.g. a temperature stability of 20 °C ± 0.1 °C all over the year. According to GUM the measurement uncertainty is calculated with the VCMM[3] software, which bases on Monte-Carlo simulations.

The shaft collet was calibrated at the reference CMM *Zeiss UPMC 1200 CARAT S-ACC* in several positions to get traceable results (figure 2). The measurement results are documented in the calibration certificate together with the feature specific extended measurement uncertainties.

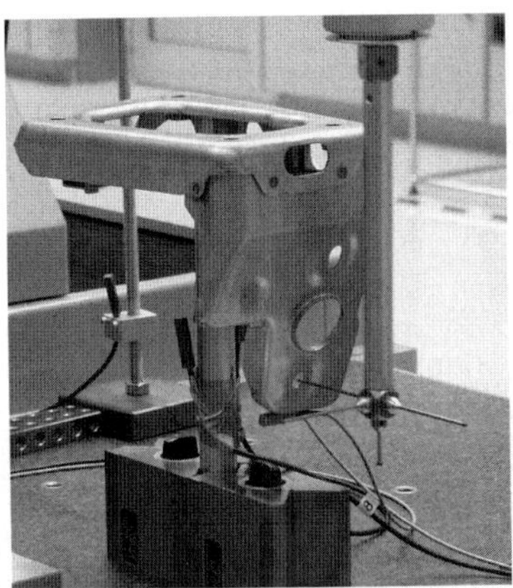

*Figure 2; DKD calibration of a shaft collet at the reference CMM*

### 4.2. Segmenting the CMM measuring range

Even to allow for a statement about the capability of the measurement process of a large CMM like a horizontal-arm CMM, its measuring volume is divided into twelve monitoring fields in (figure 3). The shaft collet was measured at six positions in the x-y-plane and at two z-levels.

---

[1] Physikalisch-Technische Bundesanstalt (PTB), the German national metrology institute
[2] Deutscher Kalibrierdienst (DKD), association of calibration laboratories
[3] Virtual Coordinate Measuring Machine (VCMM)

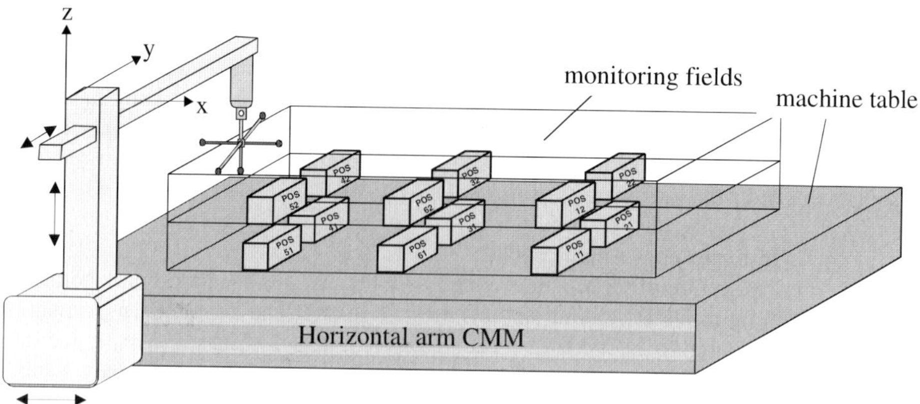



*Figure 3; Schematic representation of the sheet metal part distribution*

Since the feature specific deviations $b$ are depended on the workpiece position in the measuring volume of the CMM, individual positions of the workpiece coordinate systems have to be stored, so that the deviations $b$ can be traced back to the machine coordinate system.

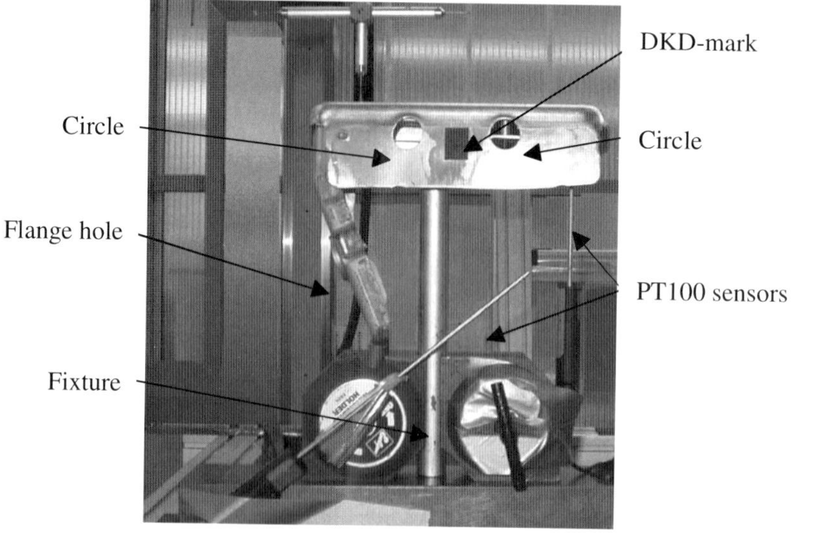



*Figure 4; Shaft collet on a horizontal-arm CMM (courtesy Audi AG)*

The shaft collet is fixed at three points with a magnet fixture and repeatedly measured at each monitoring position (figure 4). The temperature is permanently recorded with calibrated platinum resistors Pt 100. The gradients have to be considered during the subsequent uncertainty calculation.

### 4.3. Measurement strategy

The selection of the geometrical features was restricted to circles and slotted holes of different dimensions. In each position all circles and slotted holes were measured 20 times. During the experiments the same strategy was used as applied for calibration. Thus the influences of measurement force, probe characteristic, support and fixture can be disregarded. The probing strategy is accomplished in each case with and without feature related alignment (only workpiece alignment).

Due to the repeated measurements also a range of dispersion for the respective probing strategy can be approximated by means of statistical evaluation.

### 4.4. Calculation of uncertainty budget and evaluation the interim check

The calculation of the measurement uncertainty is automatically carried out software assisted by the management console using the current measurement results together with the stored experience values. As illustration of the above mentioned procedure, an actual measured value for the positional deviation of a flange hole is presented (table I). The deviations $b$ of the measurement results from the calibrated values are smaller than the extended measurement uncertainty. So it is proved that the monitored CMM is still capable to measure the shaft collet in position 51.

| workpiece: | | **shaft collet** | | position on CMM: | | | **POS 51** |
|---|---|---|---|---|---|---|---|
| feature: | | flange hole | | tolerance in mm: | | | 0,2000 |
| nominal size in mm: | | 0 | | calibrated value $x_{kal}$: | | | 0,1795 |
| tolerance in mm: | | 0,2 | | extended measurement uncertainty $U$: | | | 0,0374 |
| medial temperature T in °C: | | 22,6 °C | | *extending coefficient k:* | | | 2 |
| uncertainty of T in °C: | | 0,1 °C | | $\alpha_{steel}$: | 11,5 | $u_{\alpha}$: | 1 |

| measurement cycle no. | date: | time: | operator | calibrated workpiece | un-calibrated workpieces 1 | 2 | 3 |
|---|---|---|---|---|---|---|---|
| 1 | 2004-12-22 | 07:10 | Buchner | 0,6208 | 0,6959 | 0,2550 | 0,2584 |
| 2 | 2004-12-22 | 08:12 | Buchner | 0,6213 | 0,6955 | 0,2559 | 0,2595 |
| ⋮ | ⋮ | ⋮ | ⋮ | ⋮ | ⋮ | ⋮ | ⋮ |
| 20 | 2004-12-22 | 02:12 | Buchner | 0,6208 | 0,6961 | 0,2576 | 0,2686 |
| | | | average value: | 0,6210 | 0,6952 | 0,2567 | 0,2632 |
| | | | standard uncertainty $u_{prz}$: | 0,00038 | 0,00098 | 0,00092 | 0,00393 |
| | | | number n of workpieces: | 4 | | | |

| **symbol of influences** | $u_{kal}$ | $u_{prz}$ | $u_T$ | $u_{ws}$ | b | U | **check result** |
|---|---|---|---|---|---|---|---|
| **uncertainty** | 0,0187 | 0,0021 | 0,0000 | 0,0000 | 0,4415 | **0,4791** | **Yes** |

*Table I; Monitoring results of the shaft collet; feature: flange hole, position 51*

## 5. CONCLUSION

A procedure for monitoring of CMMs with calibrated workpieces was introduced. The monitoring is based on measurements of workpieces with known shape deviations in pre-defined monitoring fields within the measuring volume of the CMMs, wherein all relevant influence factors are considered. All measurements results are stored in a database. All unknown and known systematic deviations of the real process are represented during the experiments. Values and sensitivity coefficients of the influence factors have not to be considered separately, since a documented scenario contains the value of the influence factor together with the assigned measurement uncertainty. By means of continuously recording of the results, a statement about the long-run behavior of the CMM is possible, in reference to several input parameters. Thus, short-termed optimizing of these parameters can be avoided.

Based on the gained information the calculation of feature specific measurement uncertainties are already possible an early development stage. So the measurement uncertainty can be considered during the definition of tolerances

Determining the task specific measurement uncertainty using calibrated workpieces increases the acceptance of measurement uncertainty in industrial practice and opens new possibilities to integrate the idea of measurement uncertainty into series and pilot production processes.

## REFERENCES


**[ISO 15530-3, 2000]** ISO/TS 15530-3; "Geometrical Product Specifications (GPS) – Coordinate measuring machines (CMM): Techniques for evaluation of uncertainty of measurement – Part 3: Use of calibrated workpieces"; Geneva 2000.

**[ISO 9001, 2000]** ISO; "Quality systems – Model for quality assurance in design/development, production, installation and serving"; Geneva 2000.

**[GUM, 1995]** ISO; "Guide to the Expression of Uncertainty in Measurement"; Geneva 1995; ISBN 92-67-10188-9.

**[ISO 14253-2, 1999]** ISO/TS 14253-2; "Geometrical Product Specifications (GPS) – CMMs (CMM): Techniques for evaluation of uncertainty of measurement – Part 2: Guide to the estimation of uncertainty in GPS measurement, in calibration of measuring equipment and in product verification"; Geneva 1999.

**[ISO 10360-2, 2001]** ISO 10360-2; "Geometrical Product Specifications (GPS) – Inspection by measurement of workpieces and measuring equipment -- Part 2: Guide to the estimation of uncertainty in GPS measurement, in calibration of measuring equipment and in product verification"; Geneva 2001.

**[VDI 2617-7, 2006]** VDI/VDE 2617 Part 7; "Accuracy of coordinate measuring machines - Parameters and their checking - Estimation of measurement uncertainty of coordinate measuring machines by means of simulation"; Düsseldorf 2006.

**[ISO/CD 15530-4, 2006]** ISO/CD 15530-4; "Geometrical Product Specifications (GPS) -- Coordinate measuring machines (CMM): Technique for determining the uncertainty of measurement -- Part 4: Use of computer simulation"; Geneva 2006.

**[Weckenmann, 2001]** Weckenmann, A.; Knauer, M.; "The Influence of Measurement Strategy on the Uncertainty of CMM-Measurements"; In: *Annals of the CIRP 47 (1998) 1*; pp. 451-454; Athens 1998; ISBN 3-905 277-29-8.

# Evaluation of Geometric Deviations in Sculptured Surfaces Using Probability Density Estimation


**A. Barari*, H. A. ElMaraghy**, G. K. Knopf***

** Department of Mechanical & Materials Engineering, University of Western Ontario, London, Ontario, Canada, N6A 5B9*
*** Industrial and Manufacturing Systems Engineering, University of Windsor, Windsor, Ontario, Canada, N9B 3P4*
*Corresponding Author: hae@uwindsor.ca*



**Abstract:** The number and location of samples acquired during coordinate measurement are among the most important parameters that affect the accuracy of evaluating the actual geometric deviations and comparing them with tolerance specifications. A method based on iterative sampling of discrete 3D points is presented for evaluating actual geometry deviations of machined sculptured surfaces and for assisting the important accept/reject decisions making process. The strategy relies on estimating the density function of the geometric deviations between the actual and substitute surfaces, using the Parzen windows technique. The computed error density function is used to identify portions of the surface that require further iterative sampling of measurement points until the desired accuracy is achieved. This method may be used for assessing any geometry deviations due to systematic or non-systematic error sources. No prior assumptions about the fundamental form of the underlying probability density function of errors are required. Examples are presented to both illustrate the proposed methodology and validate its performance. The use of this method reduces inspection cost as well as the cost of rejecting good parts or accepting bad parts.

**Keywords**: Tolerances, Minimum Deviation Zone, Inspection, Adaptive Sampling.


## 1. INTRODUCTION

Effective evaluation of geometric deviations of fabricated sculptured surfaces for the purpose of determining compliance with specified form tolerances is an important metrology and inspection task. It requires the acquisition of numerous measured data points and an extensive computation of surface geometry errors to accurately characterize the inspected part geometry. Numerous research projects in the area of tolerance evaluation based on discrete measured points were undertaken. However, there is still a need for robust methods to accurately evaluate complex sculptured surfaces.

Surface inspection is traditionally accomplished by sequentially performing three major tasks: (i) measurement planning and capturing sample coordinate points; (ii) evaluation of the optimum substitute geometry; and (iii) calculation of geometric

135


*J.K. Davidson (ed.), Models for Computer Aided Tolerancing in Design and Manufacturing*, 135–146.
© 2007 Springer.

deviations. In contrast, the presented method performs the above three tasks concurrently in an adaptive closed loop with continuous feedback to each other.

The proposed method seeks to model the surface errors patterns, by studying their probability density function, in addition to supporting the part acceptance or rejection decisions. A pattern recognition technique called *Parzen windows* [Parzen, 1962] is utilized for this purpose. Studying the second order discontinuities of the density function identifies the critical portions of the surface that require further iterative sampling of measurement points. In order to validate the proposed methodology, a set of experiments is conducted and the results show a significant reduction in computation time and an improvement in the accuracy of the tolerance evaluation process.

## 2. BACKGROUND

Coordinate Measuring Machines (CMM) and/or non-contact scanning equipment are used to sample discrete points from the manufactured surfaces. The number and locations of measurements are among the most important parameters that affect the accuracy and validation of the whole inspection process. Unfortunately, no standard procedure or guideline has been developed to determine the optimum number and ideal sample location. After a surface point measurement, the conformance of the actual surface to the desired tolerance is evaluated by fitting a *substitute geometry* to the measured points. Fitting is achieved by minimizing an error function between the acquired points and the substitute geometry. Linearization and approximation methods are utilized by most fitting algorithms to simplify the problem and to speed up computation. In addition, many CMM manufacturers implement their own proprietary programs without explicitly stating the underlying assumptions and limitations of their algorithms. The discrepancies between the evaluation results from different evaluation software programs, for the same set of data points, can be as large as 50% [Yau, 1999].

Appropriate consideration of measurement planning, estimation of the best substitute geometry and uncertainty analysis as inter-related issues is essential. The work presented by Nassef and ElMaraghy shows that the performance of the error function for the best fitting, depends on the number of measured points relative to the number of points that would ideally represent the whole surface, the type of the geometric deviation and the inherent errors in the measuring system [Nassef & ElMaraghy, 1999].

### 2.1. Fitting of Substitute Geometry
Fitting substitute geometry to the data points by minimizing an error objective function is defined by the $L_p$-*norm* equation [Hopp, 1993]:

$$L_p = \left[ \frac{1}{n} \sum_i^n |r_i|^p \right]^{1/p} \tag{1}$$

where $r$ is the residual error between the $i^{th}$ measured point and the substitute geometry fitted to the $n$ data points, and $p$ is an exponent. Least square ($p=2$) and minimum deviation zone functions ($p=\infty$) are typically utilized in the fitting process. The criterion in the least square function requires that the sum of square errors be minimized. For the minimum deviation zone function, equation (1) becomes [Nassef & ElMaraghy, 1999]:

$$L_\infty = \max_i \left(|r_i|\right) \tag{2}$$

The minimum deviation zone function has received much attention in recent years [Yau, 1997] [Choi & Kurfess, 1998], because the studies show that it yields a smaller zone value than that evaluated by using a least squares fit [Choi *et al.*, 1998] [Lin *et al.*, 1995]; and it best conforms to the standard definition of the tolerance zone in ASME Y14.5. In addition, it numerically simulates the physical fitting process in traditional metrology. However, such an extreme fit has limitations on accuracy [Hocken *et al.*, 1993] [Hopp, 1993] and on stability. It is also very sensitive to number and location of the measured points. Minimization of the deviation zone, for sculptured and parametric surfaces, is highly non-linear. Since the analytical derivatives of the objective function are not usually available, a direct search method is required to solve the problem and its success is very dependent on the initial conditions [ElMaraghy *et al.*, 2004].

The least square best fit is the most likelihood estimation used to fit the substitute geometry onto a set of discrete data. Since all the measurement points contribute to the best-fit result, the substitute geometry is more stable and less sensitive to the local deviations, asperities and local surface effects. However, least square best fit is a statistical estimation rather than an exact solution and some concerns always exist regarding interpretation of its fitting results. In this work, the advantages of both minimum deviation zone and least square fitting methods are exploited.

### 2.2. Data Sampling

Research on coordinate sampling mostly focuses on the sample size and not the sample locations. Different sampling techniques, including uniform sampling, random sampling and stratified sampling [Hocken *et al.*, 1993] are studied. Extensive numerical experiments on various geometric primitives, such as line, plane, circle, sphere and cylinder conclude that an inspection result significantly varies by sampling size [Caskey *et al.*, 1991] [Hocken *et al.*, 1993]. Weckenmann *et al.* observed that the evaluation accuracy is increased by raising the sample size for circular and linear features. Furthermore, it was also concluded that different evaluation criteria call for different sampling strategies [Weckenmann *et al.*, 1991] [Weckenmann *et al.*, 1995].The distribution of the form errors on the actual geometry for flat surfaces was found to be critical for estimating the sample size as well as uncertainty of evaluation [Choi *et al.*, 1998].

An iterative sampling method [Edgeworth & Wilhelm, 1999], utilizing a CMM capable of measuring the surface normal as well as point coordinates, uses an interpolation method to decide when and where further sampling would occur.

A sampling method for inspecting sculptured surfaces was presented by Elkot and ElMaraghys [Elkot *et al.*, 2002]. A sample plan is generated based on the surface complexity and features, such as gradients, that warrant more condensed sampling. Sampling criteria include: equi-parametric, surface patch size and surface patch mean curvature. Both the number and location of sampling points were optimized and an algorithm for the automatic selection of the best sampling criterion strategy for a given surface was developed. The sampling plan is determined based on the ideal CAD geometry. However, the distribution and magnitude of surface geometry errors introduced during fabrication may require different sampling for accurately evaluating the actual surface, which is the motivation for the work presented in this paper.


3.   METHODOLOGY


If the substitute geometry is properly defined by the minimum deviation zone method, there is a unique solution that satisfies the optimality condition and is constrained by the extreme deviations. However, the extreme deviations in the sampled points are not necessarily the extreme deviations in the actual geometry. The important fact is that the uncertainty in the minimum deviation zone evaluation is closely related to the location and number of sample points [Choi *et al*, 1998]. For such a non-linear objective function, increasing the number of samples dramatically increases the computation time. Therefore, selection of sufficient number of proper sample points is very important. Also, optimization procedures to solve this problem mostly are trapped in local minima; therefore the obtained solution is very dependent on the initial condition of optimization. The methodology presented in this paper utilizes the statistical nature of the least square function to estimate the sample number and location in an iterative sampling approach.

In order to interpret the results of the least square fitting, a pattern recognition technique is employed to estimate density function of geometric deviations. Geometric deviations are calculated after each iterative fitting of substitute geometry to the updated set of samples using the least square function. The substitute geometry is defined as:

$$S_G = T(t) \times D_G \tag{3}$$

where $S_G$ is the substitute geometry and $D_G$ is the desired or nominal geometry. $T(t)$ is the rigid body transformation matrix defined by vector variables $t$, which consists of three rotation and three transformation parameters [ElMaraghy *et al.*, 2004].

Using The $L_2$-*norm* equation for least square fitting, the objective function of the optimization problem can be written as follows:

$$Obj = \min_{t} \left[ \frac{1}{n} \sum_{i}^{n} \left\| P_i - T(t) \times D_G \right\|^2 \right]^{1/2} \tag{4}$$

where the statement inside the norm sign indicates the Euclidian distance of any sample point, $P_i$, from the temporary substitute geometry. By utilizing a direct search method, the optimum vector variable $t^*$ can be found in a way that the $L_2$-*norm* equation is minimized. The nearest point of substitute geometry, $P_i^*$, to the measured point $P_i$ is called the corresponding point. The deviation of the measured points from the best substitute geometry, $e_i$, is the dot product of the vector from the corresponding point to the measured point, and the normal vector of the substitute geometry at the corresponding point, $n_i$.

$$e_i = \frac{\left(P_i - P_i^*\right) \bullet \overline{n}_i}{\left\| \overline{n}_i \right\|} \tag{5}$$

Evaluated geometrical deviations, $e_i$, are labeled data sampled from a continuous random variable, $e$. Geometrical deviations are caused by manufacturing errors and have a probability density function, $f(e)$. A variety of density functions, such as uniform, Gaussian or Beta distribution, have been suggested for manufacturing errors [Choi *et al.*, 1998][ Elkot *et al.*, 2002]. In order to avoid the limitations of such assumptions, the Parzen windows method as a nonparametric procedure is used to estimate the unknown underlying density function from the sampled data. Discontinuities in the estimated density function indicate lack of sampling in corresponding deviation regions.

### 3.1. Density function of geometrical deviations

Using equation (5), geometrical deviation in the manufactured surface can be assumed as a one-dimensional continuous random variable, $e$. Estimation of the probability density function fundamentally relies on the fact that the probability $Pr$ that a given deviation $e_i$ will fall in a region $R$ is given by:

$$Pr\left(e_i \in R\right) = \int_R f(e)de \tag{6}$$

where $f(e)$ is the probability density function of the geometrical deviation. The Parzen windows method is a nonparametric method of estimating a probability density function using kernel functions [Parzen, 1962]. Parzen windows estimate the probability density function based on the weighted average of potentially every single sample point; although, only those falling within the selected region have any significant weight. For a one-dimensional variable, regions can be represented by widows with a constant length of $h$, which are centered at any arbitrary value of $e$. The probability density function for $n$ sample points is estimated as follows [Duda & Hart, 1973]:

$$f_n(e) = \frac{1}{n} \sum_{i=1}^{n} \frac{1}{h} \varphi\left(\frac{e_i - e}{h}\right) \tag{7}$$

where $\varphi$ is the window function. A Gaussian function with a mean of zero and a variance of unity is the popular choice for the $\varphi$. In this work, since we wish to recognize discontinuities in the density function, the window function $\varphi$ is defined as follows:

$$\varphi(u) = \begin{cases} 1 & |u| \le \dfrac{1}{2} \\ 0 & Otherwise \end{cases} \tag{8}$$

This function simply counts the number of observed samples inside the window centered on any arbitrary value of $e$. Therefore the probability density function is estimated based on the ratio of samples inside the selected window to the total samples.

### 3.2. Discontinuity in the density function

The density function is calculated for $m$ windows centered with incremental distance of $h$. The first and the last windows are centered at:

$$e^1 = floor(\frac{\min_i(e_i) - \dfrac{h}{2}}{h}) \times h \tag{9}$$

$$e^m = ceil(\frac{\max_i(e_i) + \dfrac{h}{2}}{h}) \times h \tag{10}$$

where the floor and ceil functions return the largest and smallest integers which are larger and smaller than their arguments. Here, the distance between $e^1$ and $e^m$, is defined as the range of windowing which is an estimate of the minimum deviation zone of the inspected geometry. It is expected that in the scale of profile inspection, least square best fitting converges to a statistical estimation of the substitute geometry which results in the maximum density for the average observed geometric deviations the maximum density geometric deviations. Therefore, studying the gradient of the obtained density function can provide a suitable metric to evaluate efficiency of the data sampling. By studying the changes in gradient of density function, regions of the geometric deviations with the local maximum and minimum densities are recognized. The new points are sampled to validate the current estimation of the local maximum/minimum densities. Regions of the geometrical deviation corresponding to the local maximum/minimum are recognized as the critical regions.

The Hessian of the density function provides information about the changes in the gradient of density. The Hessian for a center of the window, $e^j$, is as follows:

$$f''(e^j) = \frac{2f(e^j) - f(e^{j+1}) - f(e^{j-1})}{h^2} \qquad j = 2,...,m-1 \tag{11}$$

New samples are captured from the regions that contribute in the maximum absolute Hessian value. If the maximum absolute Hessian is observed in region $e^{j*}$, two possible conditions may occurs based on the Hessian sign for the region, $e^{j*}$. Figures 1 and 2 illustrate the two possible conditions of $e^{j*}$. When the sign of Hessian is positive (Figure 1), the validation sampling is conducted for the region $e^{j*}$ as the critical region and when it is negative, $e^{j*+1}$ and $e^{j*-1}$ are critical regions selected for the validation sampling (Figure 2).

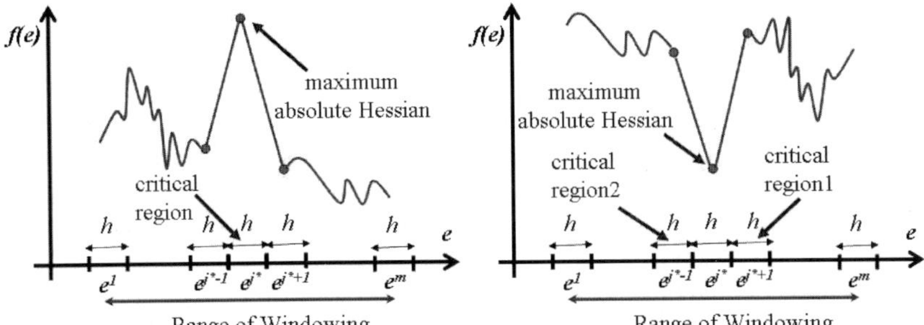


*Figure 1; Probability density function with one critical region.*

*Figure 2; Probability density function with two critical regions.*

### 3.3. Iterative sampling

Considering that a deviation value, $e_i$, is assigned to any of corresponding points, $P_i^*$, on the substitute surface, the new samples will be picked in a neighborhood of points $P_i^*$, that have critical geometric deviation. In a parametric surface, point $P_i^*$ is defined by its two associated surface parameters $u_i^*$ and $v_i^*$. By $\eta$ times random sampling in the searching zone, the zone is shrunk, until the termination criteria are satisfied. The searching zone is the area of the surface with the surface parameters between two concentric circles, centered at $(u_i^*, v_i^*)$ with two radiuses of $\gamma_i$ and $\gamma_i - \varepsilon$; where $\varepsilon$ is a constant positive value and $\gamma_i$ starts with an initial value bigger or equal to $\varepsilon$. At each new search step, $\gamma_i$ is shortened by the $\varepsilon$. In terms of the default values in the implemented algorithm, $\eta$ is equal to four and $\varepsilon$ is estimated based on the radius of the measurement probe tip. The initial value for $\gamma_i$ is half of the length of each element in the initial stratified sampling. For any new sample point:

1. If the evaluated deviation for the new sample point belongs to a critical region, sampling is terminated and the new point is added to a temporary sample set. Using the main sample set and the temporary sample set, a modified density function is generated for the new iteration.
2. If the evaluated deviation for the new sample point is not in the current range of deviation, the new point is likely to be effective data for the minimum deviation zone fitting. Therefore, the current iteration is terminated. The new point is added to the main sample set and the temporary sample set is deleted. New substitute geometry is fitted to the main sample set. A new iteration is started.
3. Otherwise, the new sample point is ignored because most likely it is not an effective data in the minimum deviation zone evaluation.

The algorithm is terminated when: i) the number of times that situation 3 in one iteration occurs more than a constant positive value, $\alpha$, and ii) when the ratio of the number of points in the temporary sample set to those in the main sample set is larger than a constant positive value, $\beta$. $\alpha$ and $\beta$ are governed by the sampling cost. The main sample set is used for minimum deviation zone fitting, using the objective function in Equation (2).

## 4. SIMULATION

An arbitrary B-spline surface created from 16 control points with an overall dimension of 900mm×600mm×200mm is modeled to simulate the inspection process. Systematic machining errors are simulated [ElMaraghy *et al.*, 2004] in order to emulate geometric deviations resulting from machining of the surface, by considering the combination of all independent errors in the elements and linkages of the machine tool using a kinematics approach. The actual independent errors used were measured in a typical vertical 3-axis machine tool [Hocken, 1980]. The non-systematic part of the machining errors are simulated by adding normally distributed error, with a mean equal to zero and standard deviation of 1 µm, to the generated surface in all three X, Y and Z directions.

### 4.1. Initial sampling

A stratified strategy for initial sampling is applied. The surface is divided uniformly into 16 elements with the same parametric area. The point chosen for each element is randomly selected within the element. Figure 3 shows samples and the evaluated minimum deviation zone, which is 0.06766 mm. There is, however, no guarantee that the extreme deviation in these samples is the same extreme deviation in simulated geometry. Figure 4 shows the estimated density function using the Parzen windows method. The evaluated deviation by least square method is 0.0912 mm, which confirms the idea of overestimation of fitting by least square method.

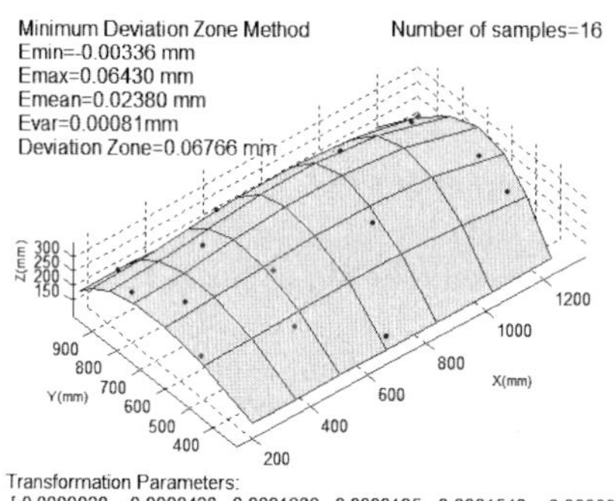

Figure 3; Inspection based on the 16 stratified samples.

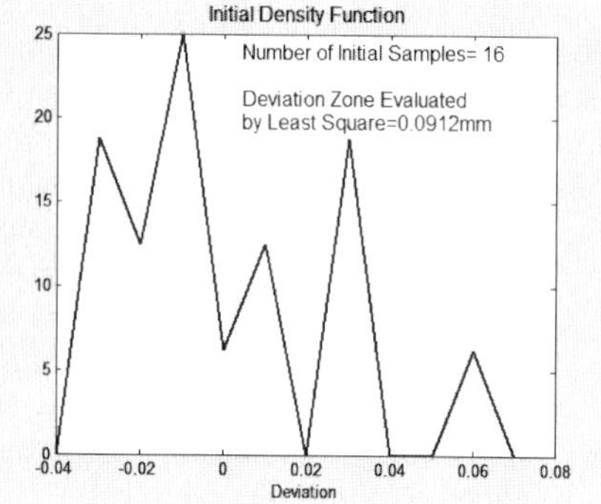

Figure 4; Density function estimated based on 16 stratified samples.

### 4.2. Iterative sampling

Although 154 points are captured during the automated sampling process, only 24 points are selected as the critical points to evaluate the minimum deviation zone. Figure

5 shows how the density function changes by iterative sampling. It can be seen that gradually over each iteration, the shape of the density function is extended, it takes a more symmetric shape and the density of samples within the average deviations is increased. The final density function, when the algorithm is terminated, is represented in figure 6. The final density function shows a characteristic behavior similar to a Gaussian distribution function with a significant maximum value around the mean and is very symmetric. The mean value is also very close to zero, which validates the high accuracy of both the sample planning as well as performed fitting.

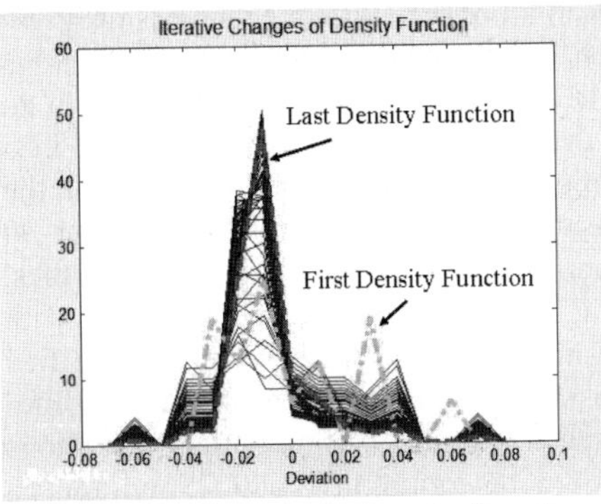


*Figure 5; Changes in density function.*

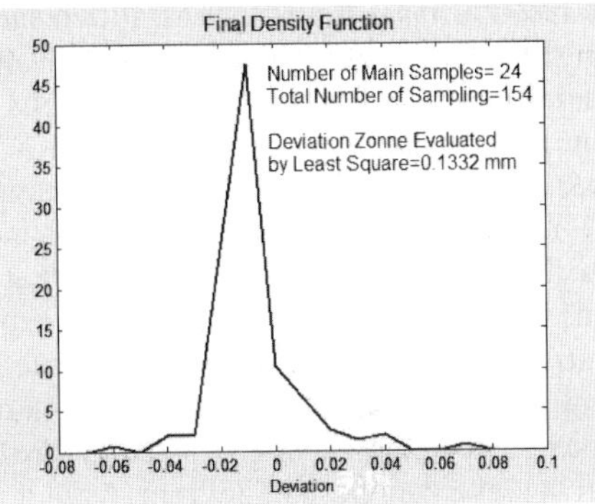


*Figure 6; Final density function.*

Figure 7 shows 24 selected samples and the evaluated minimum deviation zone as 0.10778 mm, which is almost 60% larger than what was evaluated before, based on the 16 initial samples. This shows that the points obtained by iterative sampling are more representative of the actual surface. The deviation zone by least square fitting is again overestimated and is equal to 0.01332 mm. Figure 8 shows all of the 154 samples and their corresponding evaluated deviation zone as well as the optimum transformation parameters $t^*$, which are exactly identical to those of the 24 selected samples.

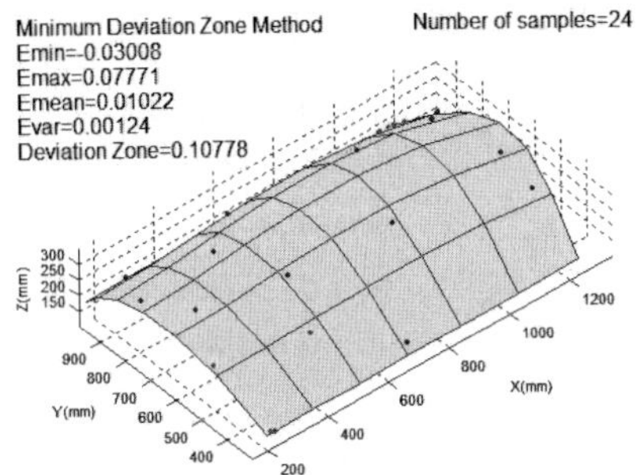


*Figure 7; Evaluation based on the 24 points selected by iterative sampling.*

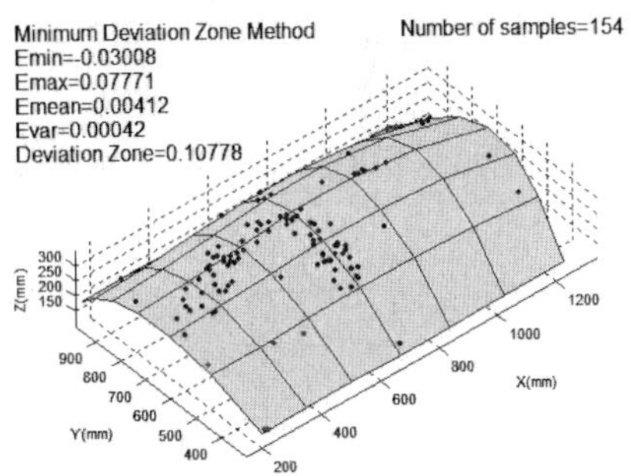


*Figure 8; Evaluation based on all of the 154 points sampled in iterative sampling.*

Experiments showed that in each iteration, by using the vector variable, $t^*$, associated with the previous substitute geometry as the initial condition of the optimization, new substitute geometry is found quickly. The computation time for the iterative sampling was almost 2.2 times longer than what was required for the initial 16 points. However, iterative sampling is still highly time efficient, since its computation time is almost 4.8 times shorter than the evaluation time based on the 154 points.

### 4.3. Accuracy and performance of the method

The result of inspection by 16 stratified samples in all of the experiments was much less than the result of iterative sampling. Evaluation based such a sample set results as an underestimation of geometrical deviations on the surface. Furthermore, it results in an improper substitute geometry that will affect the consecutive inspection process in a GD&T qualification. However, computation time for 16 points is much shorter than the time required for iterative sampling. It is observed, that the computation time for 36 stratified sampling is almost similar to that of iterative sampling with 16 initial samples, in most cases. In order to compare the inspection uncertainty of stratified sampling with iterative sampling, the simulated surface was inspected 100 times with 16 stratified sampling, 36 stratified sampling and iterative sampling. Minimum deviation zones evaluated for each sampling strategy are plotted in Figure 9. Geometrical deviations with an average of 0.05357mm with a standard deviation of 0.000075mm for 16 samples, an average of 0.06989mm with standard deviation of 0.000048mm for 36 samples and an average of 0.0907mm with standard deviation of 0.000075mm for iterative sampling are evaluated. These results indicate that, iterative sampling not only causes a more precise estimation for the geometric deviations, but also has the least standard deviation value.

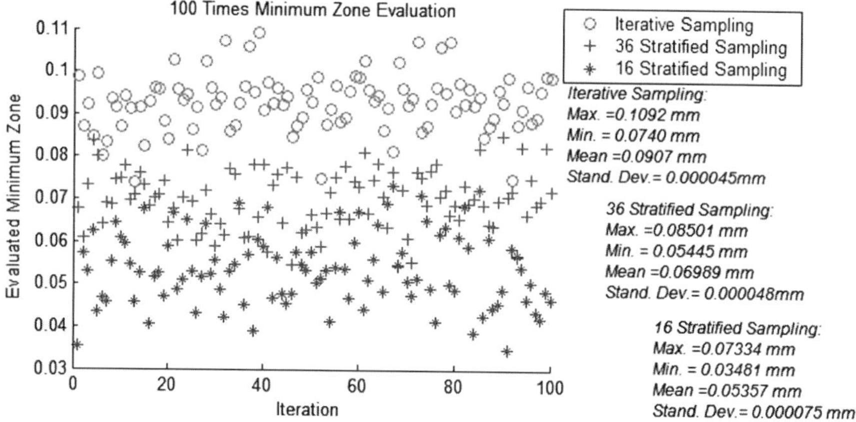


*Figure 9; Uncertainty experiment by 100X inspection with 3 sampling strategy.*

Shown in Figure 10 and 11 are histograms of a number of selected samples and total number of samples in 100 times iterative sampling, respectively.

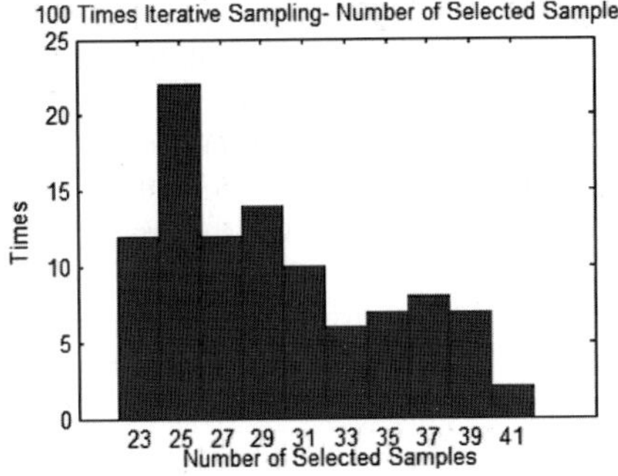


*Figure 10; Number of selected samples for 100X inspection.*

*Figure 11; Total number of samples for 100X inspection.*

## 5. CONCLUSION

An innovative online adaptive sampling method for use in an integrated environment of measurement planning, sampling and evaluation of geometric deviations was presented.

The resulting deviation zone is readily comparable to the surface form tolerances for sculptures surfaces and is used to check compliance of the fabricated surfaces with specified design tolerances.

The proposed method uses the surface errors probability density function to identify regions where more measurements should be made. It improves the accuracy of minimum zone evaluation. Also, by selecting the most useful candidates for additional measurements, this reduces the total number of required sample points and associated computation time for surface fitting and optimization.

## REFERENCES


**[Yau, 1999]** Yau, H.T.; "A model-based approach to form tolerance evaluation using non-uniform rational B-splines"; *Robotics and Computer Integrated Manufacturing*, 15, pp. 283-295; 1999

**[Hopp, 1993]** Hopp, T.; "Computational metrology"; *Manufacturing Review*, vol. 6, no. 4, pp. 295-304; 1993

**[Nassef & ElMaraghy, 1999]** Nassef, A.O.; ElMaraghy, H.A.; "Determination of best objective function for evaluating geometric deviations"; *International Journal of Advanced Manufacturing Technology*, 15, pp. 90-95; 1999

**[Yau, 1997]** Yau, H.T.; "Evaluation and uncertainty analysis of vectorial tolerances"; *Precision Engineering*, 20, pp. 123-137; 1997

**[Choi & Kurfess, 1998]** Choi, W; Kurfess, T.R.; "Uncertainty of extreme fit evaluation for three-dimensional measurement data analysis"; *Computer-Aided Design*, vol. 30, no. 7, pp. 549-557; 1998

**[Choi et al, 1998]** Choi, W; Kurfess, T.R.; Cagan, J.; "Sampling uncertainty in coordinate measurement data analysis"; *Precision Eng.*, 22, pp.153-163; 1998

**[Lin et al, 1995]** Lin, S.S.; Varghese, P.; Zhang, C.; Wang, H. B. "A comperative analysis of CMM form-fitting algorithms"; *Manufacturing Review*, vol. 8, no. 1, pp. 47-58; 1995

**[ElMaraghy et al, 2004]** ElMaraghy, H.A.; Barari, A.; Knopf, G.K.; " Integrated inspection and machining for maximum conformance to design tolerances "; *CIRP Annals - Manufacturing Technology*, vol. 53, no. 1, pp. 411-416; 2004

**[Caskey et al, 1991]** Caskey, G.; Hari, Y.; Hocken, R.; Palanvelu, D.; Raja, J.; Wilson, R.; Chen, K.; Yang, J. "Sampling techniques for coordinate measuring machines"; *Proceed. of NSF Design & Manufacturing Systems Conference*, pp. 983-988; 1991

**[Hocken et al, 1993]** Hocken, R.; Raja, J.; Babu, U.; "Sampling issue in coordinate metrology"; *Manufacturing Review*, vol. 6, no. 4, pp. 282-294; 1993

**[Weckenmann et al, 1991]** Weckenmann, A.; Heinrichowski, M.; Mordhorst, H.J.; "Design of gauges and multipoint measuring systems using coordinate measuring machine data and computer simulation"; *Precision Eng.*, 13, pp. 203-207; 1991

**[Weckenmann et al, 1995]** Weckenmann, A.; Eitzert, H.; Garmer, M.; Webert, H.; "Functionality oriented evaluation and sampling strategy in coordinate metrology"; *Precision Engineering*, 17, pp. 244-251; 1995

**[Elkott et al, 2002]** Elkott, D.; ElMaraghy, H.A.; ElMaraghy, W.H.; "Automatic sampling for CMM inspection planning of free-form surfaces"; *International Journal of Production Research*, vol. 40, no. 11, pp. 2653-2676; 2002

**[Edgeworth & Wilhelm, 1999]** Edgeworth, R. G.; Wilhelm, G.; "Adaptive sampling for coordinate metrology"; *Precision Engineering*, 23, pp. 144-154; 1999

**[Parzen, 1962]** Parzen, E.; "On estimation of a probability density function and mode"; The *Annals of Mathematical Statistics*, 33, pp 1065-1076; 1962

**[Duda & Hart, 1973]** Duda, R.O.; Hart, P. E.; "Pattern classification and scene analysis"; *John Wiley & Sons Inc.*; 1973; ISBN 0-471-22361-1

**[Hocken, 1980]** Hocken, R. J.; "Technology of Machine Tools, Vol. 5: Machine Tool Accuracy "; *Lawrence Livermore Laboratory Report, UCRL –52960-5*; 1999

# How to Automate the Geometrical Tolerances Inspection: A Reverse Engineering Approach

*M. Germani, F. Mandorli*
*Department of Mechanics, Polytechnic University of Marche,*
*Via Brecce Bianche, 60100 Ancona (ITALY)*
*www.dipmec.univpm.it*
*m.germani@univpm.it*

**Abstract:** Identifying theories and methods to link geometrical tolerances specification and inspection processes is a widely spread research topic. The growing use of virtual product models which not only represent the geometry information, but also collect attributes, notes, parameters, rules and procedures, can facilitate the digital simulation of many real processes. In the present work we are interested in using such technologies for the purpose of simulating the inspection process of manufactured products. In particular, we propose an approach for virtual inspection of geometrical tolerances based on a feature-based 3D CAD model coupled with a 3D points cloud data model. The 3D CAD model stores the tolerances specifications defined by the designer as well as the specification of the inspection methods. The points cloud model is a complete representation of the manufactured product and it is defined by the digital acquisition of the real product. The acquisition phase is performed with non-contact techniques to ensure high performances in terms of speed. The information stored in the CAD model is used to select and drive the different inspection procedures to be performed on the acquired data.

**Keywords**: Geometrical Tolerances, Automated Inspection, Feature-Based Model, Reverse Engineering

## 1. INTRODUCTION

The Geometrical Tolerances (GT) inspection of mechanical components (sheet-metal components, injection molded components, machined components, etc.) is aimed at measuring if the relevant geometrical features of a product, defined during the design phase and represented by dimensionally constrained geometries, have been manufactured with the required accuracy. International standards provide well-structured theories and methodologies to represent tolerances specifications and to perform the inspection process. The traditional approach is based on the specification of GT information on 2D drawings and on the use of Coordinate Measuring Machines (CMMs) to inspect the manufactured component.



*J.K. Davidson (ed.), Models for Computer Aided Tolerancing in Design and Manufacturing, 147–156.*
© 2007 *Springer.*

Today, Computer Aided Tolerancing (CAT) tools based on 3D CAD models, can support the identification of the best inspection strategy thanks to the computation and the visual simulation of alternative sensor probe paths, the support to measured data visualization and interpretation, etc. However, the time required by the CMM machine to measure the product, the complexity of running up the inspection set-up and the know-how required by the human operator, are still significant bottlenecks in the inspection process.

Our intent is to provide the final user with a system able to support the prescription, on the nominal (CAD) model, of the tolerances required to satisfy the assembly constraints and the functional requirements and a virtual inspection environment to automatically perform the GT inspection on the manufactured component. The main benefit of such approach will be the elimination of the physical inspection set-up and the speed-up of the inspection process.

The proposed method is based on the integration of feature-based CAD models and 3D optical digitizing systems, and it is based on the "duality principle" proposed in [ISO/TS 17450-1].

In this context, one of the objectives of our research work is to identify appropriated technologies to acquire from the manufactured product all the data required to perform the control. Because of the required speed of acquisition, we have focused our attention on optical digitizing tools. The advantage of the use of optical systems, with particular reference to the ones based on the fringe projection technology, is the high acquisition speed. On the other hand, the main drawbacks of such technology are: the difficulty to acquire high reflecting surfaces, the accuracy of the measurement that is in the range of 0.1 mm and the need of multiple acquisition for the representation of the complete object shape; however, we have experienced that in several practical applications the advantages of the use of such a tools outweigh the disadvantages.

The method we suggest is based on the use of a CAD product model that encapsulates the tolerance specifications as geometry attributes within the model data structure. The traditional information about GT specifications is integrated with the rules and the procedures to be used to verify GT according to the international standards recommendations. This enriched CAD model, which we call *Full of Information (FoI)* model, becomes the master model to drive the inspection of the measured data.

As mentioned earlier, in order to acquire the 3D shape of the manufactured object we propose to use optical acquisition systems. The output of such a systems is the virtual representation of the 3D shape given in terms of points cloud data. Appropriated algorithms are used in order to automatically separate the different surfaces on the points cloud data. The information stored in the *FoI* model is then used to drive the mapping between the specific GT specifications defined by the designer on the nominal shape and the corresponding zones on the points cloud data. Finally, on the basis of each specific type of GT, the inspection process based on the ISO standard procedures is applied to the virtual model.

The proposed procedure can be applied for the inspection of both regular geometric features, such as planes and cylinders as well as for free-form surfaces. At the present stage of our work the methodology has been tested on orientation tolerances using sheet-metal components for automotive applications. The preliminary experiments have been performed using a fringe projection reverse engineering system (Comet VarioZoom by Steinbichler Gmbh) that allows the acquisition of the component shape with the desired resolution. The first prototypal version of computer aided inspection software has been implemented using the development kit environment of a commercial CAD system (CATIA V5.12, by Dassault Systemes). The initial validation of the methodology shows promising results in terms of time savings and usability.

The present paper describes the proposed methodology and the preliminary results. Possible future developments are also discussed in the conclusions of the paper.


## 2.    PREVIOUS WORK

In these last years the ISO/TC 213 (Dimensional and Geometrical Product Specifications and Verification) has been carrying out a meaningful work to rethink procedures and standards related to the tolerancing problem. In particular, they have identified a set of rules showing the correspondence between processes of tolerance specification and tolerance inspection. This is leading to studies aimed at harmonizing these two processes, as in the case of the definition of the "duality principle" [Srinivasan, 2003]. In fact, two main operators have been determined: the specification operator and the inspection operator. Each operator contains a set of feature operations that are dual between the groups. Such one-to-one mapping provides an integrated view of the tolerancing problem.

An open issue is the development of methodologies and tools which allow the efficient application of such principles and operations. They can not disregard the tolerances representation within the design systems, in particular within the three-dimensional CAD software packages, that are used during the design intent definition. Many tolerance representation models have been studied and proposed aimed at supporting the different product development stages. As reported in literature [Kandikjian *et al.*, 2001], they can be classified as follows: documentation oriented, analysis oriented, production oriented and control oriented. From the CAD modeling viewpoint, the use of feature-based approach has been largely investigated. For example an interesting method to link dimensioning and tolerancing schemes in CAD systems, based on graph representation, has been described in [Shah *et al.*, 1998]. A feature classification (atomic, primitive and compound) has been defined as the base to implement a tolerancing module in [Gao et al., 1994].

In our approach the feature-based CAD model with tolerances is documentation oriented, since it collects the nominal geometry and the related tolerances. The analysis of data structure allows the identification of the atomic and compound features with tolerances, the resulting sub-model is integrated with relations between features and verification procedures. This information can be used to determine the *skin model*, as defined in [ISO/TS 17450-1], and to perform the comparison between the real model and the *skin model* itself.

The other part of the dual model is the virtual representation of the real object, as a set of points cloud data, once it has been digitized. Several technologies allow the acquisition of the 3D object geometry, but the optical systems, especially based on the triangulation principle, have evident advantages in terms of speed and usability. Their adoption for inspection tasks has been widely studied [Li & Gu, 2004], [Son et al, 2002]. The measurement accuracy is not comparable with CMMs but systems are in continual improvement [Prieto et al., 2003]. The connection between CAD models and 3D inspection systems has been approached to determine optimal inspection strategies [Huang & Gu, 1998], driving the scanning system for freeform surfaces and related data verification. The points cloud data analysis to perform the geometrical tolerance verification is a consolidated functionality of the most common reverse engineering commercial software systems (RapidForm by Inus Tech., Geomagic Qualify by Raindrop, Polyworks by Innovmetric), but they provide only algorithms to facilitate the feature extraction within a dense points cloud data. In fact, they are completely disjointed from the tolerance specification process and, furthermore, they require constant user interactive decisions for the inspection task and for data segmentation. An advanced and integrated solution proposed in literature [Prieto et al., 2000] is the most effective example of inspection automation, but the CAD model is used only as reference to perform the range data segmentation activity. A robust method to partition the points cloud data [Benko & Varady, 2004] is one of the main problems to be considered.

The proposed method based on optical 3D digitizing systems, generates a dense and noisy points cloud data, hence suitable filtering algorithms have to be applied. The resulting extracted data are partitioned using a methodology based on local differential properties [Germani et al., 2004] calculated directly on points cloud data. These sub-clouds are used to identify the *skin model* and to represent the real object geometry to be verified.

## 3.   GEOMETRICAL TOLERANCES INSPECTION METHODOLOGY

In this chapter we describe in more details the proposed GT inspection methodology.

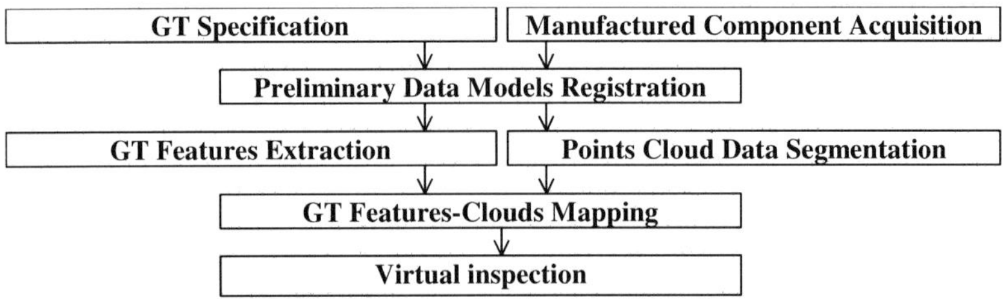

*Figure 1; Block diagram of the proposed inspection process.*

The different steps of the whole process are graphically represented in the block diagram shown in figure 1. The blocks on the left part of the diagram refer to the actions performed on the *FoI* CAD model while the ones on the right side refer to the activity performed on the points cloud data.

*GT Specification*

During this phase the designer specifies GT on the basis of required product functionalities by attaching attributes representing datum and tolerance information (GT features) to the nominal shape represented by the 3D CAD model. In order to be able to completely represent the required GT features, the *FoI* CAD model data structure must explicitly represent not only the geometric entities that are part of the object boundary, but also implicit entities such as reference points, symmetry axis, middle planes, etc. Moreover, each GT feature type is linked to an inspection method defined on the basis of a reference standard (in our work we refer to the ASME 14.5 model).

*Manufactured Component Acquisition*

The shape of the physical object to be inspected is acquired using non contact digitizing tools. The identification of the best acquisition strategy (involving decisions about the number of views and the resolution to be used) depends on the specific application. When multiple views are required, particular attention must be paid to  the process of matching the different clouds in order not to introduce additional errors on the model.

*Preliminary Data Models Registration*

The data stored in the *FoI* CAD model will be used to identify regions and entities of the digitized model to be used during the inspection procedure. In order to be able to perform this mapping the two models must be overlapped.

The model overlapping can be performed by using a best-fit method based on the ICP (Iterative Closest Points) [Besl & McKay, 1992]. The strategy to select the points to be used for the best-fit algorithm can be different according to the precision required by the specific application (interactive user selection, use of target points, etc.).

*GT features Extraction*

The *FoI* CAD model is navigated in order to extract the explicit information required for the tolerance inspection. In particular, the GT features that define the different tolerances to be inspected are collected and organized as follows:

*Tolerance(id-number, category, type, value, datum_list), method(category, type)*
  *GT_Feature(reference), GT_Feature_Type(type), CofG($G_x$, $G_y$, $G_z$), Tol_Datum(ch), Tol_Value( value);*

An identification number, a category (form, orientation, location, etc) a type (flatness, parallelism, position, etc.) a value and a datum list (list of characters; it can be an empty list) is used to explicitly represent each specific GT to be inspected. On the basis of the tolerance category and type, an inspection method, retrieved from a data base, is linked to the tolerance. For each entity that is part of the GT definition a set of information is collected: the reference to the geometric entity, the type of the geometric entity (point, axis, plane, etc.), the center of gravity of the geometric entity and, depending on the referenced entity, the text character that defines the datum or the required tolerance value (a possible instantiation of the described structure is shown in the next chapter where a complete application example is reported).

### *Points cloud Data Segmentation*

The overall points cloud data is partitioned into different zones corresponding to different form features (e.g. plane or cylindrical surfaces, etc.). Canonical surfaces can be easily identified as for free-form surfaces identification more sophisticated partitioning algorithms must be applied. At the end of this task the overall points cloud data is partitioned into several sub-clouds; each cloud has an attribute that identifies the sub-cloud type (i.e. plane, cylinder, free-form, etc.).

### *GT Features-Clouds Mapping*

In this phase the *FoI* CAD model is used to drive the identification of the sub-clouds of points that are relevant for the inspection procedures. The identification of the sub-cloud related to a particular GT feature is carried out on the basis of a minimum distance algorithm application. The centers of gravity (*CofG*) of the geometric entity representing the GT feature is compared with the center of gravity of the bounding box of all the sub-clouds of *compatible* type (i.e. plane GT features are compared to plane sub-clouds; axis GT feature are compared to cylinder sub-clouds, etc.); the sub-cloud with the minimum distance is identified to be the sub-cloud representing the GT feature on the acquired model. In case of common zones, the sub-clouds corresponding to the different regions are merged and treated as a single sub-cloud.

### *Virtual Inspection*

Once the other tasks have been successfully performed, the virtual inspection process can take place.

The appropriated inspection method is selected according to the specific GT category and type to be inspected. First, the method must describe the procedure to precisely identify the datum geometry on the sub-cloud. For each different datum and tolerance type different strategies can be identified. (The definition of the best identification strategy for all the cases is still an open issue in our research. At present we are focused on orientation tolerances with plane datum). Once the virtual datum has been identified, the method describes the procedure for the computation of the tolerance zone. Finally, the points belonging to the sub-cloud corresponding to the geometry to be inspected are checked in order to verify if they are bounded by the tolerance zone.

## 4.   PRELIMINARY RESULTS

A prototype system providing functionalities to specify and inspect orientation GT has been implemented using the software development kit of a commercial CAD system (CATIA v.5.12 by Dassault Systemes) and tested on a reference component.

The reference component is a sheet metal part for the automotive industry. It has been selected because of the following reasons: its shape combines freeform and canonical surfaces; it has a medium size; the required tolerance values are in the accuracy range of the measurement systems.

In order to be able to describe in details the different steps of the virtual inspection process, we limit our example to the parallelism required among different planar surfaces involved in the component assembly.

*GT Specification:* figure 2 shows the nominal shape of the component together with the required GT specification as modeled in the 3D CAD environment (*FoI* CAD model). The GT reference plane is defined by a common zone that is split into three separated planar surfaces (yellow planes in figure 2). The toleranced geometry is a common zone split into two planar surfaces (red surfaces in figure 2). The tolerance value is 0.5 mm.

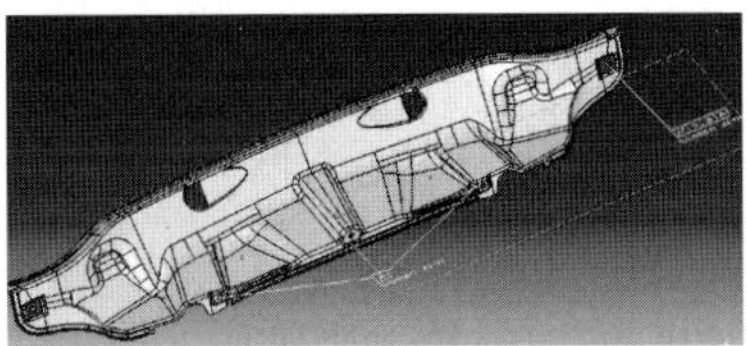

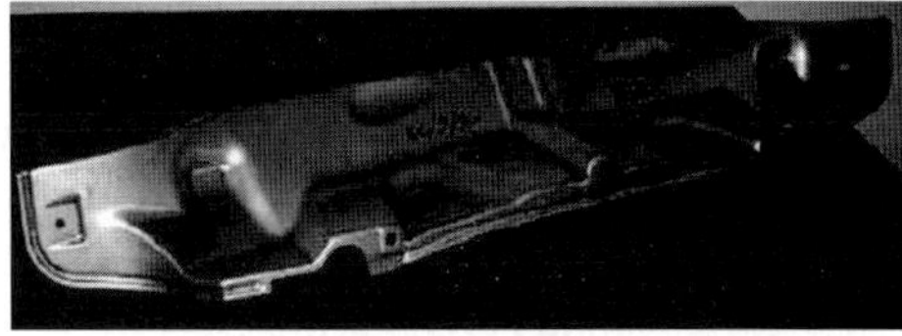

*Figure 2; Nominal shape and GT specifications*     *Figure 3; manufactured component.*

*Manufactured Component Acquisition:* the manufactured component (see figure 3) has been digitized using a 3D scanner based on fringe projection technology (COMET Vario Zoom by Steinbichler Gmbh). The clouds acquired from different points of view have been matched, filtered and sub-sampled in order to obtain the final points cloud data that is shown in figure 4.

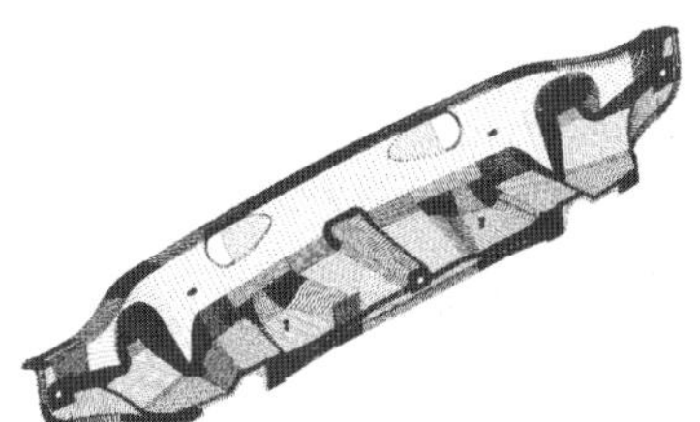

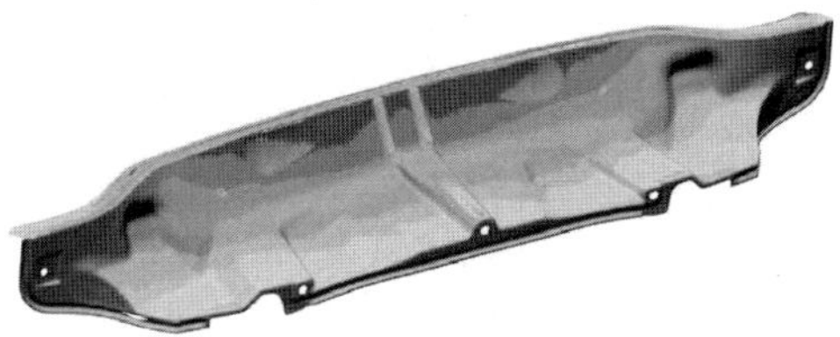

*Figure 4;    the points cloud data*          *Figure 5; the color map*

*Preliminary Data Models Registration:* the align clouds commands of the digitized shape editor module of CATIA have been used to find the relative positioning of the two models. Figure 5 shows the color map of the distances between the nominal shape and the points cloud data after the application of the align commands, as computed by the distance analysis algorithms of CATIA.

*GT Features Extraction:* the data structure of the *FoI* CAD model is navigated in order to extract and organize the information about GT specifications. By using the representation method defined in the previous chapter, the data contained in the *FoI* model of our reference example is organized as follows:

*Tolerance (1, orientation, parallelism, 0.5, (A)), method (orientation, parallelism)*
  *GT_Feature(112), GT_Feature_Type(plane), CofG (65.17 ,-275.9, 603.6), Tol_Datum (A), Tol_Value();*
  *GT_Feature(111), GT_Feature_Type(plane), CofG (65.17, -8.4, 624.1), Tol_Datum (A), Tol_Value();*
  *GT_Feature(109), GT_Feature_Type(plane), CofG (65.17, 275.9, 603.6), Tol_Datum (A), Tol_Value();*
  *GT_Feature(139), GT_Feature_Type(plane), CofG (-51.3, 602.1, 659.6), Tol_Datum (),Tol_Value(0.5);*
  *GT_Feature(387), GT_Feature_Type(plane), CofG (-51.3, -602.1, 659.6), Tol_Datum(), Tol_Value(0.5)*

*Points Cloud Data Segmentation:* the points cloud data segmentation is performed with an algorithm based on the computation of local curvature and slope on a set of cross sections extracted from the overall cloud (more details about the method can be found in [Germani et al., 2004]). The cross sections used for the segmentation of the points cloud data of our reference example are shown in figure 6.

*GT Features-Clouds Mapping:* the mapping between GT features and the sub-clouds is performed on the basis of the computation of the minimum distance between the center of gravity of the feature and the center of gravity of the bounding box of all the *compatible* sub-clouds. At the end of this process only the sub-clouds related to GT features are considered for virtual inspection. Figure 6 shows the identified sub-clouds for our example.

*Figure 6; cross sections on the points cloud data and identified sub-clouds.*

*Virtual Inspection:* the appropriated inspection method is selected on the basis of the specific GT category (orientation) and type (parallelism) to be inspected. First of all the virtual datum need to be defined. In our example the orientation tolerance to be inspected is a parallelism tolerance and the datum is a plane split into three different regions corresponding to a single *disconnected* sub-cloud. In this case the virtual datum plane is defined as follows: first, the cloud medium plane is taken as reference, then a set of points is selected as the outer points in the direction perpendicular to the medium plane. The number of selected points represents 20% of the total points belonging to the cloud. Finally, the datum plane is defined as the medium plane of the set of selected points.

Once the virtual datum plane has been defined, the tolerance is controlled by comparing the required value with the computed distance between two parallel planes, parallel to the datum plane, passing through the most external points belonging to the cloud. In our reference example, the described inspection process has reported a value of 0.55 mm, that is 0.05 mm above the allowed tolerance (see also figure 7). This result has been proved to be consistent with the result obtained with a traditional inspection process.

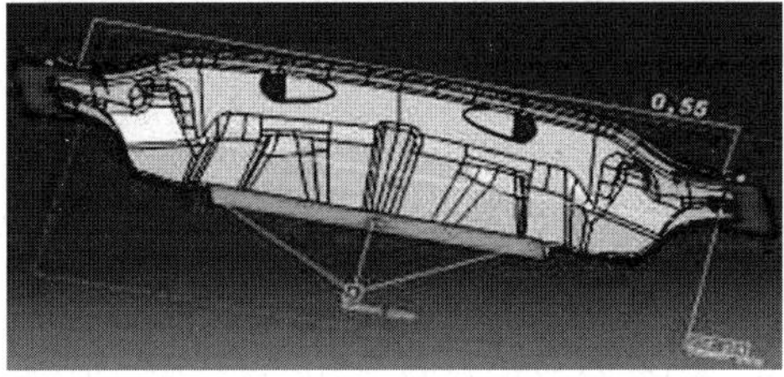

*Figure 7; Graphical representation of the inspection result.*

## 5. CONCLUSIONS AND FUTURE DEVELOPMENTS

This paper presents a methodology for the virtual inspection of GT based on the use of 3D CAD models and a 3D scanner based on fringe projection technology.

The CAD model is enriched in order to store the information about GT specifications in terms of attributes linked to the model geometrical features. The 3D digitizing system is used to perform the fast acquisition of the manufactured component shape. The proposed methodology has been tested on a real case showing promising results.

The main advantages of the proposed methodology concern the integration of design aspects with inspection aspects: once the manufactured product has been digitized, the virtual inspection can be performed in the CAD environment on the basis of procedures that are controlled by the designer. Moreover, the virtual inspection completely overcomes the need to arrange a physical inspection set-up and this will dramatically decrease the inspection costs and it will limit the problems related to misunderstandings between designers and quality control operators.

However, some aspects related to the proposed methodology need to be deeply investigated. The present limitations for a wider application of the methodology are mainly related to the acquisition phase and the definition of the virtual datum.

The acquisition process with optical systems has intrinsic limitations that under certain conditions may prevent the successful application of the methodology.

The greatest limitation is the accuracy in the measurement, that is in the range of 0.1 mm. However, while this accuracy can be acceptable for several applications, the errors introduced during the matching, filtering and sampling of the clouds have to be kept under control. This will also include a deeper knowledge and a critical analysis of the CATIA algorithms used during the preliminary data models registration phase.

At the present stage of the work we have focused our attention on orientation tolerances and we have identified different strategies to define the planar virtual datum.

More experiments are required to identify the best strategy and to extend the set of procedures to define the datum and to compute the tolerance values in case of other GT types. This will also include experiments to eventually find a cloud sampling strategy before computing the datum plane.


## REFERENCES


**[Benko & Varady, 2004]** Benko, P.; Varady, T.; "Segmentation methods for smooth point regions of conventional engineering objects"; In: *Computer Aided Design*, Vol.36, Number 6, May 2004, pp. 511-523.

**[Besl & McKay, 1992]** Besl, P.J., McKay, N.D., "A method for registration of 3-D shapes"; In: *IEEE Transactions on Pattern Analysis and Machine Intelligence*, 14(2), pp. 239 - 256, 1992.

**[Gao et al., 1994]** Gao, J., Case, K., Gindy, N.; "Geometric elements for tolerance definition in feature-based product model"; In: *Advances in Manufacturing Technology VIII*, eds. K. Case and S.T. Newmann, Taylor & Francis, September 1994, pp.264-268; ISBN 0-7484-0254-3.

**[Germani et al., 2004]** Germani, M., Corbo, P.; Mandorli, F.; "Aesthetic and functional analysis for product model validation in reverse engineering applications"; In: *Computer Aided Design*, Vol.36, Number 1, January 2004, pp. 65-74.

**[Huang & Gu, 1998]** Huang, X., Gu, P.; "CAD model based inspection of sculptured surfaces with datums"; In: *Int'l Journal of Production Research*, Vol. 36, No. 5, pp. 1351-1367.

**[ISO/TS 14750-1]** "Geometrical Product Specification (GPS) – General Concepts – Part1: model for geometrical specification and verification".

**[Kandikjian et al., 2001]** Kandikjian, T., Shah , J., Davidson, J.; "A mechanism for validating Dimensioning & Tolerancing schemes in CAD systems", In: *Computer-Aided Design*, Vol.33, Number 10, July 2001, pp. 721-737.

**[Li & Gu, 2004]** Li, Y., Gu, P.; "Free-form surface inspection techniques state of art review"; In: *Computer Aided Design*, Vol.36, Number 13, November 2004, pp. 1395-1417

**[Prieto et al., 2000]** Prieto, F., Lepage, R., Boulanger, P., Redarce, T.; "Inspection of 3D parts using high accuracy range data"; In: *Proceedings of the SPIE conference VIII Machine Vision Applications and Industrial Inspection*, pp. 82-93; San Josè 2000.

**[Prieto et al., 2003]** Prieto, F., Lepage, R., Boulanger, P., Redarce, T.; "A CAD-based 3D data acquisition strategy for inspection", In: *Machine Vision and Applications Journal*, Vol.15, Number 2, December 2003, pp-76-91.

**[Shah et al., 1998]** Shah, J., Yan, Y., Zhang, B.C.; "Dimension and tolerance modeling transformations in feature-based design and manufacturing"; In: *Journal of Intelligent Manufacturing*, Vol.9, Number 5, October 1998, pp. 475-488.

**[Son et al, 2002]** Son, S., Park, H., Lee, K.; "Automated laser scanning system for reverse engineering and inspection"; In: *International Journal of Machine Tools & Manufacture*, Vol.42, pp. 889-897.

**[Srinivasan, 2003]** Srinivasan, V.; "An integrated view of geometrical product specification and verification"; In: *Proceedings, 7th International Seminar on Computer Aided Tolerancing*, Kluwer, November 2003, pp. 1-12, ISBN 1-40201-423-6.

# A New Algorithm to Assess Revolute Surfaces through Theory of Surface Continuous Symmetry

**W. Polini*, U. Prisco**, G. Giorleo****

**Dipartimento di Ingegneria Industriale, Università di Cassino, via G. di Biasio 43,*
*03043 Cassino, Italy*

*polini@unicas.it*
***Dipartimento di Ingegneria dei Materiali e della Produzione*
*Università degli Studi di Napoli "Federico II"*
*Piazzale Tecchio 80, 80125 Napoli (Italy)*


**Abstract:** This paper presents a new approach to the evaluation of revolute surfaces based on surface invariance with regard to the rigid motions. The approach transforms the measurement data through an opportune homogeneous transformation matrix in order to minimize the distance between the cloud of measured points and the reference element of the class of revolution surfaces from which the sampling comes. The best transformation parameters are searched minimizing the distance between the cloud of measured points and a geometric element having the same geometric nature of the reference element of the class of the revolution surface. Afterwards the individuation of the best-fit set of transformation parameters, the form tolerance of the inspected feature can be assessed following the tolerance definition supplied by the GD&T Y 14.5 standard. Numerical simulations were performed in order to validate the effectiveness and the robustness of the approach. The great advantage of this new algorithm is that its formulation may be applied to assess any kind of surfaces without any adaptation effort.
**Keywords:** profile tolerance, geometric dimensioning and tolerancing (GD&T), geometric product specification (GPS), continuous subgroups of rigid motion

## 1. INTRODUCTION

Dimensional inspection of a mechanical part produces a set of Cartesian coordinates of points measured by a coordinate measuring machine (CMM) from a manufactured surface of the part. The coordinates are elaborated to yield the geometric deviations of the manufactured surface from the nominal one. Many best fitting techniques, that calculate the substitute feature, were carried out based on either direct or random search techniques, Monte Carlo simulation, simplex search and spiral search techniques [Murthy et al., 1980; Shummugam et al., 1986 and 1987; Wang et al., 1992; Samuel et al., 2003].

In recent years a new class of algorithms for the establishment of form tolerance have received much attention [Roy et al., 1992]. This new class instead of a substitute feature directly searches the minimum zone of features, for example in planarity assessment



*J.K. Davidson (ed.), Models for Computer Aided Tolerancing in Design and Manufacturing*, 157–168.
© 2007 *Springer.*

two parallel planes with the minimum distance such that they include the whole cloud of measured point. This method best conform the ISO standards that define the form tolerance by means of a tolerance zone within which the feature is to be contained. However, this kind of algorithms is very sensitive to asperities (such that caused by measure errors), while the methods minimizing an objective function are much less sensitive to asperities and mathematically well found, [Yau et al., 1996; Choi et al., 1999].

A very interesting algorithm was proposed by Yau and Menq [Yau et al. 1996]. Instead of computing the substitute geometric feature, the method inversely transforms the measured data such that the sum of the squared distances of the transformed points from the nominal geometry is minimized. The authors used a sensitivity matrix to relate transformation errors to dimensional errors. Chiabert proposed a practical methodology for the statistical recognition of 3D shapes from sparse measurements, extending the ISO/TC classification of 3D objects based on their symmetry to probability density functions so as to include the measurement process in the formalism [Chiabert et al., 2001]. Capello and Semeraro address the definition of the relationship among the estimate error of the substitute geometry, the inspection plan and the machining and measuring processes [Capello et al., 2001]. A new method is proposed, called Harmonic Fitting Method, that is valid when the least square method is used as a fitting objective.

Those traditional approaches should be integrated with the new concepts under development by the ISO/TC213 Technical Committee [Srinivans et al., 1999]. The Geometrical Product Specifications (GPS) standards are evolving towards the innovative concept of symmetry groups for feature classification. It implies a feature classification based on surface invariance with regard to the rigid motions. It is based on the twelve connected Lie subgroups of rigid motion that led to a classification of the symmetry groups of features into seven classes.

This paper presents a new approach to the evaluation of revolution surfaces based on surface invariance with regard to the rigid motions. The approach transforms the measurement data through an opportune homogeneous transformation matrix in order to minimize the distance between the cloud of measured points and the reference element of the class of revolution surfaces from which the sampling comes. The best transformation parameters are searched minimizing the distance between the cloud of measured points and a geometric element (datum) having the same geometric nature of the reference element of the class of the revolution surface. If the datum is composed by two or three geometrical element we can find the best fit as result of two fitting operations. When we consider a complex feature, like a cone or a paraboloid, the datum is composed by two reference elements. As you can verify the two reference feature must have different dimensions, i.e. there must be no homeomorphism which carries an element of the datum in the other. This characteristic gives us the possibility to find an order between the reference elements of the datum, by means of an adequate criterion based on the number of dimensions of the reference elements. This order can be used for decomposing the fitting operation in two sub-operations. The first operation is the fitting of the set of points to the higher order reference element of the datum, while the second operation is the fitting of the set of points to the remaining reference element of the datum, under the constraint given by the first operation.

Afterwards the individuation of the best-fit set of transformation parameters, the form tolerance of the inspected feature can be assessed following the tolerance definition supplied by the GD&T Y 14.5 standard.

Numerical simulations were performed in order to validate the effectiveness and the robustness of the approach. The great advantage of this new algorithm is that its formulation may be applied to assess any kind of surfaces without any adaptation effort. In the following the problem proposed algorithm is mathematically described (§2) and applied to tolerance assessment of revolute surfaces (§3). Finally, the performances of the proposed algorithm are deeply discussed by means of different sets of data (§3).

## 2. PROPOSED ALGORITHM

### 2.1. Generic approach

The mathematical basis of our works have been explained in two previous works [Carrino et al, 2002 and 2004], in this paper we put our attention about the specific variation related to the studied case.

The extracted integral feature is represented by a finite set of points, that is expressed by the (1)

$$
W_M = \left\{ \begin{array}{l} \mathbf{m}_i = \left[ m_{xi}, m_{yi}, m_{zi}, 1 \right]^T \\ : m_{xi}, m_{yi}, m_{zi} \in \Re, i = 1 ... M \end{array} \right\}, \tag{1}
$$

By applying a homogeneous coordinate transformation $\mathbf{T}(t)$ the extracted integral feature is roto-translated so that the measured points minimize their distance from the reference element. In this paper, the Roll-Pich-Yaw angles are used to describe the spatial rotation and translation, so the transformation matrix can be described as:

$$
\mathbf{T}(t) = \begin{bmatrix} C\varphi C\theta & C\varphi S\theta S\psi - S\varphi C\psi & C\varphi S\theta C\psi + S\varphi S\psi & t_x \\ S\varphi C\theta & S\varphi S\theta S\psi + C\varphi C\psi & S\varphi S\theta C\psi\gamma - C\varphi S\psi & t_y \\ -S\theta & C\theta S\psi & C\theta C\psi & t_z \\ 0 & 0 & 0 & 1 \end{bmatrix}. \tag{2}
$$

The transformation matrix is characterized by six degrees of freedom so $\mathbf{t} = [\psi, \theta, \varphi, t_x, t_y, t_z]$ is the set of transformation parameter.

The minimization of the distance between the set of measured points and the nominal surface is made by the following objective function:

$$
Q(\mathbf{t}) = \sum_{i=1}^{N} \left( \mathbf{T}(\mathbf{t}) \cdot \mathbf{m}_i - \mathbf{n}_i \right)^2, \tag{3}
$$

where $\mathbf{n}_i$ represents the nearest point of $\mathbf{T}(\mathbf{t}) \cdot \mathbf{m}_i$ on the reference element, point chosen in according of geometrical definition of distance.

Equation (3) represents the standard expression of the objective function. The expression itself is sufficient to give us an acceptable solution of the problem, but the resultant algorithm is relatively inefficient. In order to improve the speed of the algorithm it is useful the concept of equivalence class, based on the Lie subgroups. Speaking shortly, the concept of equivalence class implies that the feature is invariant for certain class of translations or rotations. So there are some parameters in the (3) that have no influence on the value of the objective function. This parameters can be simply set to zero, and the objective function can be rewritten without these parameters.

### 2.2. Revolute surfaces

The revolute class of surfaces encompass all feature invariant under coaxial rotation displacement, but not invariant under translation along axis. The datum of the feature is constituted by the axis of the feature, which is unique, and by a point along the axis, that can be chosen between the infinite points of the axis.

Our algorithm is composed by two optimization routines, which are executed sequentially in order to find the optimized set of parameters.

The first step of the algorithm is the minimization of the mean distance between the measured point set and the axis of the feature, while the second step is the minimization of the vertical distance, measured along the z axis, between the measured points and the nominal surface. For sake of simplicity the axis of the feature is chosen coincident with the z axis.

The nearest point to the ith actual measurement $\mathbf{m_i}$ on the z axis is naturally: $\mathbf{n_i} = [0, 0, n_{zi}, 1]^T$. Then, the generic objective function, equation (3), becomes for the first step:

$$Q(\mathbf{t}) = \sum_{i=1}^{N} \left( \begin{bmatrix} C\theta & S\theta S\psi & S\theta C\psi & t_x \\ 0 & C\psi & S\psi & t_y \\ -S\theta & C\theta S\psi & C\theta C\psi & 0 \\ 0 & 0 & 0 & 1 \end{bmatrix} \cdot \mathbf{m_i} - \begin{bmatrix} 0 \\ 0 \\ n_{zi} \\ 1 \end{bmatrix} \right)^2 . \tag{4}$$

The equation (4) can be simplified turning the attention to the fact that the only need is to minimize the distance of the first and the second transformed coordinate from z axis, never minding of the third component of $\mathbf{m_i}$. Thus, the only first two components of the vector in (4) are considered, obtaining the following objective function:

$$Q(\psi, \theta, t_x, t_y) = \sum_{i=1}^{N} \left[ C\theta \cdot m_{xi} + S\theta S\psi \cdot m_{yi} + S\theta C\psi \cdot m_{zi} + t_x \right]^2 +$$
$$\sum_{i=1}^{N} \left[ C\psi \cdot m_{yi} + S\psi \cdot m_{zi} + t_y \right]^2 . \tag{5}$$

The method of Levenberg-Marquardt is applied to the numerical optimisation of this equation. Then, the best-fit parameters for the equation (5), is $\bar{t} = \left| \overline{\psi}, \overline{\theta}, 0, \overline{t_x}, \overline{t_y}, 0 \right|$.

The second step of our algorithm is the minimization of the vertical distance between the measured points and the nominal surface. The result to this step is the parameter $\bar{t_z}$, that combined with the parameters found in the previous step give us the final best fit parameter $\bar{t} = \left[ \bar{\psi}, \bar{\theta}, 0, \ \bar{t}_x, \bar{t}_y, \bar{t}_z \right]$.

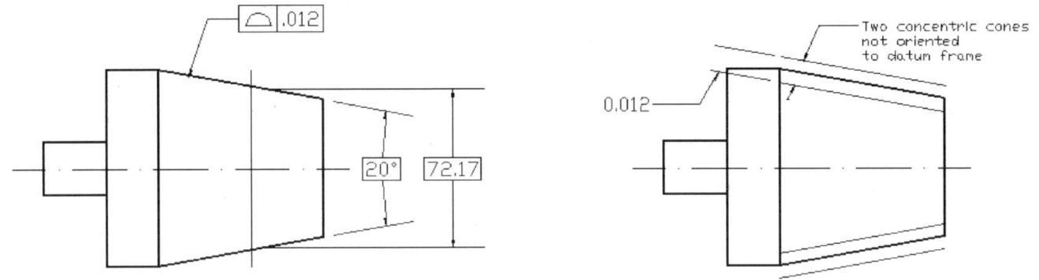


*Figure 1: Definition of tapering tolerance*

Afterwards the individuation of the best-fit set of transformation parameters the tolerance assessment of the feature can be assessed in according with the standard tolerance definition (see Figure 1): *a profile tolerance specifies a tolerance zone bounded by parallel features, identical to the nominal feature, within which the surface must lie.* This definition can be easily expressed for the fitted feature in the appropriate reference system.


## 3. APPLICATION EXAMPLES

The above algorithms was implemented on a Athlon Thunderbird, 512 MB, 1000 MHz machine, using the Mathematica 5.0 package. The proposed method was tested to verify its correctness using datasets created by numerically simulating actual rotational features, specifically 6 truncated cones and 6 truncated paraboloids. The simulated features were generated expressing the surfaces in their parametric form. The equation of a generic surface in parametric form is:

$$\begin{cases} x = x(u,v) \\ y = y(u,v) \\ z = z(u,v) \end{cases} \qquad (6)$$

Naturally, for a parametric representation of a surface, "other parameters" are hidden inside equation (6) besides the independent variable $u$ and $v$ (i.e. the radius for parametric equation of a sphere). For a revolute surface one of the two parameters belongs to the interval $[0, 2\pi]$. Be $v$ this parameter. Then, it is possible to write: $v \in [0, 2\pi]$. The simulated surface was generated sampling points from two perfect surfaces characterised by two different value of the "other parameters". This simulation procedure of datasets was implemented following the well-established Carr and Ferreira method [Carr et al, 1995]. The difference between these two parameters is the tolerance measured horizontally, i.e., along the x-y plane. The resulting form tolerance, defined according the ASME Y 14.5-1994 standard, of the generated revolute surfaces can be calculated by simple calculations. The results are reported in the fifth column of Tables

1 and 2. The last column, sixth, of Table 1 and 2, entitled Form deviation (proposed algorithm) $t_p$, reports the form error of the tested datasets calculated using the proposed algorithm. Comparison of column sixth with column fifth gives the assessment capability of the proposed algorithm.

For the conical frustum the datasets were extracted by the parametric equation of the corresponding cone, see figure 2 below.

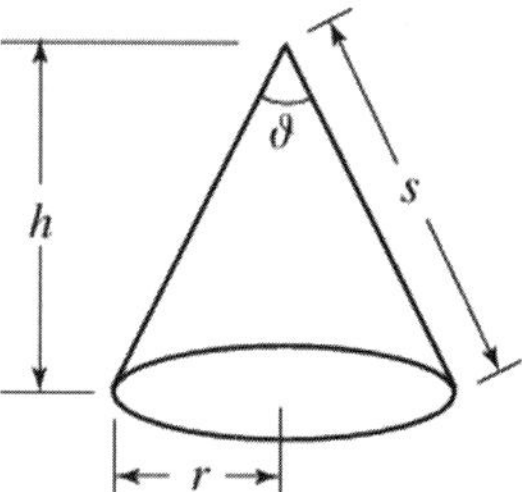

*Figure 2: Representation of a cone used to generate the conical frustrum*

The set of parametric equations, explicating equation (6) for a cone and used to generate the conical frustum, is:

$$\begin{cases} x = \dfrac{h-u}{h}\, r \cdot \cos v \\[2mm] y = \dfrac{h-u}{h}\, r \cdot \sin v \;, \\[2mm] z = u \end{cases} \qquad (7)$$

where $h$ and $r$ are cone height and base radius, respectively. The cone base radius corresponds to the base radius of the truncated cone and it is reported in Table 1. The height of the conical frustum, reported in the forth column of Table 1, is the height of the original cone truncated by a given quantity: $h - diff$. This difference represents one of the two limits of the definition interval of $u$: $u \in [0, h - diff]$. The "other parameters" present in the set (7) are $h$ and $r$. Varying randomly these parameters between two sets, differing just for a factor of scale, so that the similarity of the two surfaces is preserved, for every point sampled, it is possible to build datasets of points of known form tolerance. The resulting form error can be easily calculated and it is reported in Table1.

The set of parametric equations, explicating equation (6) for a paraboloid, is:

$$\begin{cases} x = a\sqrt{\dfrac{u}{h}}\, \cos v \\[2mm] y = a\sqrt{\dfrac{u}{h}}\, \sin v \;. \\[2mm] z = u \end{cases} \qquad (8)$$

This is the parametric form of the paraboloid having radius *a* at height *h*. *a* and *h* are reported in Table 2 as max radius and height, respectively, see Figure 3. As well as for the truncated cone, paraboloid datasets are obtained varying the "other parameters" *a* and *h*, so that every sampled point is randomly extracted from two different perfect surfaces of known distance. The resulting form error can be easily calculated and it is reported in Table 2 as form error.

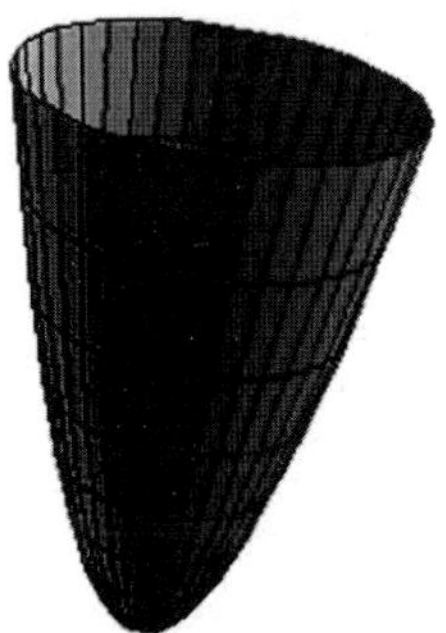

*Figure 3: Representation of a paraboloid used to generate the dataset*

Moreover, each dataset was randomly oriented into the coordinate system. The datasets 6 of the truncated cone and the dataset 5 of the paraboloid are reported in Appendix.
Figures 4 and 5, that was taken from Mathematica®, show the representation of the two step of the optimization process.
The obtained results are higher than the value of the form tolerance in all cases, since the adopted approach is based on least squares principle, even if they are enough close to the form tolerance.

*Table 1. Results for the six datasets of truncated cone*

| Series | N° of points | Base radius | Height | Form Tolerance $t_r$ | Form Deviation (proposed algorithm) $t_p$ |
|---|---|---|---|---|---|
| 1 | 24 | 5 | 6 | 0.0447214 | 0.0565002 |
| 2 | 24 | 5 | 6 | 0.0894427 | 0.0967931 |
| 3 | 24 | 8 | 11 | 0.0941176 | 0.105575 |
| 4 | 24 | 20 | 6 | 0.0447214 | 0.0592273 |
| 5 | 48 | 8 | 11 | 0.0447214 | 0.0611619 |
| 6 | 60 | 5 | 6 | 0.0447214 | 0.0578712 |

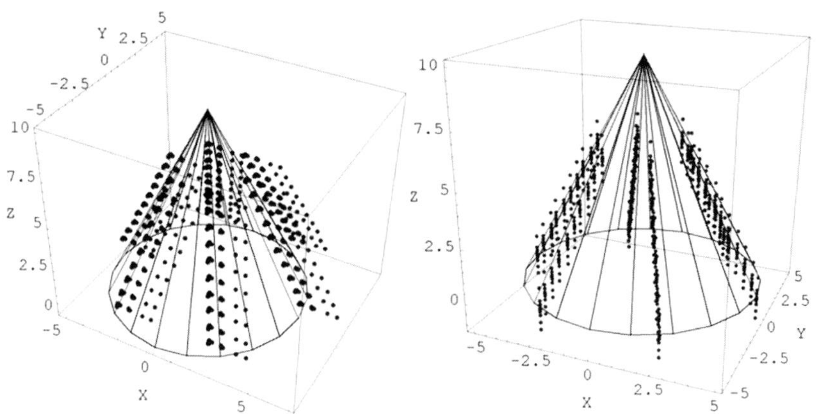

*Figure 4. Fitting of the points to the nominal surface for a truncated cone*

*Table 2. Results for the six datasets of paraboloid*

| Series | N° of points | Max radius | Height | Form Tolerance $t_r$ | Form Deviation (proposed algorithm) $t_p$ |
|---|---|---|---|---|---|
| 1 | 24 | 5 | 10 | 0.1 | 0.1344782 |
| 2 | 24 | 5 | 10 | 0.05 | 0.0636638 |
| 3 | 24 | 8 | 15 | 0.05 | 0.0609883 |
| 4 | 24 | 20 | 10 | 0.01 | 0.0141566 |
| 5 | 48 | 10 | 10 | 0.05 | 0.0696541 |
| 6 | 160 | 8 | 15 | 0.001 | 0.0013479 |

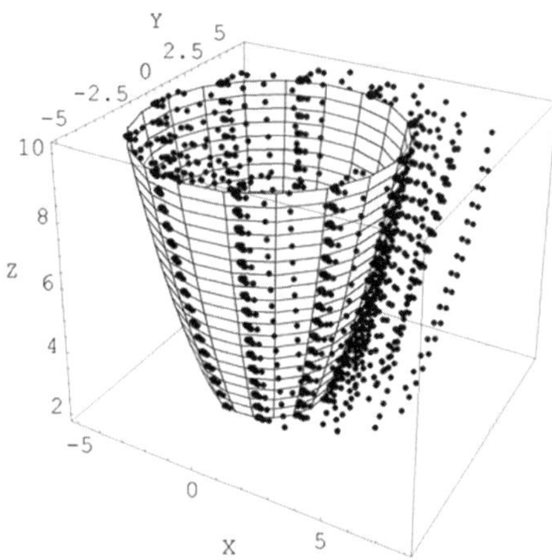

*Figure 5. Fitting of the points to the nominal surface for a paraboloid*

## 4. CONCLUSION

The present work shows an approach to assess profile tolerance related to revolute surfaces that is both general and easy to implement. The proposed approach is based on the new concepts under development by the ISO/TC213 Technical Committee. The approach transforms the measurement data through an opportune homogeneous transformation matrix in order to minimize the distance between the cloud of measured points and the reference element of the class of surfaces from which the sampling comes. The best transformation parameters are searched by fitting the set of points to the higher order reference element of the datum and, then, by fitting the set of points to the remaining reference element of the datum, under the constraint given by the first operation. The results indicate that the proposed algorithm provides accurate and quick assessment.

## ACNOWLEDGEMENTS

The work has funded partially by the Italian M.I.U.R. (Italian Ministry of University and Research).

## REFERENCES

**[Capello et al., 2001]**  Capello, E.; Semeraro, Q.; "The Harmonic Fitting Method for the Assessment of the Substitute Geometry Estimate Error. Part I: 2D and 3D Theory", *Int. J. machine Tools and Manufacture*, 41, pp. 1071-1102; 2001.

**[Carr et al., 1995]**  Carr, K.; Ferreira, P; "Verification of form tolerances. Part I: Basic Issues, Flatness and Straightness", *Precision Engineering*, 17, pp. 131-143; 1995.

**[Carr et al, 1995]**  Carr, K.; Ferreira, P; "Verification of form tolerances. Part II: Cylindricity and Strightness of a median line", *Precision Engineering*, 17, pp. 144-156; 1995.

**[Carrino et al., 2002]**  Carrino, L.; Giorleo, G.; Polini, W.; Prisco, U.; "Automatic Technique Based on Symmetry Group Classification To Asses Form Tolerance", In: *Proc. of the 3rd CIRP International Seminar on Intelligent Computation in Manufacturing Engineering*, pp. 409-414; Ischia, Italy, 2002.

**[Carrino et al., 2004]**  Carrino L.; Coticelli F.; Giorleo G.; Polini W.; Prisco U.; "Cylindricity assessment by means of symmetry groups classification theory", In: *Proceedings of 4th CIRP International Seminar on Intelligent Computation in Manufacturing Engineering*, pp. 237-242; Sorrento, Italy, 2004.

**[Chiabert et al., 2001]** Chiabert, P.; Costa, M.; Pasero, E.; "Detection of Continuous Symmetries in 3d Objects from Spars Measurements trough Probabilistic Neural

Networks", In: *Proc. of IEEE international Workshop on Virtual and Intelligent Measurement Systems*, pp. 104-110; Budapest Hungary, 2001.

**[Choi et al., 1999]**	Choi, W.; Kurfen, T.R.; "Dimensional Measurement Data Analysis, Part I: A Zone Fitting algorithm", *Journal of Manufacturing Science and Engineering, Transaction of ASME*, 121, pp. 238-245; 1999.

**[Murthy et al., 1980]** Murthy, T.S.R.; Abdin, S.Z.; "Minimum Zone Evaluation of Surfaces", *Int. J. Machine Tools and Manufacture*, 20, pp. 123-136; 1980.

**[Wang, 1992]** Wang, Y.; "Application of Optimisation Techniques to Minimum Zone Evaluation of Form Tolerances", In: *Quality Assurance Through integration of Manufacturing Processes and Systems*, Edited by A.R. Thangaraj, ASME, pp. 15-28; 1992.

**[Roy et al., 1992]**	Roy, U.; Zhang, X.; "Assessment of a Pair of Concentric Circles with the Minimum Radial Separation for Assessing Roundness Errors", *Computer Aided Design*, 24, pp. 161-168; 1992.

**[Samuel et al., 2003]**	Samuel, G.L.; Shunmungam, M.S.; "Evaluation of circularity and spheriticy from coordinate measurement data", *J. of Materials Processing Technology*, 139, pp. 90-95, 2003.

**[Shunmugam et al., 1986]**	Shunmugam, M.S.; "On Assessment of Geometry Errors", *Int. J. Production Research*, 245, pp. 413-425; 1986.

**[Shunmugam et al., 1987]**	Shummugam, M.S.; "New Approach for the Evaluating Form Errors of Engineering Surfaces", *Computer Aided Design*, pp. 368-374; 1987.

**[Srinivasan et al., 1999]** Srinivasan, V; "A Geometrical Product Specification Language Based on Classification of Symmetry Groups", *Computer Aided Design*, 31, pp. 659-668; 1999.

**[Yau et al., 1996]**	Yau, H.-T.; Meng, C.-H.; "A Unified Least-Square Approach to the Evaluation of Geometric Errors Using Discrete Measurement Data", *Int. J. Mach. Tools and Manufact.*, 36, pp. 1269-1290; 1996.

APPENDIX

| Conical frustrum dataset: 6 | | | Parabolid dataset: 5 | | |
|---|---|---|---|---|---|
| X | Y | Z | X | Y | X |
| 7.12623062 | 0.24885346 | 0.75160440 | 8.5477214 | 2.1495057 | 1.8504715 |
| 4.37827810 | 4.48565926 | 0.84234309 | 7.1592119 | 5.3433292 | 1.8737314 |
| -0.61563203 | 4.31126808 | 1.01684058 | 3.9193621 | 6.6197416 | 1.9298858 |
| -2.96146784 | -0.10341674 | 1.10408932 | 0.7260321 | 5.2310376 | 1.9860401 |
| -0.36031024 | -4.43200406 | 1.01858555 | -0.4719975 | 1.9920659 | 2.0079353 |
| 4.73347809 | -4.25412505 | 0.84059812 | 0.8383408 | -1.2031221 | 1.9860401 |
| 6.80989688 | 0.23780684 | 1.42973078 | 4.0768259 | -2.4013513 | 1.9298858 |
| 4.28893014 | 4.27456638 | 1.51279158 | 7.2152802 | -1.0365115 | 1.8746964 |
| -0.41584158 | 4.02939382 | 1.67728454 | 10.0965772 | 2.1765411 | 3.4236758 |
| -2.60528834 | -0.09097867 | 1.75871670 | 8.1940001 | 6.3866184 | 3.4555978 |
| -0.13375231 | -4.04858606 | 1.67728454 | 3.9219815 | 8.0696771 | 3.5296421 |
| 4.52723047 | -3.88582096 | 1.51442023 | -0.2886968 | 6.2385524 | 3.6036864 |
| 6.40700205 | 0.22373744 | 2.11088178 | -2.0431853 | 1.9646407 | 3.6356084 |
| 4.11200430 | 3.89865859 | 2.18649736 | -0.1406082 | -2.2454366 | 3.6036864 |
| -0.21605114 | 3.74751956 | 2.33772851 | 4.1326622 | -4.0002150 | 3.5296421 |
| -2.24910884 | -0.07854061 | 2.41334409 | 8.3420887 | -2.0973706 | 3.4555978 |
| 0.00522775 | -3.82998296 | 2.33924082 | 11.2635217 | 2.1969102 | 5.0035473 |
| 4.37394434 | -3.60232273 | 2.18649736 | 9.1122759 | 7.3092371 | 5.0395343 |
| 6.09732686 | 0.21292335 | 2.78877550 | 3.9275286 | 9.3518860 | 5.1293984 |
| 4.01640078 | 3.67579321 | 2.85717851 | -1.1318410 | 7.0803445 | 5.2183886 |
| -0.01626069 | 3.46564530 | 2.99817248 | -3.1542906 | 1.9452463 | 5.2552495 |
| -1.89292934 | -0.06610255 | 3.06797147 | -1.0030448 | -3.1670807 | 5.2192624 |
| 0.18799674 | -3.52897241 | 2.99956846 | 4.1804666 | -5.1389260 | 5.1293984 |
| 4.26302740 | -3.38666920 | 2.85717851 | 9.2410721 | -2.9381881 | 5.0404082 |
| 5.86089569 | 0.20466699 | 3.46410992 | 12.3978567 | 2.2167100 | 6.5839881 |
| 3.88013610 | 3.37640662 | 3.52937198 | 9.8597071 | 8.0549141 | 6.6265064 |
| 0.14469133 | 3.24596202 | 3.65989610 | 3.9386189 | 10.3165209 | 6.7291547 |
| -1.53674984 | -0.05366449 | 3.72259886 | -1.8999334 | 7.8496488 | 6.8318030 |
| 0.37076573 | -3.22796186 | 3.65989610 | -4.1611644 | 1.9276713 | 6.8730709 |
| 4.06737208 | -3.03532628 | 3.53065163 | -1.6946368 | -3.9117830 | 6.8318030 |
| 5.54456196 | 0.19362037 | 4.14223630 | 4.2264514 | -6.1733897 | 6.7291547 |
| 3.74387143 | 3.07702004 | 4.20156545 | 10.0650037 | -3.7065177 | 6.6265064 |
| 0.38332020 | 2.90189679 | 4.31906042 | 13.3459909 | 2.2332598 | 8.1676796 |
| -1.18057034 | -0.04122642 | 4.37722625 | 10.5269572 | 8.7175485 | 8.2149032 |
| 0.55353473 | -2.92695131 | 4.32022373 | 3.9505008 | 11.2358084 | 8.3289110 |
| 3.91408595 | -2.75182806 | 4.20272876 | -2.5340634 | 8.4895675 | 8.4429189 |
| 5.16830131 | 0.18048106 | 4.82245665 | -5.0519051 | 1.9121233 | 8.4888650 |
| 3.60760675 | 2.77763345 | 4.87375891 | -2.3060477 | -4.5734426 | 8.4429189 |
| 0.55133375 | 2.67090605 | 4.98055137 | 4.2704087 | -7.0917026 | 8.3289110 |

| Conical frustrum dataset: 6 | | | Parabolid dataset: 5 | | |
|---|---|---|---|---|---|
| X | Y | Z | X | Y | X |
| -0.82439084 | -0.02878836 | 5.03185363 | 10.7549729 | -4.3454617 | 8.2149032 |
| 0.73630372 | -2.62594076 | 4.98055137 | 14.2103993 | 2.2483481 | 9.7528327 |
| 3.79257672 | -2.51921335 | 4.87375891 | 11.1360376 | 9.3199372 | 9.8043335 |
| 4.85862613 | 0.16966696 | 5.50035037 | 3.9625397 | 12.1460994 | 9.9286673 |
| 3.44631982 | 2.43115690 | 5.54688303 | -3.0539469 | 9.0191427 | 10.0520745 |
| 0.75465496 | 2.38337807 | 5.64087901 | -5.8585798 | 1.8980428 | 10.1031914 |
| -0.52147972 | -0.01821047 | 5.68834232 | -2.8071326 | -5.1208390 | 10.0520745 |
| 0.94409496 | -2.27784024 | 5.63994836 | 4.3128984 | -7.9259367 | 9.9286673 |
| 3.63575982 | -2.23006140 | 5.54595238 | 11.3306955 | -4.8740623 | 9.8052602 |
| 4.54895094 | 0.15885287 | 6.17824409 | | | |
| 3.33507740 | 2.17886029 | 6.21814584 | | | |
| 0.95797618 | 2.09585008 | 6.30120665 | | | |
| -0.11203184 | -0.00391224 | 6.34110840 | | | |
| 1.12373617 | -1.98271593 | 6.30039232 | | | |
| 3.47894293 | -1.94090945 | 6.21814584 | | | |
| 4.27922704 | 0.14943390 | 6.85474182 | | | |
| 3.19881272 | 1.87947370 | 6.89033931 | | | |
| 1.16129739 | 1.80832210 | 6.96153428 | | | |
| 0.20419638 | 0.00713069 | 6.99713177 | | | |
| 1.28461070 | -1.72290911 | 6.96153428 | | | |
| 3.30094143 | -1.61783515 | 6.89103730 | | | |

# Statistical Modelling of Geometrical Invariant Sampled Sets


***P. Chiabert, M. De Maddis***
*DISPEA – Politecnico di Torino,*
*C.so Duca degli Abruzzi 24 – 10129 Torino (Italy)*
*paolo.chiabert@polito.it*



**Abstract:** In 2001 Mario Costa proposed a methodology for the statistical identification of three dimensional shapes based on the synthesis of two mathematical tools: the classification of geometrical shapes according to their invariant properties under the action of rigid motions and the Parzen's method for the non-parametric estimation of Probability Density Functions from a finite number of sampled points. This paper provides an extensive experimental test on the original algorithms developed by Mario Costa by analyzing a number of cases which differ on size and surface classes. From a conceptual point of view, the non-parametric, model-independent estimation of the Probability Density Function of a set of points closes the loop composed of design, manufacturing and inspection activities along the product development process. It seems therefore possible to adopt the proposed methodology to provide an unambiguous description of product morphology.

**Keywords**: geometrical invariant set, statistical analysis, shape recognition


## 1.  GEOMETRICAL PRODUCT SPECIFICATIONS

Since 1995 researchers involved in geometric dimensioning and tolerancing declared the need of parallel investigations on specification and verification activities performed on the geometrical controls of manufactured parts.

In 1995 International Standard Organization (ISO) published a technical report [ISO/TR14638, 1995] on the Geometrical Product Specification giving an overview of the international standardization including the existing and future standards. On the basis of such document the ISO Technical Committee 213 started the Geometrical Product Specifications and Verification (GPS) project, aiming to a new technological system in the field of geometrical control of manufactured parts.

Some of the results obtained within the GPS project are the conceptual background which justifies the applicability of the proposed methodology from a mathematical as well as from an industrial point of view.

### 1.1.  The GPS project

The scope of GPS project is the standardization in the field of macro and micro geometry specifications and the related verification principles, measuring equipment




*J.K. Davidson (ed.), Models for Computer Aided Tolerancing in Design and Manufacturing,* 169–178.
© 2007 *Springer.*

and calibration requirements including the uncertainty of dimensional and geometrical measurements. Such an ambitious goal required solid foundations able to support a new complete and coherent building for geometrical specifications in the age of Computer Aided Design. The foundations of GPS Project consist of several concepts which are resumed in the following:

- the **GPS matrix** assures the completeness in the standardization of geometrical controls. The General GPS matrix consists of individual chains of standards related to specific controls along the design and verification phases in the product development process;
- the **mathematical** approach avoids ambiguity of standards as well as successful implementation in computer based technology, thus reducing the discretionality of human interaction;
- the **uncertainty** is a concept largely accepted in the verification phase [ISO/GUM, 1993] and it is now extended to the specification phase. The uncertainty management is a tool to explain the discrepancies between the intended functionalities, geometrical specifications and physical measurements.

### 1.2. Classification of Euclidean surfaces

A GPS mathematical concept is the classification of three-dimensional surfaces based on their invariance properties [Srinivasan, 1999]. It relies on the definition of connected Lie subgroups of $T(3) \times SO(3)$ - the group of rigid motions - where $T(3)$ [$SO(3)$] denotes the group of translations [rotations] in $\Re^3$, respectively. Actually only seven subgroups do leave invariant some proper subset of $\Re^3$, thus obtaining the classes listed in Table I.

| Class $C_i$ | Surface $S \subseteq \Re^3$ | Rigid motion group $G_i$ | Reference element | Parameter |
|---|---|---|---|---|
| $C_1$ | Sphere | $SO(3)$ | Point | Radius (ro) |
| $C_2$ | Cylinder | $T(1)xSO(1)$ | Straight line | Radius (ro), Height (z) |
| $C_3$ | Plane | $T(2)xSO(1)$ | Plane | Length (x), Wide (y), Height (z) |
| $C_4$ | Helical | $T(1)xSO(1)$ pitch $\varepsilon$ | Helix | Radius (ro), Height (z), Pitch ($\varepsilon$) |
| $C_5$ | Revolution | $SO(1)$ | Point, Straight line | Profile (ro(z)), Height (z) |
| $C_6$ | Prismatic | $T(1)$ | Straight line, Plane | Profile (x,y), Height (z) |
| $C_7$ | Complex | $I_3$ | Point, Straight line, Plane | None |

*Table I; Classes of invariant surfaces in $\Re^3$*

Given any $S \in \Re^3$, let: $Aut(S)$ be the group of automorphisms of $S$, i.e. $Aut(S)=\{g \in T(3) \times SO(3): gS=S\}$; $Aut_0(S)$ be the connected component of $Aut(S)$ that contains the identity rigid motion $I3$. Set $S$ is assigned to class $C_i$ if and only if $Aut_0(S)=G_i$. According to Table I, seven semi-parametric models can describe all elementary surfaces in the Euclidean space. Moreover, each semi-parametric model has an intrinsic Euclidean reference system which localizes the surface in the space. The MRGE (Minimum Reference Geometric Element), is composed of a set of elements (point, line and plane) derived from the semi-parametric model associated to the surface. The classification proposed in Table I can also be applied to the description of invariant surfaces building up any complex mechanical model. Prof. A. Clement adopted such a classification to develop a new tolerancing approach based on TTRS (Topologically and Technologically Related Surfaces), which reduces to only 28 situations all the possible combinations of elementary surfaces describing mechanical parts. [Clement *et al.*, 1994].

### 1.3. Duality between Specification and Verification phases

Another result of mathematization in GPS project is the recognition that different geometrical models can be used to describe a mechanical product along its lifecycle and that relationships among the different models should be identified in order to assure a unique description of the part.

In the GPS project there exist two different contexts: specification and verification.

In the specification phase designers describe product shapes by means of nominal surfaces. In order to assure the product functionality, designers use a virtual defective model (the skin model), to establish the admissible limits on the variability of non-ideal surfaces in manufactured parts.

In the verification phase, metrologists extract information from the manufactured parts in order to identify non-ideal surfaces, the physical alter-ego of the skin model, and to establish their congruence with the specifications imposed by geometrical tolerancing.

The gap existing between the specification and verification phases is solved in the GPS project by means of duality principle: the sets of operations used in the specification phase to address variability limits are in biunivocal relationship with the sets of operations used in the verification phase to identify the element subject to the specification and to evaluate its conformity.


## 2. THE IDENTIFICATION OF PRODUCT SHAPE


The paper proposes a probabilistic approach to the recognition, segmentation and characterization of geometrical shapes based on a data-driven methodology applied to a set of measured points. The use of statistical empirical tools offers many advantages in term of clearness, reliability and simplicity thus providing a unifying approach to the geometrical control of product shape. As a matter of fact, statistical control techniques are widely adopted in verification and manufacturing phases and their extension to the design phase could improve quality and reduce costs.

Other interesting methodologies [Fischer *et al.*, 2003], [Gelfand *et al.*, 2004] are mainly focused on the inspection phase and do not achieve a mathematical universal description of 3D shapes which could be adopted in the verification phase and in the design phase too. Moreover, these methodologies do not exploit the advantages of the duality

principle (thus preventing a direct matching of inspection results with skin model specifications) and do not provide a direct identification of parameters and MRGEs.
The final goal of the proposed approach is the metrological resolution of a complex part. In a previous paper [Costa *et al.*, 2002] the authors demonstrated the feasibility of the proposed approach by analyzing the composed axial part illustrated in Figure 1.

*Figure 1; Composed axial surface:  cone, cylinder, sphere*

The methodology recognizes the axial symmetry of the object and uses in the opposite direction the method proposed by Prof. A. Clement, searching for the eight binary compositions of elementary surfaces which create an axial surface.
In this example, the algorithm recognizes a cylinder and an axial surface as the most likelihood decomposition of the original object. Figure 2 illustrates the cylinder's parameters and the PDF of the residual axial surface.

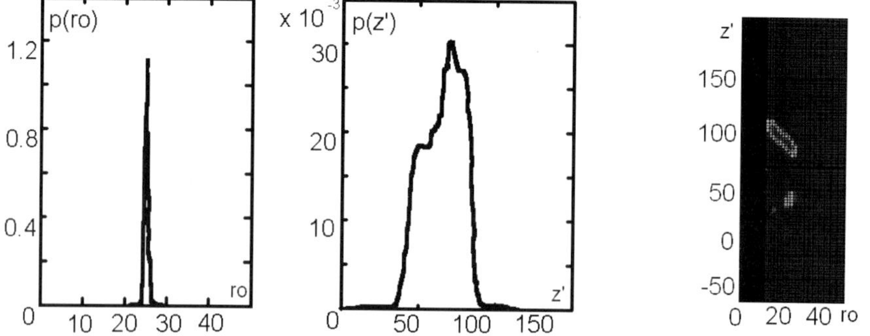

*Figure 2; PDFs from cylindrical surface (left and center) and from residual axial surface (right)*

Further decompositions could be applied on the residual axial surface, but it would require criteria which are not available in the classification of invariant surfaces.
The relevance of the proposed methodology relies in its objectivity: the measured points are partitioned among the invariant surfaces identifiable on the object. This partitioning does not require any human support, even if any a-priori information can be used to speed up the cognitive process. In addition, the process becomes faster and faster as it recognizes and extracts new invariant surfaces.

## 3.  THE STATISTICAL APPROACH

In a nutshell the proposed approach consists of the statistical modelling of the seven classes of symmetry. For each class $C_i$ $i=1,\ldots7$, a semi-parametric model $M_i$ and a set of reference parameters $\Omega_i$ are created in order to produce a PDF which is invariant under the action of the group of symmetry $G_i$. Furthermore, the invariance is enforced by replacing the set of measured points $D=\{(x_1,y_1,z_1),(x_2,y_2,z_2),\ldots,(x_n,y_n,z_n)\}$ with his projection on the quotient set $\Re^3/E_i$ being $E_i$ the set of equivalent points with respect to the rigid motions $g\in G_i$. This projection is performed by a suitable parametric function $V_i(\cdot;\Omega_i)$.

$$\tilde{p}\left(x, y, z \mid M_i, \Omega_i, V_i(D;\Omega_i)\right) \tag{1}$$

For any set of points in $\Re^3$, it is possible to build the seven PDFs representing the classes of symmetries. In order to identify which PDF provides the best description of the measured points it is necessary to compare their performances by computing their likelihood by means of the following expression:

$$\hat{L}(M_i) = \prod_{j=1}^{n} \tilde{p}\left(x_j, y_j, z_j \mid M_i, \Omega_{ij}, V_i(D_j;\Omega_{ij})\right) \tag{2}$$

where $D_j=D\setminus\{(x_j,y_j,z_j)\}$ $\forall j\in\{1,\ldots,n\}$ and $\Omega_{ij}$ is the maximum likelihood estimate of Minimum Reference Geometric Elements $\Omega_i$ based on $D_j$, namely:

$$\Omega_{ij} = \arg\max_{\Omega_i} \tilde{p}\left(D_j \mid M_i, \Omega_i, V_i(D_j;\Omega_i)\right) \tag{3}$$

An exhaustive formulation of the seven PDFs is illustrated in [Chiabert *et al.*, 2003], with a detailed description of the semi-parametric model $M_i$, the set of reference parameters $\Omega_i$ and the projecting function $V_i(\cdot;\Omega_i)$.

## 4.  THE EXPERIMENTAL TESTS

The strength of the proposed methodology relies on its ability to identify a surface according to the seven classes of symmetries available in the Euclidean space. The experimental test involves seven simple surfaces and demonstrates the ability of algorithms in classifying each surface by identifying its symmetries.

Some considerations on the sampling methodology have to be made before analyzing experimental results.

Algorithms work on sets of points extracted from analysed surfaces and such points should be uniformly sampled from the whole surface in order to avoid that local characteristics override the global behaviour. For instance, sampling a part of a sphere provides an axial surface overriding the global spherical behaviour.

Another interesting consideration invests the role of sampling noise. The symmetries of the underlying surface are retained in the set of sampled points if and only if the sampling noise does not introduce distortions. A measured point $P\in\Re^3$ is a noisy version of some unknown point $Q$ belonging to the surface $S$; this is mathematically translated in the convolution of $p_n(P\,|\,Q)$, the conditional PDF that accounts for the noise, and $p_S(Q)$, the probabilistic version of the indicator function on $S$:

$$\forall P \in \mathfrak{R}^3,\; \hat{p}_S(P) = \int_{\mathfrak{R}^3} p_n(P\,|\,Q)\,p_S(Q)\,d^3Q \tag{4}$$

It can be demonstrated [Costa *et al.*, 2001] that $\hat{p}_S(P)$ shows the same symmetries of the underlying surface $p_S(Q)$ if $p_n(P\,|\,Q)$ is actually a function of $\|P{-}Q\|$. This is what applies, for instance, to the familiar homoscedastic, isotropic gaussian noise.

### 4.1. Structure of the experimental test

Six simple invariant surfaces (the helix is neglected because is less relevant for the paper's scope) are illustrated in Table II, with their characteristics.

| $C_1$-Sphere | $C_2$-Cylinder | $C_3$-Plane | $C_5$-Cone | $C_6$-Prism | $C_7$-Complex |
|---|---|---|---|---|---|
|  |  |  |  |  |  |
| C (20,30,40) R=50mm S=31416mm$^2$ | R=50mm H=100mm S=31416mm$^2$ | L=124mm H=70mm S=30572mm$^2$ | R$_{min}$=40mm R$_{max}$=70mm H=86mm S=31476mm$^2$ | H=96mm S=31402mm$^2$ | S=34253mm$^2$ |

*Table II; Surfaces analyzed in the experimental test*

Each surface listed in Table II has been uniformly sampled and the set of sampled points has been perturbed with a noise having a Normal distribution N(0;0,001). The original size of the sampled set is about two thousand points, but subsets of different size are randomly extracted and analyzed.

### 4.2. Results of the experimental test

The paper focuses its analysis on classes of symmetry which have not been tested in other works. The analysis of spherical, cylindrical and complex surfaces, already tested in [Chiabert *et al.*, 2001], is limited to a brief summary while major attention is reserved to planar, axial and prismatic surfaces.

$C_1$ - **Spherical surface.** The test on spherical surface provides robust results with just a limited number of points. The numerical results from a sampled set of 10 points demonstrate that the spherical class wins the ranking with a log-likelihood value of -6,07, a large advantage over the axial class which has a log-likelihood value of -13.44. The PDF representing the spherical radius *ro* is roughly centered on the value of 50mm while the set of reference parameters identifying the centre of the sphere is $\Omega_1$ ={19,99985, 30,00002, 40.00023}.

$C_2$ - **Cylindrical surface.** The test on the cylindrical surface provides similar good results when applied to a sampled set of 10 points. Algorithms identify the cylindrical class as the optimal description of the sampled points with a log-likelihood value of -11,00. The prismatic class holds the second position with a log-likelihood of -13,23.

The PDFs representing cylinder radius and height are compatible with the values of the nominal cylinder. The set of reference parameters identifying the cylinder axis shows some discrepancy, because $\Omega_2=\{\alpha=26{,}79582°, \beta=0{,}00002°, a=-0{,}00282\ b=0{,}00024\}$ while the nominal values should be $\{\alpha=30°, \beta=0°, a=0, b=0\}$.

$C_7$ - **Complex surface.** The complex surface is easily identified by the proposed algorithms when the sampled set has more than 15 points. The log-likelihood of the complex model is -13,69 and slowly decreases when the number of sampled points increases. The complex model is less efficient than any other invariant model but its role is relevant within the classification procedure: thanks to its slow convergence, it is possible to state that any model having a log-likelihood worse than the complex model is not compatible with the sampled set. Moreover, when the number of sampled point tends to be infinite, the log-likelihood of the complex model approaches the value of the invariant class corresponding to the symmetries exhibited by the sampled surface.

The reference parameter set $\Omega_7$ of the complex class is empty.

$C_3$ - **Planar surface.** The planar surface has been investigated in more detail to exploit the potentiality of the proposed approach. Table III reports the results from the ranking of the seven classes over the planar surface with different size of the sampled set.

| n=10 | | n=12 | | n=15 | | n=25 | | n=50 | | n=75 | | n=100 | | n=200 | |
|---|---|---|---|---|---|---|---|---|---|---|---|---|---|---|---|
| 3 | -5,26 | 3 | -5,48 | 3 | -5,11 | 3 | -5,20 | 3 | -5,19 | 3 | -5,10 | 3 | -5,09 | 3 | -5,04 |
| 7 | -15,20 | 6 | -5,49 | 6 | -13,51 | 6 | -13,54 | 4 | -5,32 | 6 | -13,22 | 6 | -13,09 | 6 | -12,97 |
| 1 | -15,43 | 7 | -14,96 | 1 | -14,41 | 5 | -14,36 | 2 | -5,32 | 7 | -13,93 | 7 | -13,88 | 7 | -13,72 |
| 6 | -15,80 | 1 | -15,27 | 7 | -14,44 | 7 | -14,37 | 6 | -13,47 | 2 | -14,26 | 2 | -14,16 | 2 | -14,12 |
| 5 | -16,84 | 5 | -15,57 | 2 | -14,50 | 2 | -14,38 | 7 | -14,18 | 5 | -14,33 | 5 | -14,16 | 5 | -14,26 |
| 4 | -20,20 | 4 | -16,05 | 5 | -14,79 | 1 | -14,55 | 5 | -14,58 | 1 | -14,56 | 4 | -14,56 | 4 | -14,47 |
| 2 | -26,97 | 2 | -19,97 | 4 | -14,83 | 4 | -14,67 | 1 | -14,76 | 4 | -14,61 | 1 | -14,59 | 1 | -14,57 |

*Table III; Ranking of invariant classes $C_i$ i=1,...7, on the planar surface with different size of sampled set*

The planar model shows the highest log-likelihood among the seven classes, and there is no doubt that the set of points shows a planar symmetry. All other classes have lower log-likelihood values, with two noteworthy exceptions with sets of 12 and 50 points maybe related to singularities in the sampled sets.

It is important to observe that the prismatic class holds often the second position while the complex class holds the third position. This fact agrees with the classification of invariant surfaces: the planar class differs from the prismatic class thanks to a tighter factorization of the PDF that accounts for the additional independence between x and y.

Table III also highlights the good performance of the planar class and shows the decreasing distance from the complex model. As previously stated, with an infinite number of sampled points the complex class and the more efficient planar model would provide the same results.

The set of reference parameters identifying the normal vector to the plane is $\Omega_3=\{-3.009926, 1.000002, 3.009915\}$. Figure 3 shows the factorized PDFs representing the

                    P. Chiabert and M. De Maddis

distribution of 200 sampled points under the hypothesis of planar symmetry. The nominal dimensions are easily extracted from the three pictures.

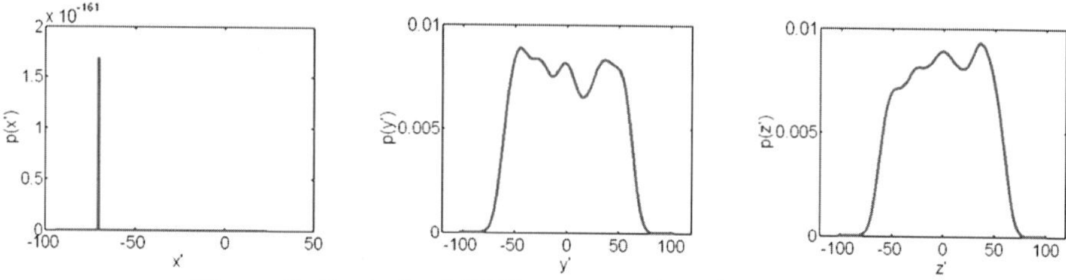

*Figure 3; Factorized PDFs representing planar symmetry*

$C_5$ - **Axial surface.** Also the axial surface is easily identified by the proposed algorithms. Table IV shows the preeminence of the axial class over any other solution. The second position is hold by the spherical class, and the third position by the cylindrical class.

| n=10 | | n=12 | | n=15 | | n=25 | | n=50 | | n=75 | | n=100 | | n=200 | |
|---|---|---|---|---|---|---|---|---|---|---|---|---|---|---|---|
| 5 | -10,60 | 5 | -10,54 | 5 | -10,11 | 5 | -10,19 | 5 | -10,20 | 5 | -9,71 | 5 | -9,48 | 5 | -9,57 |
| 2 | -13,97 | 1 | -13,86 | 1 | -13,42 | 1 | -13,72 | 1 | -13,77 | 1 | -13,52 | 1 | -13,23 | 1 | -13,46 |
| 1 | -14,06 | 2 | -14,47 | 2 | -14,27 | 2 | -13,97 | 2 | -14,02 | 2 | -13,80 | 7 | -13,84 | 2 | -13,77 |
| 6 | -14,20 | 3 | -14,72 | 4 | -14,32 | 4 | -14,07 | 6 | -14,13 | 7 | -14,02 | 4 | -14,02 | 7 | -13,80 |
| 3 | -14,23 | 6 | -14,78 | 3 | -14,54 | 6 | -14,23 | 4 | -14,23 | 4 | -14,07 | 6 | -14,03 | 4 | -14,04 |
| 7 | -14,45 | 7 | -14,86 | 6 | -14,60 | 7 | -14,41 | 7 | -14,24 | 6 | -14,14 | 3 | -14,21 | 6 | -14,05 |
| 4 | -16,86 | 4 | -14,86 | 7 | -14,63 | 3 | -14,53 | 3 | -14,36 | 3 | -14,48 | 2 | -14,32 | 3 | -14,25 |

*Table IV; Ranking of invariant classes $C_i$ i=1,…7, on the axial surface with different size of sampled set*

This fact can be easily explained by looking at the shape of the cone, whose generatix is roughly described by the PDF depicted in Figure 4. A minimum radius of 40mm, a maximum radius of 70mm and a height of 86mm make the cone more similar to a sphere than a cylinder. The reference parameter set $\Omega_5$={-0,00020, 0,08721, 0,01075, -0,01258} identifies the axis of symmetry resultant from a sampled set of 100 points which is used to define the PDF in Figure 4.

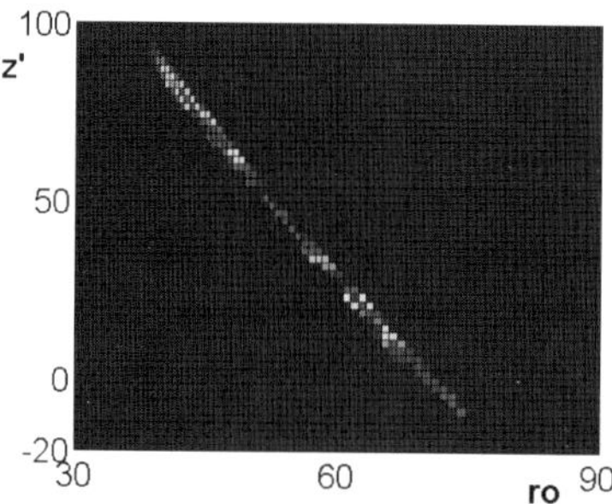

*Figure 4; Factorized PDF representing axial symmetry*

$C_6$ - **Prismatic surface.** The experimental test on the prismatic surface presents some problems which are presumably related to the nominal shapes. In Table V the prismatic class is correctly identified with a small number of points. Increasing the number of points other classes, the cylindrical and the axial, provide a better log-likelihood.

| n=10 | | n=12 | | n=15 | | n=25 | | n=50 | | n=75 | | n=100 | | n=200 | |
|---|---|---|---|---|---|---|---|---|---|---|---|---|---|---|---|
| 6 | -13,89 | 6 | -14,12 | 6 | -14,00 | 6 | -13,43 | 2 | -13,26 | 6 | -13,18 | 5 | -13,80 | 5 | -13,61 |
| 5 | -13,97 | 2 | -14,45 | 2 | -14,06 | 3 | -13,73 | 6 | -13,33 | 4 | -13,19 | 3 | -13,91 | 7 | -13,64 |
| 3 | -14,09 | 3 | -14,57 | 3 | -14,27 | 2 | -13,82 | 4 | -13,34 | 2 | -13,30 | 7 | -13,98 | 3 | -13,68 |
| 7 | -14,29 | 7 | -14,85 | 4 | -14,31 | 5 | -13,92 | 5 | -13,59 | 5 | -13,74 | 6 | -14,14 | 6 | -13,84 |
| 1 | -14,33 | 1 | -15,11 | 5 | -14,42 | 7 | -14,06 | 3 | -13,86 | 3 | -13,80 | 1 | -14,15 | 4 | -13,84 |
| 2 | -14,90 | 4 | -15,43 | 7 | -14,52 | 1 | -14,14 | 7 | -14,08 | 7 | -13,89 | 4 | -14,23 | 2 | -14,02 |
| 4 | -17,22 | 5 | -29,02 | 1 | -14,74 | 4 | -14,28 | 1 | -14,10 | 1 | -14,19 | 2 | -14,26 | 1 | -14,16 |

*Table V; Ranking of invariant classes $C_i$ i=1,...7, on the prismatic surface with different size of sampled set*

In order to explain this abnormal situation it is possible to observe that differences in the log-likelihood of the classes are very small, less than 1% and that the nominal prismatic shape is not too different from cylindrical or axial symmetries.

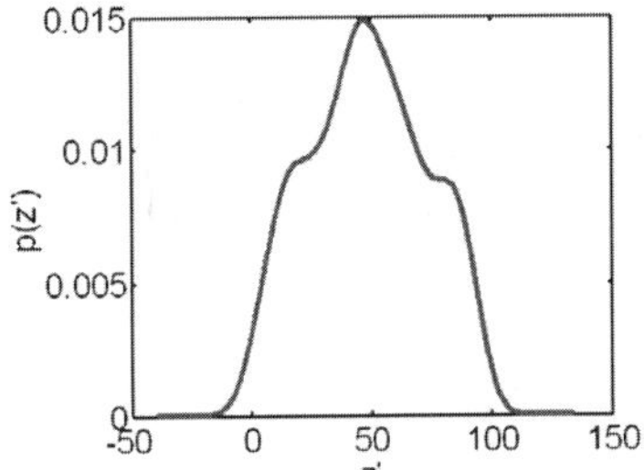
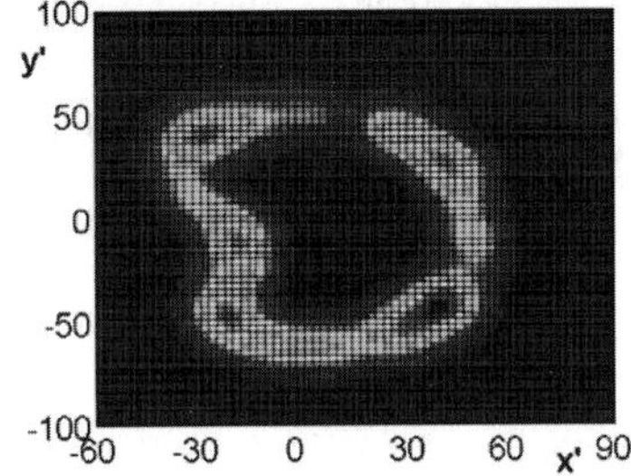

*Figure 5; Factorized PDFs representing prismatic symmetry*

Figure 5 shows the best results of the PDFs which represent the height of the prism (on the left) and the normal section (on the right).

The reference parameter set $\Omega_6$={-0,00160, -0,00177} defines the direction of the prism resultant from a sampled set of 75 points.


## 5.  CONCLUSION


The paper illustrates some experimental results derived by the application of the methodology proposed by Prof. Mario Costa for the recognition of symmetries in set of points sampled from mechanical surfaces.

The algorithms seem to be robust in the identification of symmetries but new investigations should solve some of the problems highlighted in the text. Moreover, new experimental tests could discover unexplored capability of the proposed methodology.

Authors are gratefully to Prof. Mario Costa for his overwhelming enthusiasm and his beautiful mind. Yesterday he was a star on the earth, today he is a star in the heaven.


REFERENCES

**[Clement et al., 1994]** Clement, A.; Riviere, A.; Temmerman, M.; "Cotation Tridimensionelle des Systemes Mecaniques"; PYC Edition, Ivry sur Seine, 1994.

**[Chiabert et al., 2001]** Chiabert, P.; Costa M.; "Probabilistic Evaluation of Invariant Surfaces through the Parzen's method", In: *Proceedings of the 7th CIRP International Seminar on Computer Aided Tolerancing*, pp. ; Cachan 2001

**[Chiabert et al., 2003]** Chiabert P.; Costa M.; "Statistical Modelling of Nominal and Measured Mechanical Surfaces", In: *Journal of Computing and Information Science in Engineering*, Vol.3, No.1; 2003.

**[Costa et al., 2001]** Chiabert, P.; Costa, M.; Pasero, E.; " Detection of Continuous Symmetries in 3D Objects from Sparse Measurements through Probabilistic Neural Networks"; In: *Proceedings of the IEEE International Workshop on Virtual and Intelligent Measurement Systems*, Budapest (Hungary) 2001

**[Costa et al., 2002]** Chiabert P.; Costa M.; "Probabilistic description of mechanical surfaces"; In: *Proceedings of 3rd CIRP International Seminar on Intelligent Computation in Manufacturing Engineering*, pp. 479-484; Ischia (Italy), 2002;

**[Fischer et al., 2003]** Azernikov, S.; Miropolsky, A.; Fischer, A.; "Surface Reconstruction of Freeform Objects Based on Multiresolution Volumetric Method"; In: *Journal of Computer and Information Science and Engineering*, Vol. 3, No. 1; 2003

**[Gelfand et al., 2004]** Gelfand, N.; Guibas L. J.; "Shape Segmentation Using Local Slippage Analysis"; In: *Proceedings of the Second Symposium on Geometry Processing*, Nice (France), 2004;

**[ISO/GUM, 2000]** "Guide to the expression of uncertainty in measurement"; ISO International Standard

**[ISO/TR14638, 1995]** "Geometrical Product Specification – Masterplan"; ISO International Standard

**[Srinivasan, 1999]** Srinivasan, V.; "A Geometrical Product Specification Language Based on a Classification of Symmetry Groups"; In: *Computer-Aided Design*, Vol.31, No.11, pp.659-668, 1999

# Simulation of the Manufacturing Process in a Tolerancing Point of View: Generic Resolution of the Positioning Problem


**F. Villeneuve, F. Vignat**
*University of Grenoble, Laboratory 3S,*
*Domaine Universitaire, BP53, 38041, Grenoble cedex 9, France*
*frederic.vignat@inpg.fr, francois.villeneuve@inpg.fr*



**Abstract:** To verify the capacity of a manufacturing process to realize corresponding parts it is necessary to simulate the defects that it generates and to analyze the correspondence of produced parts with the functional tolerances. In a previous paper, a method was proposed to simulate the process and to determine its effect on the manufactured part in term of surfaces deviations. To do this, a model of the part named MMP (Model of the Manufactured Part) is virtually manufactured which allows stacking the defects created at each set-up of the process. This paper presents a generic method to determine the parameters of the MMP, especially those concerning the positioning deviation torsor of the part in the part-holder. This set of hierarchically arranged functions and constraints describing the manufactured part is then used in every case of tolerances analysis

**Keywords**: manufacturing process, simulation, positioning deviation,


## 1. INTRODUCTION

In previous papers [Villeneuve et al., 2001] [Villeneuve et al., 2004] the way to simulate a process in a geometrical defects point of view was already presented. In this method, the defects are divided into two categories, the positioning deviation (variation of the position of the nominal part / machine-tool) on one hand and the manufacturing deviations (variation of the position of surfaces realized in the set-up relative to the machine-tool) on the other hand. The positioning deviation is due to the combination of the deviations of the assembled pairs of surfaces and the links between these surfaces. The manufacturing deviations are due to the deviations of the surfaces swept by the cutting-tools relative to the machine.

Determining the positioning deviation is detailed in this paper. How to analyze the consequences of defects on functional tolerances has been presented in [Vignat et al., 2005].

Some papers already presented the way to simulate assembly operations for functional tolerance analysis. [Sodhi et al., 1994] define 5 types of relations between the assembled parts (contacts, attachments, assembly dimensions, enclosures and




*J.K. Davidson (ed.), Models for Computer Aided Tolerancing in Design and Manufacturing, 179–189.*

alignments) and determine the relative position of the parts using a contrained optimization algorithm. [Anselmetti et al., 2003] assembles nominal part and study the relative surfaces position variation due to defects within the tolerance zone. [Thiebaut, 2001] and [Dantan, 2000] describe surface defects and links between surfaces by a small displacement torsor, define 3 types of contact (floating, slipping and fixing) and determine the extreme position in every possible configuration. In this paper a similar method is applied but the use of a constrained optimization algorithm exempts us enumerating whole configurations. Another original aspect is that the textual expression of the Model of Manufactured Part (MMP) contains all the information required for the tolerance analysis.

After a reminder of the whole method, the positioning resolution is presented in 3 points: first, the global determination of the positioning deviation torsor, second, the determination of the driving link parameters, third the generic method of resolution of the conditions of contact which allows taking into account a great number of positioning cases (isostatic, hyperstatic, floating...).

The set of hierarchically arranged functions and constraints obtained are describing the MMP and can then be used in every cases of tolerances analysis. It is shown that this method can also be applied to the resolution of the problem of assembly between the part to be controlled and a virtual control gauge.

## 2.   REMINDER OF THE METHOD

The method exposed in [Villeneuve et al., 2004] is based on the determination and the analysis of the deviations of the surfaces of a part relative to their nominal position (nominal part). The method is developed in two steps:

- The first step determines the effect of the process in term of deviation of the surfaces of the part. At the end of this step, a model of the manufactured part with deviation defects is generated (it will be called MMP for Model of Manufactured Part).
- The second one consists in analyzing, using the MMP, the consequences of the process on the respect of the functional tolerances (analysis) and transferring these tolerances in manufacturing tolerances.

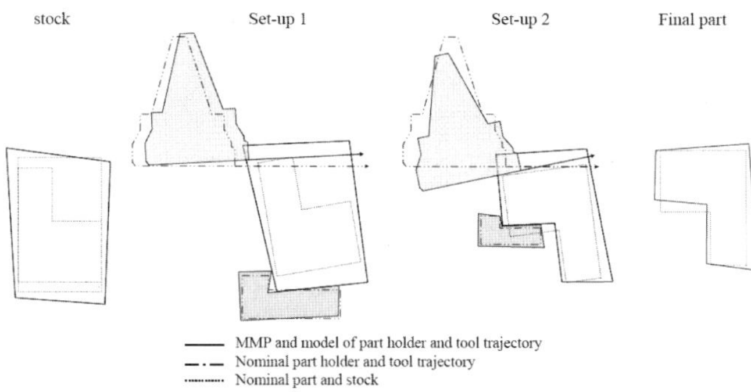


*Figure 1; process deviations*

The deviations of the surfaces of the MMP are expressed relative to their nominal position in the part (nominal part). These deviations are expressed by a small displacement torsor plus the variations of the intrinsic characteristics of the surfaces (diameter for a cylinder or a sphere). The domain of variation of the parameters characterizing these deviations represents the 3D capabilities of the manufacturing means (machine-tool, cutting tools, part holders) used during the machining process.

For the first step, we assume that a process generates two independent defects, the positioning deviations (due to the deviations of the part-holder surfaces and of the Model of Workpiece (MWP) surfaces) and the machining deviations (deviation of a realized surface relative to its nominal position in the machine) as represented figure 1.

The deviation of each realised surface can be determined relative to the nominal part by expressing for each set-up:

$$\text{Deviation}_{\text{part,surface}} = \underbrace{-\,\text{Deviation}_{\text{machine,part}}}_{\text{positionning}} + \underbrace{\text{Deviation}_{\text{machine,surface}}}_{\text{machining}} \tag{1}$$

Or, expressing by torsors:        $T_{P,Pi} = -T_{Sj,P} + T_{Sj,Pi}$

At each set-up, the positioning deviation $T_{Sj,P}$ is function of the part-holder defects, the links between the part holder and the workpiece positioning surfaces and the defects of the workpiece positioning surfaces that have been machined in the previous set-up. The determination of the positioning deviation of the workpiece is the object of the present paper.

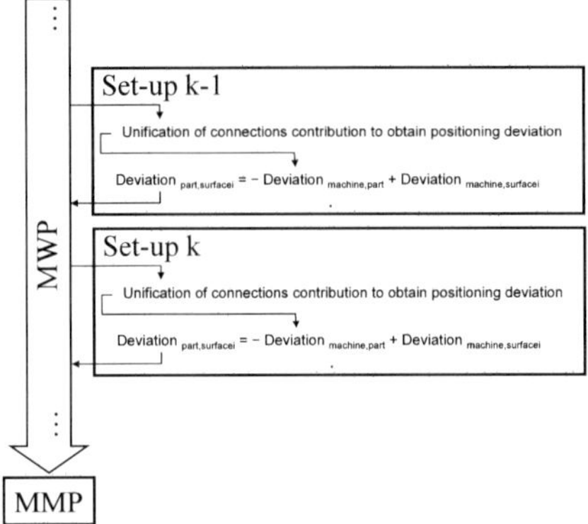



*Figure 2; generation of the MMP*

Each set-up simulation defines a MWP (Model of Workpiece) built around the nominal model of the part. For set-up n, the MWP built at the end of set-up n-1 is virtually positioned on the part-holder of set-up n to obtain the positioning deviation in set-up n. (fig. 2). This operation must be repeated for each set-up until the final part is virtually manufactured.

The defects collected in the MMP at each set-up are classified in 3 categories:

- Parameters of deviations of the surfaces of the part-holder (DH).
- Parameters of deviations of the manufactured surfaces (surfaces swept by the cutting-tools) relative to the machine-tool (DM).
- Link parameters between the surfaces of the part and of the part-holder (LHP) connected by elementary connections.

Furthermore, the MMP includes the deviation torsor of each realised surfaces relative to the nominal part obtained by formula (1).


## 3.   POSITIONING RESOLUTION

The positioning deviation of the MWP relative to the machine tool is determined by "unification" of the elementary connections treated on a hierarchical basis (primary, secondary...). Every elementary connection (planar, centering, punctual...) links a surface of the workpiece (MWP) and a surface of the part-holder. The contacts between these surfaces are of two types: floating or slipping. The resolution of the problem ("unification" following the rules of assembly) describes the 6 components of position of the workpiece in the assembly space. These components are functions of the parameters of defects of the connected surfaces and the links between these surfaces.

| Type of connection | Coordinate system | Link torsor |
|---|---|---|
| Plane-Plane | | $T_{HkSj,Pi} = \begin{Bmatrix} lrx_{kSj} & Ultx_{kSj} \\ lry_{kSj} & Ulty_{kSj} \\ Ulrz_{kSj} & ltz_{kSj} \end{Bmatrix}_{O_{kSj},X_{kSj},Y_{kSj},Z_{kSj}}$ |
| Cylindrical | | $T_{HkSj,Pi} = \begin{Bmatrix} lrx_{kSj} & ltx_{kSj} \\ lty_{kSj} & lty_{kSj} \\ Ulrz_{kSj} & Ultz_{kSj} \end{Bmatrix}_{O_{kSj},X_{kSj},Y_{kSj},Z_{kSj}}$ |

*Table 1; some elementary connection links torsors*

### 3.1. First step: determining the positioning deviation of the workpiece in set-up Sj

$T_{Sj,HkSj}$ is the torsor modeling the deviations of the surfaces of the part-holder

$T_{P,Pi}$ is the torsor modeling the deviations of the surfaces of the workpiece

$T_{HkSj,Pi}$ is the torsor modeling the link between the surfaces of the part-holder and the workpiece (table 1 and figure 3).

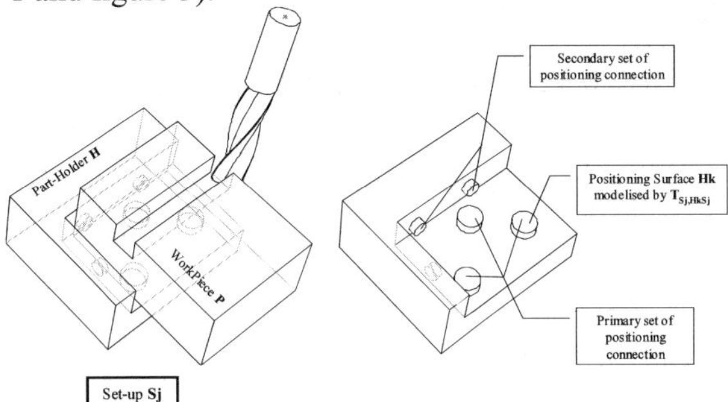


*Figure 3; positioning connections workpiece/part-holder*

The goal is to determine $T_{Sj,P}$ that is the torsor modeling the positioning deviation of the workpiece relative to the machine tool in set-up Sj. Each positioning surface HkSj of the part-holder connected to a surface Pi of the workpiece contributes to position it. All positioning connections contribute to a unique positioning deviation of the workpiece. So $T_{Sj,P}$ coming from connection n is the same than $T_{Sj,P}$ coming from connection m. Expressing the equality between all the expressions of $T_{Sj,P}$ is called "unification". So, in set-up Sj, for positioning connection n and m:

$$\boxed{Pos[n]=Pos[m]=T_{Sj,P}=T_{Sj,HkSj}+T_{HkSj,Pi}-T_{P,Pi}}$$

For each connection, the link torsor includes parameters determined (lr or lt) and undetermined (Ulr or Ult). The "unification" allows establishing the relations between the undetermined parameters and the determined one. Moreover, if the positioning is over-constrained, the "unification" gives the compatibility equations between the determined parameters. These relations are obtained using a Gauss Pivot method as described in [Thiebaut, 2001].

### 3.2. Second step: determining the "driving link parameters"

The part-holders give generally a over-constrained positioning. In such a case, a hierarchical order (primary, secondary …) is defined between the different set of connections. This hierarchy allows determining the main link parameters defining the positioning. We call them "driving link parameters". For a complete positioning, there are at most 6 independent parameters. To position a revolution part, 5 parameters are sufficient.

To determine the "driving link parameters", we use the compatibility equations determined at the first step. The algorithm consists in:

- Retain the link parameters of the primary set of positioning connections
- Eliminate the link parameters of the secondary set of positioning connections, which are expressed according to the first reserved parameters in the compatibility equations.
- Repeat it until the last connection

### 3.3. Third step: determining the positioning constraints

At this stage, we know the expression of $T_{Sj,P}$ function of the defect parameters of the connected surfaces and the "driving link parameters".

According to the type of link (floating or slipping), these link parameters (LHP) will be free (floating contact) or constrained to guarantee the contact Part-Holder/Workpiece (slipping contact). Furthermore, all the link parameters (LHP) are constrained by the conditions of contact, which are conditions of non-penetration workpiece/part-holder (GapHP > 0).

This thus obliges us to envisage 2 solving procedures:

- Slipping contact. We have to apply a positioning function. This function is generally defined by the displacement of a point of the workpiece along a direction appropriated to the contact that has to be maintained. The displacement function has to be maximized, respecting the non-penetration conditions between workpiece and part-holder. These non-penetration conditions are applied at all potential contact points of the boundary of the surface. In some cases, that implies to discretize the edges of the contact zone.

  In some cases, a simple numerical solution exists. For example, a primary plane/plane connection implies that the link parameters have to be set to zero:

$$T_{HkSj,Pi} = \left\{ \begin{matrix} 0 & Utx_{kSj} \\ 0 & Uty_{kSj} \\ Urz_{kSj} & 0 \end{matrix} \right\}_{Point,Base}$$

- Floating contact. The parameters of the floating link only have to respect the non-penetration condition between workpiece and part-holder.

All of these data, set-up by set-up, are compiled in a table describing the Model of Manufactured Part (MMP).

## 4.   CASE STUDY

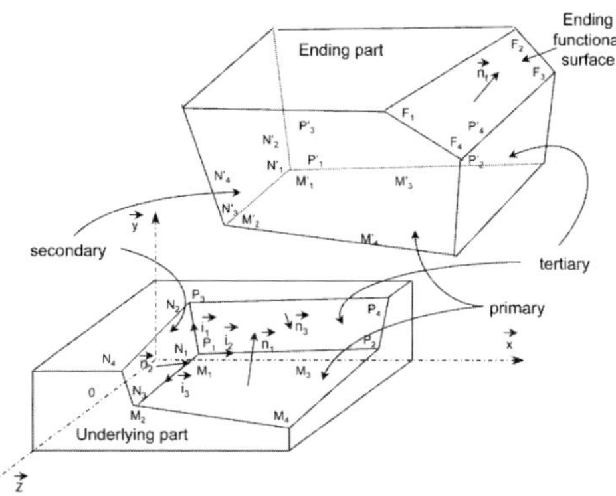


*Figure 4; Plane-plane-plane over-constrained positioning [Anselmetti et al., 2003]*

### 4.1. Over-constrained positioning

We apply our method on a first example that was proposed by [Anselmetti et al., 2003] (Figure 4).

In set-up 1, the three positioning surfaces of the set-up 2 are machined on the workpiece. For a better readability, it is assumed that the first set-up gives a perfect workpiece.

In set-up 2, the primary positioning connection is between part-holder plane 1S2 and workpiece plane 1, the secondary one between 2S2 and 2 and the tertiary one between 3S2 and 3.

Each positioning surface of the part-holder is deviated:

$$T_{S2,HIS2} = \left\{ \begin{matrix} rx_{IS2} & 0 \\ ry_{IS2} & 0 \\ 0 & tz_{IS2} \end{matrix} \right\}_{O_{IS2},X_{IS2},Y_{IS2},Z_{IS2}}$$

Each connection generates positioning link parameters:

$$T_{HIS2,PI} = \left\{ \begin{matrix} lrx_{IS2} & Ultx_{IS2} \\ lry_{IS2} & Ulty_{IS2} \\ Ulrz_{IS2} & ltz_{IS2} \end{matrix} \right\}_{O_{IS2},X_{IS2},Y_{IS2},Z_{IS2}}$$

Using the formula $T_{Sj,P} = T_{Sj,HkSj} + T_{HkSj,Pi} - T_{P,Pi}$ the positioning deviation due to one set of connection is determined as:

$$T_{S2,P} = \left\{ \begin{matrix} lrx_{IS2} + rx_{IS2} & -20lry_{IS2} - 20ry_{IS2} - 25Urz_{IS2} + Utx_{IS2} \\ Urz_{IS2} & 25lrx_{IS2} + 40lry_{IS2} + ltzlS2 + 25rx_{IS2} + 40ry_{IS2} + tz_{IS2} \\ -lry_{IS2} - ry_{IS2} & -20lrx_{IS2} - 20rx_{IS2} + 40Urz_{IS2} - Uty_{IS2} \end{matrix} \right\}$$

The positioning is obtain by 3 sets of connection. So we obtain:

- 3 expressions of  positioning deviation of the part

186                              F. Villeneuve and F. Vignat

- 18 link parameters of which 9 are undetermined

Unifying these 3 positioning deviations gives 12 equations.

$$\begin{pmatrix} 0 & 0 & 0 & -0.707 & 0 & 0 & 0 & 0 & 0 \\ 1 & 0 & 0 & -0.707 & 0 & 0 & 0 & 0 & 0 \\ 0 & 0 & 0 & 0 & 0 & 0 & 0 & 0 & 0 \\ -25 & 1 & 0 & 17.7 & 0 & 0.707 & 0 & 0 & 0 \\ 0 & 0 & 0 & -17.7 & 0 & -0.707 & 0 & 0 & 0 \\ 40 & 0 & -1 & 3.5 & 1 & 0 & 0 & 0 & 0 \\ 0 & 0 & 0 & 0 & 0 & 0 & 0 & 0 & 0 \\ 1 & 0 & 0 & 0 & 0 & 0 & -0.707 & 0 & 0 \\ 0 & 0 & 0 & 0 & 0 & 0 & -0.707 & 0 & 0 \\ -25 & 1 & 0 & 0 & 0 & 0 & -17.7 & -1 & 0 \\ 0 & 0 & 0 & 0 & 0 & 0 & 28.28 & 0 & -0.707 \\ 40 & 0 & -1 & 0 & 0 & 0 & -28.28 & 0 & 0.707 \end{pmatrix} \begin{pmatrix} Urz_{1S2} \\ Utx_{1S2} \\ Uty_{1S2} \\ Urz_{2S2} \\ Utx_{2S2} \\ Uty_{2S2} \\ Urz_{3S2} \\ Utx_{3S2} \\ Uty_{3S2} \end{pmatrix} =$$

$$\begin{pmatrix} lrx_{1S2} - 0.707lry_{2S2} + rx_{1S2} + 0.707ry_{2S2} \\ -0.707lry_{2S2} - 0.707ry_{2S2} \\ lrx_{2S2} - lry_{1S2} + rx_{2S2} - ry_{1S2} \\ 22.5lrx_{1S2} - 20lry_{1S2} + 17.7lry_{2S2} - 0.707ltz_{2S2} + 22.5rx_{2S2} - 20ry_{1S2} + 17.7ry_{2S2} - 0.707tz_{2S2} \\ 25lrx_{1S2} - 17.5lrx_{2S2} + 40lry_{1S2} + 17.7lry_{2S2} + ltz_{1S2} - 0.707ltz_{2S2} + 25rx_{1S2} - 17.5rx_{2S2} + 40ry_{1S2} + 17.7ry_{2S2} + tz_{1S2} - 0.707tz_{2S2} \\ -20lrx_{1S2} - 28.3lry_{2S2} - 20rx_{1S2} - 28.3ry_{2S2} \\ lrx_{1S2} - lrx_{3S2} + rx_{1S2} - rx_{3S2} \\ -0.707lry_{3S2} - 0.707ry_{3S2} \\ -lry_{1S2} + 0.707lry_{3S2} - ry_{1S2} + 0.707ry_{3S2} \\ -20lry_{1S2} + 14.142lry_{3S2} - 20ry_{1S2} + 14.142ry_{3S2} \\ 25lrx_{1S2} + 2.5lrx_{3S2} + 40lry_{1S2} - 28.3lry_{3S2} + ltz_{1S2} - 0.707ltz_{3S2} + 25rx_{1S2} + 2.5rx_{3S2} + 40ry_{1S2} - 28.3ry_{3S2} + tz_{1S2} - 0.707tz_{3S2} \\ -20lrx_{1S2} + 22.5lrx_{3S2} - 28.3lry_{3S2} - 0.707ltz_{3S2} - 20rx_{1S2} + 22.5rx_{3S2} - 28.3ry_{3S2} - 0.707tz_{3S2} \end{pmatrix}$$

Resolving the system gives 3 compatibility equations (degree of over-constraint = 3)

$$\begin{pmatrix} lrx_{2S2} - lry_{1S2} + rx_{2S2} - ry_{1S2} \\ lrx_{1S2} - lrx_{3S2} + rx_{1S2} - rx3S2 \\ lrx_{1S2} + lry_{1S2} + 1.414lry_{2S2} - 1.414lry_{3S2} + rx_{1S2} + ry_{1S2} + 1.414ry_{2S2} - 1.414ry_{3S2} \end{pmatrix} = \begin{pmatrix} 0 \\ 0 \\ 0 \end{pmatrix}$$

and the expression of the 9 undetermined link parameters function of the link and defects parameters.

$$\begin{pmatrix} Urz_{1S2} \\ Utx_{1S2} \\ Uty_{1S2} \\ Urz_{2S2} \\ Utx_{2S2} \\ Uty_{2S2} \\ Urz_{3S2} \\ Utx_{3S2} \\ Uty_{3S2} \end{pmatrix} = \begin{pmatrix} lry_{1S2} + ry_{1S2} - \sqrt{2}lry_{3S2} + \sqrt{2}ry_{3S2} \\ 25lrx_{1S2} + 5lrx_{2S2} + 45lry_{1S2} + 35.36lry_{2S2} - 35.36lry_{3S2} + ltz_{1S2} - \sqrt{2}ltz_{2S2} + 25rx_{1S2} + 5rx_{2S2} + 45ry_{1S2} + 35.36ry_{2S2} - 35.36ry_{3S2} + tz_{1S2}\sqrt{2} - 2tz_{2S2} \\ -5lrx_{1S2} - 25lrx_{3S2} - ltz_{1S2} + \sqrt{2}ltz_{3S2} - 5rx_{1S2} - 25rx_{3S2} - tz_{1S2} + \sqrt{2}tz_{3S2} \\ -lry_{2S2} - \sqrt{2}lrx_{1S2} + \sqrt{2}rx_{1S2} - ry_{2S2} \\ -20lr_{x1S2} - 25lrx_{3S2} - 40lry_{1S2} - 24.75lry_{2S2} + 56.57lry_{3S2} - ltz_{1S2} + \sqrt{2}ltz_{3S2} - 20rx_{1S2} - 25rx_{3S2} - 40ry_{1S2} - 24.75ry_{2S2} + 56.57ry_{3S2} - tz_{1S2} + \sqrt{2}tz_{3S2} \\ 24.75lrx_{2S2} - 56.57lry_{1S2} - \sqrt{2}ltz_{1S2} + ltz_{2S2} + 24.75rx_{2S2} - 56.57ry_{1S2} - \sqrt{2}tz_{1S2} + tz_{2S2} \\ -lry_{3S2} + \sqrt{2}lry_{1S2} + \sqrt{2}ry_{1S2} - ry_{3S2} \\ 25lrx_{1S2} + 5lrx_{2S2} + 15lry_{1S2} + 35.36lry_{2S2} + 3.54lry_{3S2} + ltz_{1S2} - \sqrt{2}ltz_{2S2} + 25rx_{1S2} + 5rx_{2S2} + 15ry_{1S2} + 35.36ry_{2S2} + 3.54ry_{3S2} + tz_{1S2} - \sqrt{2}tz_{2S2} \\ -35.36lrx_{1S2} - 3.54lrx_{3S2} - \sqrt{2}ltz_{1S2} + ltz_{3S2} - 35.36rx_{1S2} - 3.54rx_{3S2} - \sqrt{2}tz_{1S2} + tz_{3S2} \end{pmatrix}$$

The choice of the driving link parameters is obtained classifying the 9 link parameters taking into account the hierarchy of connections:

LHP[1]={$lrx_{1S2}$, $lry_{1S2}$, $ltz_{1S2}$}, LHP[2]={$lrx_{2S2}$, $lry_{2S2}$, $ltz_{2S2}$}

LHP[3]={$lrx_{3S2}$, $lry_{3S2}$, $ltz_{3S2}$}

The first 3 parameters are kept. Then, using compatibility equations step by step, we keep those that are not linked to the former one. We thus obtain:

LHP[1]={$lrx_{1S2}$, $lry_{1S2}$, $ltz_{1S2}$}, LHP[2]={$lry_{2S2}$, $ltz_{2S2}$}, LHP[3]={$ltz_{3S2}$}

and

$$T_{S2,P} = \left\{ \begin{matrix} lrx_{1S2} + rx_{1S2} & \begin{aligned} &-25lrx_{1S2} - 45lry_{1S2} - 35.35lry_{2S2} - ltz_{1S2} + 1.414ltz_{2S2} \\ &-25rx_{1S2} - 45ry_{1S2} - 35.35ry_{2S2} - tz_{1S2} + 1.414tz_{2S2} \end{aligned} \\ lrx_{1S2} + 1.414\,lry_{2S2} + rx_{1S2} + 1.414\,ry_{2S2} & \begin{aligned} &25lrx_{1S2} + 40lry_{1S2} + ltz_{1S2} + 25rx_{1S2} + 40ry_{1S2} + tz_{1S2} \end{aligned} \\ -lry_{1S2} - ry_{1S2} & \begin{aligned} &-10lrx_{1S2} + 56.57lry_{1S2} - ltz_{1S2} + 1.414ltz_{2S2} \\ &-10rx_{1S2} + 56.57ry_{1S2} - tz_{1S2} + 1.414tz_{3S2} \end{aligned} \end{matrix} \right\}$$

The last step consists in determining the constraints applied on the positioning. Assuming that the 3 connections are constituted of slipping contact implies:

For the primary connection LHP [1]={$lrx_{1S2}$, $lry_{1S2}$, $ltz_{1S2}$}={0,0,0}

For the secondary connection, it is necessary to apply a positioning function such as the displacement of the workpiece in direction of the second plane is maximum respecting a non-penetration constraint for each potential point of contact. So to maximize $_{-ltz_{2S2} - tz_{2S2}}$

respecting:

$$-3.54\text{lry}_{1S2} - 25\text{lry}_{2S2} + \text{ltz}_{2S2} + 3.54\text{rx}_{2S2} - 3.54\text{ry}_{1S2} \geq 0$$
$$3.54\text{lry}_{1S2} - 25\text{lry}_{2S2} + \text{ltz}_{2S2} - 3.54\text{rx}_{2S2} + 3.54\text{ry}_{1S2} \geq 0$$
$$-3.54\text{lry}_{1S2} + 25\text{lry}_{2S2} + \text{ltz}_{2S2} + 3.54\text{rx}_{2S2} - 3.54\text{ry}_{1S2} \geq 0$$
$$3.54\text{lry}_{1S2} + 25\text{lry}_{2S2} + \text{ltz}_{2S2} - 3.54\text{rx}_{2S2} + 3.54\text{ry}_{1S2} \geq 0$$

For the tertiary connection, the same principle is applied.

All of the positioning determination is resumed in a table (table 2) which is the textual expression of the MMP.

| Set-up 2 | | | |
|---|---|---|---|
| **Part Holder** | | | |
| **Surface** | **Surface defects DH** | | **Constraints CH** |
| Plane 1S2 | $\text{rx}_{1S2}$, $\text{ry}_{1S2}$, $\text{tz}_{1S2}$ | | Position and orientation relative to nominal |
| Plane 2S2 | $\text{rx}_{2S2}$, $\text{ry}_{2S2}$, $\text{tz}_{2S2}$ | | Position and orientation relative to nominal |
| Plane 3S2 | $\text{rx}_{3S2}$, $\text{ry}_{3S2}$, $\text{tz}_{3S2}$ | | Position relative to nominal |

**Assembly**

| Part-holder Surface | Link parameters LHP | Type of contact | Positioning function to maximize | Non penetration condition | solution | Part Surface |
|---|---|---|---|---|---|---|
| Plane 1S2 | $\text{lrx}_{1S2}$ $\text{lry}_{1S2}$ $\text{ltz}_{1S2}$ | Slipping | $-\text{ltz}_{1S2}-\text{tz}_{1S2}$ | $25\text{lrx}_{1S2}+20\text{lry}_{1S2}+\text{ltz}_{1S2} \geq 0$<br>$-25\text{lrx}_{1S2}+20\text{lry}_{1S2}+\text{ltz}_{1S2} \geq 0$<br>$25\text{lrx}_{1S2}-20\text{lry}_{1S2}+\text{ltz}_{1S2} \geq 0$<br>$-25\text{lrx}_{1S2}-20\text{lry}_{1S2}+\text{ltz}_{1S2} \geq 0$ | 0<br>0<br>0 | Plane 1 |
| Plane 2S2 | $\text{lry}_{2S2}$ $\text{ltz}_{2S2}$ | Slipping | $-\text{ltz}_{2S2}-\text{tz}_{2S2}$ | $-3.54\text{lry}_{1S2}-25\text{lry}_{2S2}+\text{ltz}_{2S2}+3.54\text{rx}_{2S2}-3.54\text{ry}_{1S2} \geq 0$<br>$3.54\text{lry}_{1S2}-25\text{lry}_{2S2}+\text{ltz}_{2S2}-3.54\text{rx}_{2S2}+3.54\text{ry}_{1S2} \geq 0$<br>$-3.54\text{lry}_{1S2}+25\text{lry}_{2S2}+\text{ltz}_{2S2}+3.54\text{rx}_{2S2}-3.54\text{ry}_{1S2} \geq 0$<br>$3.54\text{lry}_{1S2}+25\text{lry}_{2S2}+\text{ltz}_{2S2}-3.54\text{rx}_{2S2}+3.54\text{ry}_{1S2} \geq 0$ | none | Plane 2 |
| Plane 3S2 | $\text{ltz}_{3S2}$ | Slipping | $3.54\text{lrx}_{1S2}-15.91\text{lry}_{1S2}-22.5\text{lry}_{2S2}-\text{ltz}_{3S2}+3.54\text{rx}_{1S2}-15.91\text{ry}_{1S2}-22.5\text{ry}_{2S2}-\text{tz}_{3S2}$ | $10.61\text{lrx}_{1S2}+101.41\text{lry}_{1S2}+20\text{lry}_{2S2}+\text{ltz}_{3S2}+10.61\text{rx}_{1S2}+3.54\text{rx}_{3S2}+101.41\text{ry}_{1S2}+20\text{ry}_{2S2}-20\text{ry}_{3S2} \geq 0$<br>$-17.68\text{lrx}_{1S2}-101.41\text{lry}_{1S2}-20\text{lry}_{2S2}+\text{ltz}_{3S2}-17.68\text{rx}_{1S2}+3.54\text{rx}_{3S2}-101.41\text{ry}_{1S2}-20\text{ry}_{2S2}+20\text{ry}_{3S2} \geq 0$<br>$17.68\text{lrx}_{1S2}+101.41\text{lry}_{1S2}+20\text{lry}_{2S2}+\text{ltz}_{3S2}+17.68\text{rx}_{1S2}-3.54\text{rx}_{3S2}+101.41\text{ry}_{1S2}+20\text{ry}_{2S2}-20\text{ry}_{3S2} \geq 0$<br>$-10.61\text{lrx}_{1S2}-101.41\text{lry}_{1S2}-20\text{lry}_{2S2}+\text{ltz}_{3S2}-10.61\text{rx}_{1S2}-3.54\text{rx}_{3S2}-101.41\text{ry}_{1S2}-20\text{ry}_{2S2}+20\text{ry}_{3S2} \geq 0$ | none | Plane 3 |

**Machining**

| Surface | Machining defects DM | Constraints CM | $T_{P,Pi}$ |
|---|---|---|---|
|  |  |  |  |

*Table 2; Set-up 2 manufacturing table*

### 4.2. Floating positioning

A second example (figure 5) has been developed in [Vignat et al., 2005]. Here is presented the determination of the positioning deviation of the workpiece presented figure 5 for the case of a plane/centering/locating positioning in set-up 2 of the process plan. Assuming that:

- In set-up 1, surfaces plane 1, cylinders 2 and 3 are machined with defects.
- In set-up 2, the primary positioning connection is between part-holder plane 1S2 and workpiece plane 1, the secondary one between cylinder 2S2 and cylinder 2 and the tertiary one between locating 3S2 and cylinder 3.

In the same way than the previous example, the driving link parameters are obtained:

LHP $[1]=\{lrx_{1S2}, lry_{1S2}, ltz_{1S2}\}$, LHP $[2]=\{ltx_{2S2}, lty_{2S2}\}$, LHP $[3]=\{lty_{3S2}\}$

And the positioning deviation torsor of the workpiece:

$$T_{S2,P} = \left\{ \begin{array}{ll} lrx_{1S2} - rx_1 + rx_{1S2} & -5lry_{1S2} + ltx_{2S2} - 5ry_1 - 5ry_{1S2} - tx_2 + tx_{2S2} \\ lry_{1S2} + ry_1 + ry_{1S2} & 5\,lrx_{1S2} + 2lty_{2S2} - lty_{3S2} - 5rx_1 + 5rx_{1S2} - 2ty_2 + 2ty_{2S2} + ty_3 - ty_{3S2} \\ 1/50\left(-lty_{2S2} + lty_{3S2} + ty_2 - ty_{2S2} - ty_3 + ty_{3S2}\right) & 75lry_{1S2} + ltz_{1S2} + 75ry_1 + 75ry_{1S2} + tz_1 + tz_{1S2} \end{array} \right\}$$

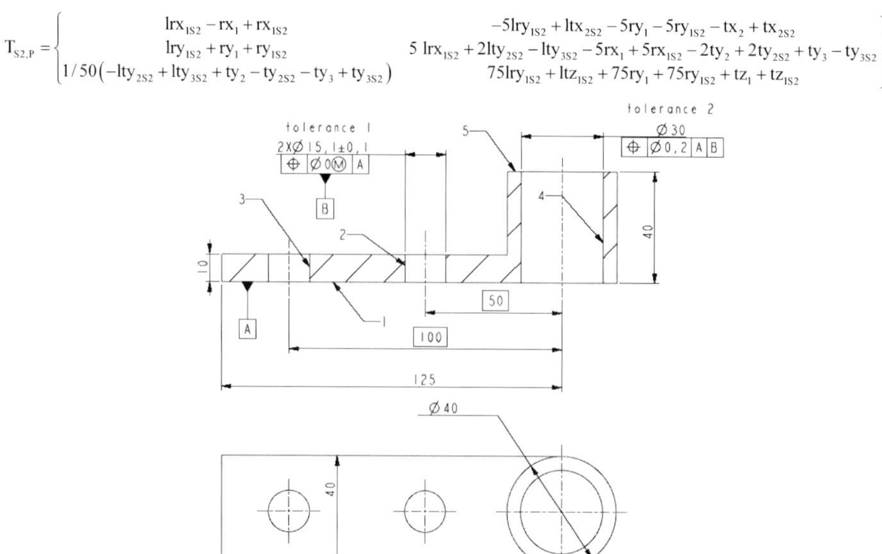


*Figure 5; definition drawing of the second example part*

The last step consists in determining the constraints applied on the positioning. Assuming that the primary connection is constituted of a slipping contact implies LHP $[1]=\{lrx_{1S2}, lry_{1S2}, ltz_{1S2}\}=\{0,0,0\}$

For the secondary connection, which is of floating type, the only constraint is to respect a non-penetration constraint for each potential point of contact. This constraint is expressed by the signed distance between each potential point of contact part holder/workpiece. The signed distance has to remain positive.

In the case of hole/pin assembly it could be expressed with 2 non-linear inequation (top and down) or with n linear inequation if the top and down circles are discretized.

See an example in the following inequalities where $R_2$ and $R_{2S2}$ are respectively the nominal radius of hole 2 and pin 2S2, and $ra_2$ and $r_{2S2}$ the radius variations.

$$-R_2 + R_{2S2} - r_{2S2} + ra_2 - \sqrt{\left(lty_{2S2} + 5\left(rx_1 - rx_{1S2} - rx_2 + rx_{2S2}\right)\right)^2 + \left(ltx_{2S2} + 5\left(ry_1 + ry_{1S2} + ry_2 - ry_{2S2}\right)\right)^2} > 0$$

$$-R_2 + R_{2S2} - r_{2S2} + ra_2 - \sqrt{\left(lty_{2S2} - 5\left(rx_1 - rx_{1S2} - rx_2 + rx_{2S2}\right)\right)^2 + \left(ltx_{2S2} - 5\left(ry_1 + ry_{1S2} + ry_2 - ry_{2S2}\right)\right)^2} > 0$$

## 5.  CONCLUSION

Some papers already presented the way to simulate assembly operations for functional tolerance analysis [Sodhi et al., 1994] [Thiebaut, 2001] [Dantan, 2000] [Anselmetti et al., 2003]. The present paper applies this approach to contribute to the generation of the parameters of the MMP (Model of the Manufactured Part) whose concept was introduced in [Villeneuve et al., 2004]. More precisely, the way to find the positioning parameters of a workpiece in a set-up is detailed.

A positioning function is introduced in order to make the resolution more generic. It allows to consider all the cases existing for a workpiece/part-holder assembly, over-constrained or isostatic, with floating or slipping contacts.

The set of hierarchically arranged functions and constraints obtained contributes to describe the MMP and can be used in every case of tolerances analysis. This method can also be applied to the resolution of the problem of assembly between the part to be controlled and a virtual control gauge. A textual formulation of the workpiece positioning is described represented by a table. This formulation provide a precise description of the assembly operation and give all the information necessary to perform the analysis step without recompile a simulation one. The way to analyze functional tolerances and the ability of a manufacturing process to realize suitable part was proposed in [Vignat et al., 2005].

## REFERENCES

**[Anselmetti et al., 2003]** Anselmetti B., Mejbri H., Mawussi K.: "Coupling experimental design—digital simulation of junctions for the development of complex tolerance chains", Computers *in Industry*, 50, pp. 277–292, 2003

**[Dantan, 2000]** Dantan J.Y.; "Synthèse des specifications géométriques : modélisation par calibre à mobilités internes" *PhD thesis*, University of Bordeaux, France, 2000

**[Sodhi et al., 1994]** Sodhi R., Turner J. U., Relative positioning of variational part models for design analysis, *Computer Aided Design*, volume 26, #5, May 1994

**[Thiebaut, 2001]** Thiebaut, F., "Contribution à la définition d'un moyen unifié de gestion de la géométrie réaliste basée sur le calcul des lois de comportement des mécanismes", *PhD thesis*, ENS de Cachan, France, 2001

**[Vignat et al., 2005]** Vignat F., Villeneuve F.: "Simulation of the manufacturing process (2): Analysis of its consequences on a functional tolerance", *Proceedings of the 9$^{th}$ CIRP International Seminar on Computer Aided Tolerancing*, Tempe (US), 10-11 April 2005

**[Villeneuve et al., 2001]** Villeneuve, F.; Legoff, O.; Landon, Y.; "Tolerancing for manufacturing: a three dimensional model"; *International Journal of Production Research*, Vol. 39, N°8; pp. 1625-1648, 2001

**[Villeneuve et al., 2004]** Villeneuve F., Vignat F.: "Manufacturing process simulation for tolerance analysis and synthesis", in: *CD Proceedings of IDMME 04*, Bath (UK), 5-7 April 2004

# Surface Best Fit: Generated Distribution of the Real Manufacturing Process


S. Aranda*,  J. M. Linares *, J. M. Sprauel*, P. Bourdet**

* EA(MS)$^2$, I.U.T., Avenue Gaston Berger - F13625 Aix en Provence cedex 1
** LURPA, ENS Cachan, Avenue du président Wilson - F94230 Cachan
aranda@iut.univ-aix.fr



**Abstract:** In the last seminar of the CIRP, a new concept based on a statistical approach has been presented. It allows evaluating the uncertainties of Coordinate Measuring Machine measurements, accounting for the measuring process. The implication of these evaluated uncertainties in the verification of specifications has been also discussed and the example of a coaxiality specification has been taken. The basis of this statistical approach presented at the CIRP seminar in 2001, had then shown that the use of an association criterion must be linked to an assumption concerning the probability density of the measured points around the associated surface. So the legitimate question that remains in suspends, is to know what kind of distribution is generated by the real manufacturing process. The aim of this paper is to present a machining model able to predict the probability density of the measured points in the case of end milling.



**Keywords:** end milling, signature, prediction, Z-map, best fit.


## 1. INTRODUCTION

The Coordinate Measuring Machine (CMM) has brought a significant change on the industrial control practices. It has allowed accessing to larger information than with the classical metrology approach. But, at the opposite of this last one, the CMM cannot measure directly the geometrical dimensions of the part (corresponding to ISO or ANSI geometrical specifications). The acquisition of the surfaces results in a series of set of points. Each set of points represents a sample of the surface where it has been acquired. It contains however both information about the spatial localization of the surface and its intrinsic geometrical characteristics.

In order to access the geometrical dimensions of the part, the sets of points must be interpreted by a specific algorithm. The first step of this process consists in describing each set of points by a mathematical model matching with the expected surface. In the second step, the parameters of the mathematical model are then optimized to define the best surface fitting the acquired points. The particular model obtained after the best fit is performed, is called the derived element of the surface. These two steps transform the real surfaces into mathematical objects easy to handle. The geometrical specifications




*J.K. Davidson (ed.), Models for Computer Aided Tolerancing in Design and Manufacturing,* 191–200.
© 2007 *Springer.*

can then be checked by applying vectorial geometrical operations to the derived elements previously obtained. The best fit methods have a great impact in this process. In fact their capacity to extract informations from the set of points, and to integrate them in the derived element parameters, strongly influences the verification of the geometrical specifications.

Today, Tchebitchev criterion (infinite norm) and least squares criterion (Euclidian norm) are the most employed best fit practices. The first one presents the advantage of calculating the derived element leading to the evaluation of the least form defect, while the second one offers excellent computation stability. By formulating the least square criterion on a purely statistical point of view, some works [Sprauel et al., 2001] demonstrated that it is possible to calculate the uncertainties of the parameters of the derived element. They proved also that the use of the least squares criterion is equivalent to the likelihood criterion when the distribution of the points around the mean surface is Gaussian. It consists in performing a maximization of the statistical information included in the set of points. This work was validated in the case of high quality surfaces.

In its continuation, a propagation method has been developed [Bachman et al., 2004] to evaluate the impact of the uncertainties calculated for each derived element on any geometrical dimension which may be computed. Applying the method to different specifications, the authors showed that some of them may become very difficult to verify. It is the case, in particular, when the tolerances are small and tend then towards the value of the propagated uncertainties. Two ways can be investigated to remove this indetermination:

- changing the control process to minimize the propagation of uncertainties,
- reducing the uncertainties of the parameters of the derived element by improving the best fit method.

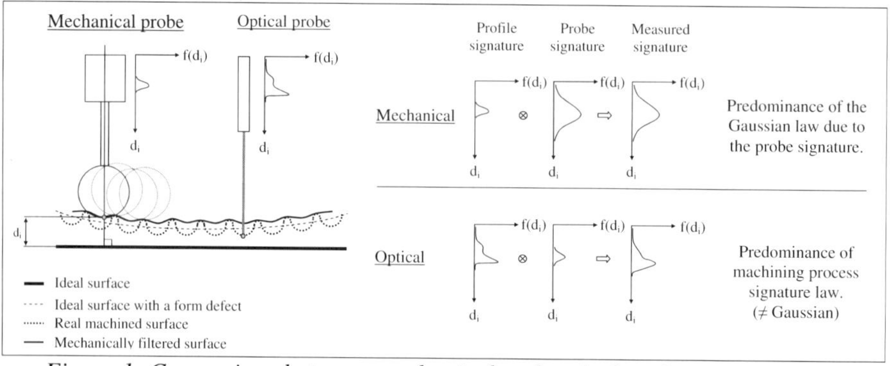


*Figure 1; Comparison between mechanical and optical probe measurements*

The recent evolution of the probes, and more exactly the apparition of optical sensors, offers an advisability of exploring the second way. The high resolution of such instrument (10 nm) associated to its low uncertainty (0.2 µm at three standard deviations) permits to get the surface with a high definition. However, the mechanical filtering carried out by the radius of a classical probe does no longer exist for optical sensors. In consequence, all types of geometrical defects are included in the measured coordinates (form, undulation and roughness). They represent the trace, or signature, left by the machining process and are almost fixed by the selection of the cutting parameters. They correspond to latent information that can be extracted from the points sample and included in the derived element definition.

Henke et al., [Henke et al., 1999] have integrated the first and second order defects directly in the definition of the mathematical model. They proposed to superimpose to the model of a perfect cylinder the cumulated effect of most classical machining defects (conicity, filet, trilobing…). The defects are described by polynomial Fourier series and all theirs possible combinations are investigated by a special algorithm. For each combination, a maximum probability is determined and used to select the combination leading to the most consistent model. The optimum defines the best fit of the analyzed surface. Choi et al., [Choi et al., 1998] have presented an alternative approach of the problem that does not require cataloging all the different kinds of defects. They used a beta distribution to reproduce the repartition of the distances between the mathematical model and the infinite set of points of the real surface. This repartition corresponds to the process signature. In assuming that the measured points form a representative sample of the analyzed surface, it characterizes also the distribution of the best fit residues. This condition is however not satisfied when the points are acquired with an automatic cycle using coordinates evenly distributed [Weckenmann et al., 1995]. The beta distribution was selected because it allows accounting for dissymmetrical distributions (case of turned and milled surfaces) and may also emulate a normal distribution (surfaces of high quality), depending on its parameters. A special algorithm has been developed to define the best fit of the analyzed surface, computing simultaneously the most probable parameters of the derived element and the beta distribution of the residues.

Another approach can be investigated, because the signature of surface is completely defined by the machining process and the cutting parameters employed. In consequence, the probability density of the points around the surface can be predicted, prior to the calculation of the derived surface. The best fit of the surface can be performed, using the likelihood criterion with the specific distribution found. The main aim of this paper is answering to the first step of this approach that is to predict the density probability of the process signature. It is a required preliminary for considering the best fit approach mentioned above. To illustrate this approach, the End milling process has been chosen.

The next section will present a review of existing end milling simulation methods and models. These models are however not satisfactory compared to our objectives. In the third part, an adapted solution based on a Z-map technique [Maeng et al., 2003], is therefore proposed. It proceeds by an analytical resolution of the end milling simulation, in order to avoid any statistical bias. Some simulated results will then be compared to a real machining case. The conclusion will finally state about the interest of this model.

## 2. LITERATURE REVIEW: END MILLING SIMULATION

End milling is one of the most employed matter removing processes and a lot of models have been developed to characterize it. The aims of these works are various and concern essentially roughness determination, cutting force estimation, chip geometry prediction and form defect evaluation. It is difficult to classify the different studies but three main groups can be defined:
- experimental approaches,
- kinematic and dynamic approaches,
- artificial intelligence approaches.

Many researchers have focused theirs works on roughness optimization, using experimental studies and artificial intelligent approaches. The surface response methodology (SRM) was employed to evaluate the impact of feed rate, cutting speed and depth of cut, on the roughness. However, it is difficult to enhance the provided models with mechanical knowledge and to integrate non-contiguous parameters, like the machining path, for example. In fine, it can be concluded that these approaches have little or no general applicability.

Kinematic and dynamic approaches tried answering to the above mentioned limitations. This category includes methods based on machining theory and mechanical knowledge. Only few works are based on a complete analytical resolution of the problem because of the complexity of the studied models. The problem is thus generally discretized and treated by numerical algorithms. Classically, the cutting movements are discretized in time. The cutting edge is divided into small segments. The workpiece is represented as a Z-Map object (figure 2).

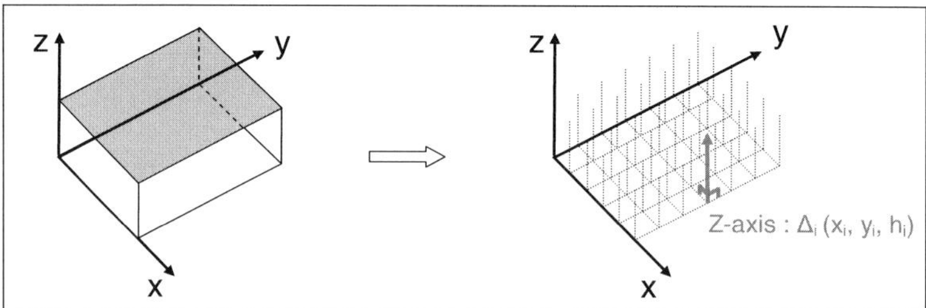


*Figure 2; Z-map representation*

The Z-Map representation consists in discretizing a volume into a set of vectors. These vectors are called z-axis and written $\Delta_i$. They are parallel to direction z of the coordinate system and characterized by three variables. The two first ones ($x_i$ and $y_i$) indicate the coordinate of the intersection between the z-axis and the x/y plane. These coordinates are generally chosen in order to fit the nodes of a grid. The third variable $h_i$ corresponds to the norm of the z-axis. It is used to store the distance between the upper surface and the x/y plane along the z-axis. So, this representation is very useful to predict the geometry of the machined surface and more particularly the cut height for some selected $x_i$ and $y_i$ coordinates.

However, it must be underlined that this representation is restricted to cases where the orientation of the normal to the surface is ranged. Nevertheless, most common milling operations satisfy this condition and lot of the recent machining simulation models are based on this method or its derivates (extended Z-map, moving node extended Z-Map).

The kinematic and dynamic approaches can be distributed in three sub groups: geometrical oriented approaches, cutting forces oriented approaches and vibration oriented approaches.

Maeng [Maeng et al., 2003] proposes a geometrical oriented model, based on the z-map method. The first step consists in computing the trace left by the extreme point of the milling tool during its displacement. This trace is expressed as a system of equations representing the geometrical surface envelope generated by the movement of the cutting edge profile. Once this operation performed, the intersection between this envelope and each z-axis is computed and associated coordinates $h_i$ are stored. These two steps then are repeated for each trajectory of the milling strategy employed. The resulting machined surface is finally represented by the minimum value of h found for each z-axis considering the whole displacements of the tool. This method is implemented by Baek [Baek et al., 2001] to predict the surface roughness. A complete analytical description and resolution of the problem is used for that purpose. He starts by performing a complete analysis of the different profiles left in the part, by the cutting edges, during one revolution of the milling tool. Each profile is represented by an equation. The profiles are then compared and only the portions that lead to the maximum machined depth are kept. The final result is expressed in the form of a function defined by intervals. Anyway, this study is seriously limited because only the machining profile of the middle cutting plane can be computed whereas it is not representative of the whole generated surface.

## 3. PROPOSED MODEL

### 3.1. Representation of the part

The signature corresponds to the distribution of the real points of the machined surface around to the nominal profile. It represents, in consequence, a statistical estimation of the point population and on only has a sense if the population is built randomly. This is not the case with a classical Z-Map representation were all the z-axes are defined by a grid of constant step along the x and y directions. Such grid construction may result in intensifying the contribution of some characteristically sub-groups of the surface point's population and therefore introduce a statistical bias [Weckenmann et al., 1995]. In first approach, increasing the number of z-axis can easily help to reduce this bias. But in fact, it is not a valuable solution for two reasons. First, it will never totally remove it and second it will dramatically increase the calculation cost. For as much, the z-map representation of the workpiece has not to be rejected but simply extended. In the extended model that is proposed here, the z-axis are not constructed on a grid, but theirs intersections with the x/y plane ($x_i$ and $y_i$), are built randomly (figure 3). Practically, theses two coordinates are generated by a Monte Carlo method, assuming a uniform surface probability density. Such procedure allows reaching a reliable representation of the population of the surface points with a limited number of points. From here, the z-axis coordinates will be expressed in the coordinate system $R_p(O_p, x_p, y_p, z_p)$ of the part.

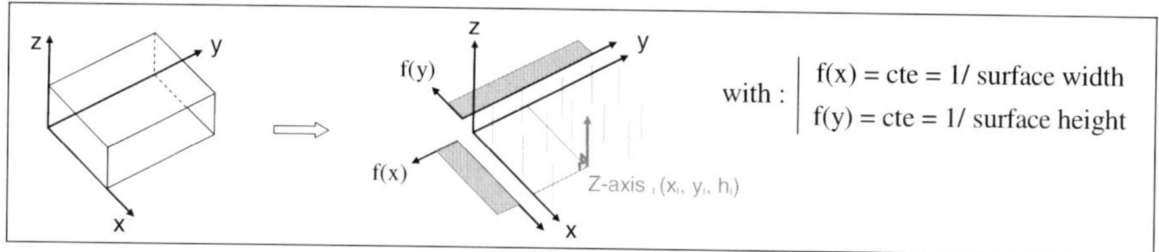


*Figure 3; Proposed workpiece representation*

### 3.2. Tool representation

First, a coordinate system $R_t(O_t, x_t, y_t, z_t)$ is associated to the tool. Vector $z_t$ corresponds to the tool axis, direction $x_t$ is located in the bottom plane of the tool and vector $y_t$ is defined by the Vectorial product $z_t \wedge x_t$ (figure 4a). The tool is then divided into a set of independent cutting edges and a coordinate system $R_e^k(O_e, x_e^k, y_e^k, z_e)$ is associated to each of them. Parameter k represents the edge number. Direction $z_e$ is taken collinear to the tool axis. Both points $O_t$ and $O_e$ are equivalent. These two elements are the same for each cutting edge. The axis $x_e^k$ is located in the bottom plane $O_t x_t y_t$ of the tool and rotated of an angle $\psi$ to $x_t$. Finally, the geometry of a cutting edge is described in its corresponding plane $O_e x_e^k z_e$, by a function $f(x)$ that will associate to each range $x \in [r_{min}, r_{max}]$ a corresponding coordinate z. For most tools the function f can be defined by three intervals as shown in figure 4a.

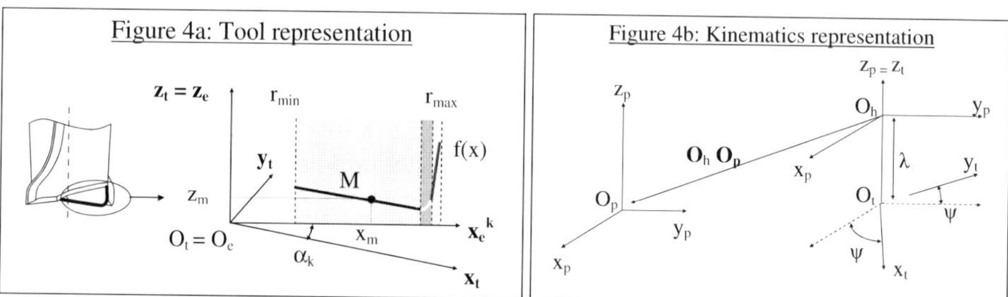


*Figure 4; Representation of the cutting edge and the machining kinematics.*

### 3.3. Kinematics of the milling machine

The link between the tool and the coordinate system of the part is realized by the kinematics of the milling machine (figure 4b). It can be divided into two simultaneous movements. The first is the displacement of the milling machine's head. It is described by the vector $O_f O_h$ that depends on the machining trajectory, the cutting feed and the time parameter t. Depending on its expression, different kinds of trajectories can be simulated. While this paper focuses on G1 interpolation, the vector becomes:

$$O_f O_h = g_x(t) \cdot x_p + g_y(t) \cdot y_p + g_z(t) \cdot z_p \quad \text{with} \quad g_i(t) = b_{i1} \cdot t + b_{i0}$$

The second movement corresponds to the rotation of the tool. It is defined by an angle $\theta$ and expressed in the head coordinate system $R_h(O_h, x_p, y_p, z_p)$. Both vectors $z_t$ and $z_h$ are identical. The angles $(x_p, x_t)$ and $(y_p, y_t)$ are thus both equal to $\theta$ which is time dependend. For a milling operation with constant feed and cutting velocity, the expression of $\theta$ is $\theta(t) = \omega \cdot t$, where $\omega$ is the spindle rotation speed.

At last, vector $O_h O_t$ expresses the height between the driven points and the bottom of the milling tool from where the cutting edge geometries are defined.

### 3.4. Global expression of a cutting edge point.

The whole kinematical chain presented in the previous paragraph, can be globally expressed using a set of transformation matrixes including both rotation and translation associated to each change of coordinate system. The final expression for this chain is:

$$O_p M = V(t, x_m, k), \quad x_m \in [r_{min}, r_{max}], \quad t \in [t_{min}, t_{max}], \quad k = 1 \ldots n$$

$$\underbrace{\begin{bmatrix} x \\ y \\ z \\ 1 \end{bmatrix}}_{O_pM} = \underbrace{\begin{bmatrix} 1 & 0 & 0 & -g_x(t) \\ 0 & 1 & 0 & -g_y(t) \\ 0 & 0 & 1 & -g_z(t) \\ 0 & 0 & 0 & 1 \end{bmatrix}}_{\text{Part displacement}} \cdot \underbrace{\begin{bmatrix} \cos(\omega \times t) & \sin(\omega \times t) & 0 & 0 \\ -\sin(\omega \times t) & \cos(\omega \times t) & 0 & 0 \\ 0 & 0 & 1 & \lambda \\ 0 & 0 & 0 & 1 \end{bmatrix}}_{\text{Tool rotation and position}} \cdot \underbrace{\begin{bmatrix} \cos(\alpha_k) & \sin(\alpha_k) & 0 & 0 \\ -\sin(\alpha_k) & \cos(\alpha_k) & 0 & 0 \\ 0 & 0 & 1 & 0 \\ 0 & 0 & 0 & 1 \end{bmatrix}}_{k^{\text{th}} \text{ edge orientation}} \cdot \underbrace{\begin{bmatrix} x_M \\ 0 \\ f(x_M) \\ 1 \end{bmatrix}}_{O_eM}$$

The resulting expression allows calculating the coordinates, in the part, of any cutting edge point as a function of the kinematical and geometrical parameters t, $x_M$ and k. While varying these variables, a set of three-dimensional cycloidal surfaces is obtained.

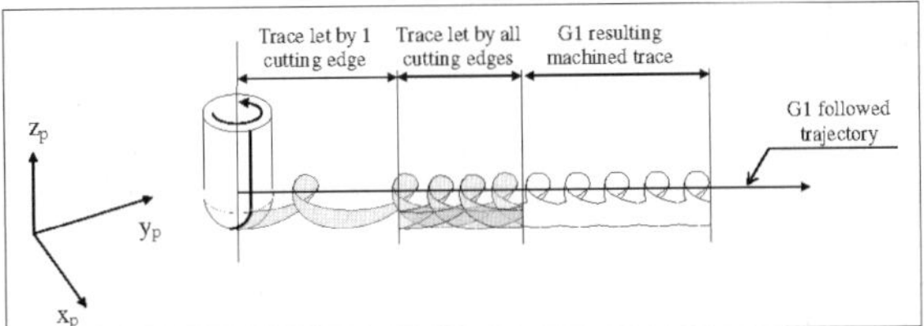


*Figure 5; Cutting edge trace in the coordinate system of the part*

The machined profile results from the superimposition of such surfaces obtained for the studied trajectory and the employed cutting parameters (figure 5).

### 3.5. Simulation of the machining process and signature prediction

The simulation of the process consists in finding all the intersections between any z-axis ($\Delta_i$) of the part, and the traces left by the cutting edges. A lot of intersections are generally found (figure 6). If one or more of them exist, the z-axis corresponds to a machined point. The intersections between the z-axis $\Delta_i$ and the cutting edges for one trajectory correspond to the set of solutions $S_j(t, k, x_M)$ of following system of trigonometric equations:

$$\begin{cases} x_i = O_p P \cdot x_p = V(t, x_m, k) \cdot x_p \\ y_i = O_p P \cdot y_p = V(t, x_m, k) \cdot y_p \end{cases} \quad \text{with: } x_m \in [r_{min}, r_{max}], \ t \in [t_{min}, t_{max}], \ k = 1 \ldots n$$

This system has been solved analytically by using a resolution method by intervals. For each z-axis $\Delta_i$ the solutions $S_j(t, k, x_M)$ provide new possible values $h_i^j = V(t, x_m, k) \cdot z_p$. They are all compared to the current one and if a deeper depth is found and the difference $h_i - h_i^j$ is larger than the minimum cutting chip thickness, the variable $h_i$ is updated. Finely, the simulated surface is obtained by repeating this operation for each z-axis and for all the trajectories of the machining process (figure 6). The resulting simulated signature left on the surface, by the machining process, corresponds simply to the repartition of the points of the calculated surface around the ideal feature.

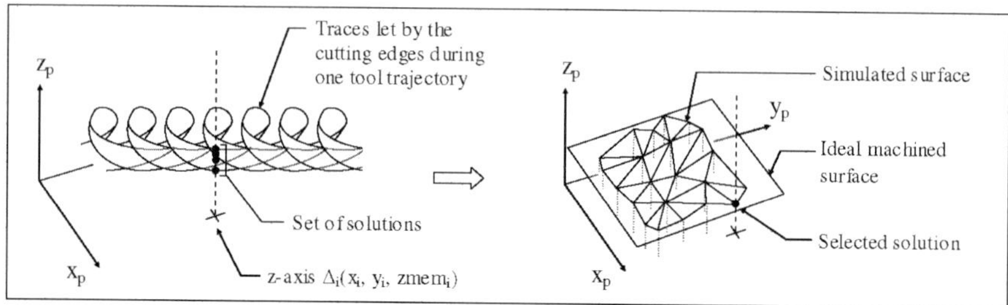


*Figure 6; Simulation principle*

## 4. MODEL VALIDATION

In order to test the suggested model, it has been confronted with a real sample. Figure 7 presents a summary of the validation method. From one side, a real sample as been machined. The employed cuttings parameters were: feed $f_d = 0.2$ mm/tooth and cutting speed Vc = 100 m/min. The employed tool was a mono-bloc carbide milling tool with a diameter of about 10 mm and two cutting edges. This sample has been measured using an optical probe. The set of surface points, thus obtained, has been used to define the real surface signature. On the other hand, the model has been implemented on Microsoft Vb.net programming framework. A simulated surface was thus computed. Such calculations cannot be compared directly to measured coordinates because this latter data includes random perturbations of the measurement process. For that reason, a flat standard of very high quality was measured to characterize the signature left by the optical probe. This information was finally added to the theoretical distributions to simulate the effect of acquisition.

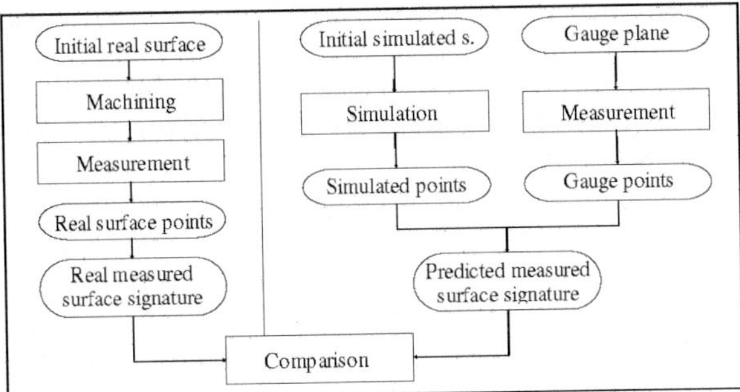


*Figure 7; Model verification*

Figure 8 shows the real and the simulated surface signatures obtained. The left hand side of the figure shows the predicted and the real textures realized by a spiral cycle. The different texture zones of the machined surface are accurately described by the model. So the model can be advantageously used to predict the texture or roughness induced by given cutting parameters, tool geometries and machining strategies. The right hand side figure shows the predicted and the real signatures obtained. Both curves have been centered and reduced in order to be correctly compared. The predicted and the real signatures are in good agreement although if some differences can be noticed. They can be partly explained by the plastic deformation and shearing induced by cutting or geometrical defects of the milling tool.

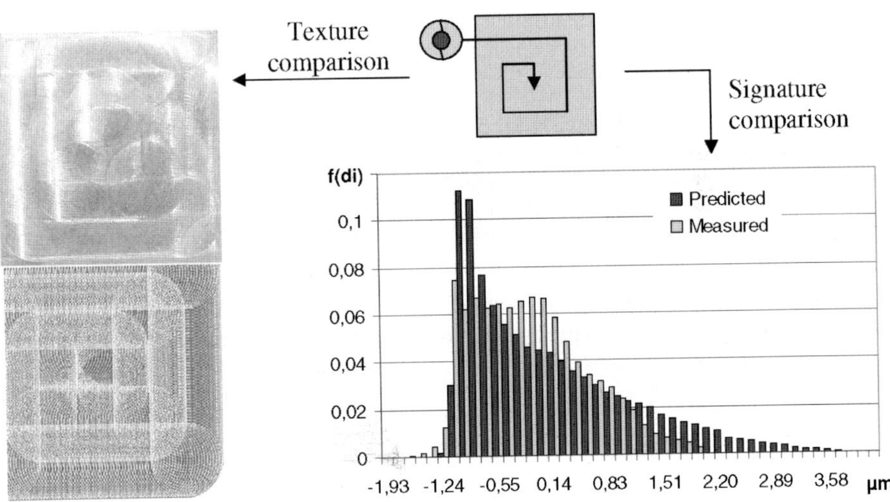


*Figure 8; Simulated and real signatures*

## 5. CONCLUSION

In this paper a model has been presented to predict the signature of the end milling machining process. This model differentiates from the previous approaches by the efforts made to reduce the statistical bias in the derived signature. It permits also integrating the cutting edges geometry. Therefore, a quasi completely analytical resolution of the cut heights leads to low computing time. The signature of the surface has first been defined as the repartition of the points around the ideal feature. This concept has been introduced to extend the best fit criteria, to the analysis of surfaces acquired with optical probes. The milling simulation model integrates the influence of the cutting parameters, the cutting edges geometry and also the strategy employed in machining. The results show that the model predictions are suitable for the followed objective and demonstrate that the signature left by an end milling operation does not match with a Gaussian or uniform distribution. Future work will develop a new best fit approach that accounts for the surface signatures left by the machining processes. An inverse method will then be used to retrieve most machining conditions from the set of measured coordinates.

## REFERENCES

**[Bachman et al., 2004] J. Bachmann, J.M. Linares, J.M. Sprauel, P. Bourdet,** "Aide in decision making: Contribution to uncertainties in 3D measurements", *International Journal of Precision Engineering, Elsevier Publisher,* pp.78-88, Vol.28 n°1, 2004,

**[Sprauel et al., 01] J.M. Sprauel, J.M. Linares, P. Bourdet,** "Contribution of non linear optimization to the determination of measurement uncertainties", *7th CIRP Seminar on Computer Aided Tolerancing , ENS Cachan France,* pp.285-292, April 2001.

**[Henke et al., 99] R.P. Henke, K.D. Summerhays, J.M. Balwin, J.M.Cassou, C.W. Brown,** "Methods for evaluation of systematic geometric deviations in machined parts and their relationships to process variables", *Precision eng.,* pp.273-292, Vol.23, 1999.

**[Choi et al., 98] W. Choi, T. Kurfess, J. CAGAN,** "Sampling uncertainty in coordinates measurement data analysis", *Precision eng.,* pp.153-163, Vol.22, 1998.

**[Weckenmann et al., 1995] A.Weckenmann, H.Eitzert, M.Garmer, H.Weber,** "Functionality-oriented evaluation and sampling strategy in coordinate metrology ", *Precision engineering,* pp.244-252, Vol.17, 1995.

**[Maeng et al., 2003] S.R. Maeng, N. B, S.Y. Shin, B.K. Choi,** "A Z-map update method for linearly moving tools", *Computer-Aided Design,* pp.995-1009, Vol.35, 2003.

**[Baek et al., 2001] D.K. Baek, T.J. Ko, H.S. Kim,** "Optimization of federate In a face milling operation using a surface roughness model", *International Journal of Machine and tools & manufacture,* pp.451-462, Vol.41, 2001.

# Position Deviation of a Holes Pattern
# Due to Six-Point Locating Principle


**W. Polini*, G. Moroni****

**Dipartimento di Ingegneria Industriale, Università di Cassino, via G. di Biasio 43,
03043 Cassino, Italy, Phone: +3907762993679*

*polini@unicas.it*

***Dipartimento di Meccanica, Politecnico di Milano, Piazza Leonardo da Vinci 32,
20133 Milano, Italy, Phone: +390223994762*

*giovanni.moroni@polimi.it*



**Abstract:** This work shows a statistical method to predict the position of the locators around a plate to minimise the deviation of drilled holes pattern from their nominal position due to the inaccuracies on locators position. The statistical method adopts 3-2-1 locating principle and models the position of each locator by a Gaussian probability density function and, consequently, calculates the probability the drilled holes pattern falls inside the location tolerance, that is centred around each hole's nominal position. Starting from this statistical method, we introduce a very simple rule in locator configuration that gives quite interesting suboptimal solutions.

**Keywords**: fixture design, 3-2-1 locating principle, holes pattern, location tolerance, statistical positioning


## 1.  INTRODUCTION

A machining fixture controls the position and the orientation of the workpiece reference frame with respect to the machine one. The reference frame in a machining fixture is defined by the locators. During machining, the tool path is defined with respect to this workpiece reference frame. Ideally, the locators make point contact with the workpiece and the position and orientation of the workpiece reference frame with respect to the machine one is perfectly accurate. However, in reality the geometry and the position of the locators are imperfect and the reference frame they produce has position and orientation errors with respect to the machine reference frame. This misalignment produces geometrical errors in the features machined on the workpiece, e.g. location error of a holes pattern.

The existing research provides essential steps towards the design of locators placement. The more largely used formalism is based on the screw theory due to its compactness and its general applicability [Ohwovoriole et al., 1981]. The screw theory allows to identify both the causes of a wrong positioning of the part and the possible corrective actions. Bourdet and Clement used the displacement screw vector to mathematically describe the misalignment between part and machine [Bourdet et al., 1974]. They extended this work by developing a model to determine the nominal positions of locators which minimise the magnitude of the screw displacement vector [Bourdet et al.,

201


*J.K. Davidson (ed.), Models for Computer Aided Tolerancing in Design and Manufacturing, 201–211.*
© 2007 *Springer.*

1988]. Weill connected the screw displacement vector to the geometric variation of critical part features and minimised this geometric function [Weill et al., 1991]. More recent studies deal with robust design of fixture configuration, by analysing the influence of workpiece surface errors and fixture set-up errors on the stability of part [Cai et al. 1997], developing algorithms for workpiece localization [Chu et al., 1998], or employing the screw parameters associated with TTRS to determine the position uncertainty of a part [Desrochers et al., 1997]. Very interesting is the work of Choudhuri that presents a method for modelling and analysing the impact of a locator tolerance scheme on the potential datum related geometric errors of linear machined features [Choudhuri et al., 1999].

This work shows a method to design the optimal locators positioning that minimises the position error of a holes pattern due to the inaccuracy of the fixture configuration.

In a previous paper [Armillotta et al., 1996], a method for checking deterministic positioning from locator configuration was proposed. Besides detecting positioning incorrectness, it derives an explanation of singularity reasons, in order to ease the redesign of a wrong fixture. A following study extended the previous method in order to highlight quasi-singularity conditions, where part inaccuracies are likely to result in excessive geometric errors on machined features [Armillotta et al., 1999]. Then, a statistical method was proposed in order to integrate the approaches based on deterministic positioning [Armillotta et al., 2003]. It uses an analytical approach to define the probability density function of positions of machined features as a function of the inaccuracy of the locators scheme. It considers only 2D parts, such as plates. The probability density function of machined feature positions was used in a following work to define the optimal locator positions, minimizing the machining inaccuracy [Moroni et al., 2003]. Finally, the position deviation of a hole due to the inaccuracy of all the six locators of the 3-2-1 locating scheme was estimated by a Monte Carlo simulation approach [Giusarma et al., 2004].

This work aims to investigate on how locators configuration affects the drilling of a holes pattern. It considers how deviation on fixturing elements propagate on location tolerance of a holes pattern. The 3-2-1 locating principle has been adopted. The position of each locator is represented by a Gaussian probability density function and, consequently, the probability the holes pattern falls inside the location tolerance, centred around each hole nominal position, is estimated as the product of the probabilities due to each hole. The probability each hole falls inside the location tolerance is estimated by means of an analytical expression. The optimal positioning of the locators is designed by minimising the deviation in holes pattern positioning during drilling due to locators inaccuracy. The proposed method may be applied to a holes pattern with any material conditions by considering an appropriate location tolerance zone. 2D parts have been considered as application examples. The present work considers error free starting workpiece and machine tool. In the following the optimal locators positioning method is deeply described (§2), some application examples are discussed and a very simple suboptimal configuration rule is introduced (§3).

## 2.   OPTIMAL LOCATORS CONFIGURATION

The case study is represented by a plate with a holes pattern to be drilled, as shown in Figure 1. A RFS location tolerance specify the position of holes pattern. The position of the workpiece is determined by two locators on the primary datum and one on the secondary datum. Each locator has coordinates related to the part nominal reference frame, represented by the following three terns of values:

$$\mathbf{p_1}(x_1,y_1,z_1); \ \mathbf{p_2}(x_2,y_2,z_2); \ \mathbf{p_3}(x_3,y_3,z_3) \tag{1}$$

The proposed approach considers the uncertainty source in the positioning error of the drilled holes pattern due to the variance in the positioning of the locators. It aims to minimise the machining uncertainty due to this source. It neglects the tool positioning error or the geometric deviations on datum elements.

The six coordinates of the locators (1) in the machine reference frame $XOY$ are considered distributed according to a gaussian probability density function with mean equal to the nominal position of locators and standard deviation $\sigma$:

$$x_1 \approx N\left(x_{1n} = 0, \sigma^2\right) \qquad\qquad y_1 \approx N\left(y_{1n}, \sigma^2\right)$$

$$x_2 \approx N\left(x_{2n}, \sigma^2\right) \qquad\qquad y_2 \approx N\left(y_{2n} = 0, \sigma^2\right) \tag{2}$$

$$x_3 \approx N\left(x_{3n}, \sigma^2\right) \qquad\qquad y_3 \approx N\left(y_{3n} = 0, \sigma^2\right)$$

with $x_{in}$, $y_{in}$ nominal values of the locator coordinates in the machine reference frame (MRF).

The perturbed part reference frame (PRF) $X'O'Y'$ is related to the actual position of the locators. In particular, the X' axis is the straight line passing through the actual position of locators $\mathbf{p_2}$ and $\mathbf{p_3}$, while the Y' axis is perpendicular to X' axis and passes through the actual position of $\mathbf{p_1}$. The nominal coordinates of the centre of the generic (i) hole of the pattern are $\mathbf{c_i}(x,y)$, while their real values are function of locator probability density function $\mathbf{c'_i}(x',y')$ in the PRF.

To determine the effect of locators' deviation from their nominal position on the position of the i-hole of the pattern, the probability density function of the centre $\mathbf{c'_i}(x',y')$ of the drilled hole in the PRF has been calculated as a function of the probability density function of the locators (2). The probability density functions of the coordinates x' and y' of the centre of the i-hole of the pattern depend only on the gaussian distribution of the coordinate $x_1$, $y_2$ and $y_3$ of the locators, as demonstrated in [Armillotta et al., 2003]:

$$f_{\mathbf{c'_i}}(x',y') = \frac{|x_{3n} - x_{2n}|}{2 \cdot \pi \cdot \sqrt{(y-y_{1n})^2 + (x-x_{2n})^2 + (x-x_{3n})^2}} \cdot$$

$$\cdot \exp\left\{ -\frac{1}{2} \cdot \left( \frac{(x'-x)^2 \cdot \sigma_{y'}^2 \cdot (x_{3n}-x_{2n})^2}{(y-y_{1n})^2 + (x-x_{2n})^2 + (x-x_{3n})^2} + \right.\right.$$

$$\left.\left. \frac{2 \cdot (x'-x) \cdot (y'-y) \cdot (y-y_{1n}) \cdot [(x-x_{2n})+(x-x_{3n})]}{(y-y_{1n})^2 + (x-x_{2n})^2 + (x-x_{3n})^2} + \frac{(y'-y)^2 \cdot \sigma_{x'}^2 \cdot (x_{3n}-x_{2n})^2}{(y-y_{1n})^2 + (x-x_{2n})^2 + (x-x_{3n})^2} \right) \right\}$$

$$(3)$$

with

$$\sigma_{x'} = \frac{\sqrt{2 \cdot (y-y_{1n})^2 + (x_{3n}-x_{2n})^2}}{|x_{3n}-x_{2n}|} \tag{4}$$

$$\sigma_{y'} = \frac{\sqrt{(x-x_{2n})^2 + (x-x_{3n})^2}}{|x_{3n}-x_{2n}|} \tag{5}$$

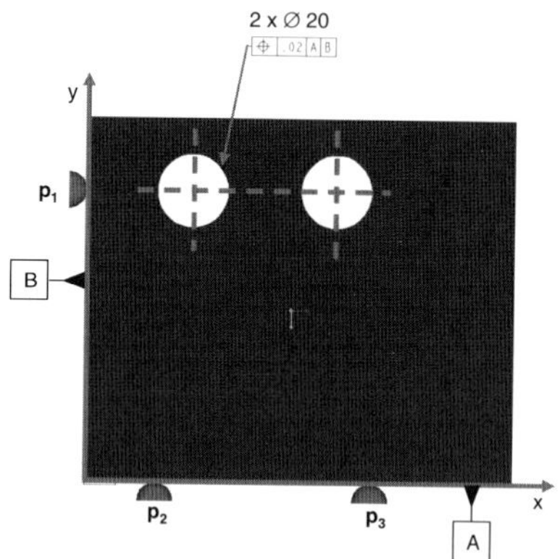


*Figure 1. 2-hole plate*

The probability the i-th hole of the pattern falls inside the location tolerance ($tz_i$) is calculated by solving the following integral:

$$\iint_{tz_i} f_{\mathbf{c'_i}}(x',y') \cdot dx' \cdot dy' \tag{6}$$

while the probability the pattern of holes falls inside the location tolerance is given by the following product:

$$\prod_{i=1}^{n}\left( \iint_{tz_i} f_{c'_i}\left(x',y'\right)\cdot dx'\cdot dy' \right) \tag{7}$$

with n equals to the number of holes constituting the pattern. The optimal locators positioning problem consists in defining the locators' position for which the probability of success, i.e. the probability that the actual position of the holes pattern due to fixturing error is inside the tolerance zone centred around the nominal position of each hole, is maximum. Therefore, the problem consists in finding the values of the $x_2$, $x_3$ and $y_1$ coordinates that guarantee equation (7) achieves the maximum value. Additional contraints to the optimization problem are linked to the coordinates of the locator that should be positive and smaller than the length of the plate sides.

The optimal part locators positioning may be mathematically represented as:

$$\max_{x_2,x_3,y_1} \prod_{i=1}^{n}\left( \iint_{tz_i} f_{c'_i}\cdot dx'\cdot dy' \right) \tag{8}$$

subject to:

$$0 \leq x_2 < x_3 \leq x_{max} \tag{9}$$

$$0 \leq y_1 \leq y_{max} \tag{10}$$

where $x_{max}$ and $y_{max}$ represent the length of the plate sides. Conditions (9) and (10) refer to the constraint due to the length of the plate sides to locators' positions. Condition (9) simplifies the mathematical problem by considering only one of the two symmetric positions of $\mathbf{p_2}$ and $\mathbf{p_3}$ locators on the primary datum.

## 3. APPLICATION EXAMPLES

The proposed approach was applied to a plate characterised by different patterns of two or four holes. It was assumed that the coordinates of the locators along the $X$ and $Y$ axes of the MRF are distributed according to a Gaussian probability density function, with mean values equal to the nominal positions and standard deviations equal to 0.01 mm. The tolerance zone of the holes pattern was squared with 0.02 mm side. The position of the locators that minimises eq. (8), under the constraints (9)-(10) was identified for the different cases. Starting from the optimal solutions we have defined and applied two simple locators positioning rules that we call "barycentre method" and " maximum distance method".

### 3.1 Two holes pattern: case 1

A first plate (120 mm x 100 mm) was taken into account with 2-holes pattern, as shown in Figure 1. Nine 2-holes patterns were considered, for each of them the optimal position of the locators was determined, as shown in Table 1. The optimal position of the locator $p_1$ on the secondary datum is always coincident with the Y-coordinate of the pattern barycentre. The optimal position of the locators $p_2$ and $p_3$ is a quasi-symmetric configuration with respect to the X-coordinate of the pattern barycentre.

*Table 1. Optimal position of 2-1 pins of the considered 2-hole plates*

| Case | hole 1 | hole 2 | Proposed approach | | | | Barycentre method | | | | Maximum distance | | | |
|---|---|---|---|---|---|---|---|---|---|---|---|---|---|---|
| | | | $p_1$ | $p_2$ | $p_3$ | Probability | $p_1$ | $p_2$ | $p_3$ | Probability | $p_1$ | $p_2$ | $p_3$ | Probability |
| 1.1 | 40, 70 | 80, 70 | 0, 70 | 0, 0 | 120, 0 | 0.3136 | 0, 70 | 0, 0 | 120, 0 | 0.3136 | 0, 70 | 0, 0 | 120, 0 | 0.314 |
| 1.2 | 20, 80 | 60, 80 | 0, 80 | 0, 0 | 102, 0 | 0.3013 | 0, 80 | 0, 0 | 80, 0 | 0.2939 | 0, 70 | 0, 0 | 120, 0 | 0.299 |
| 1.3 | 60, 20 | 100, 20 | 0, 20 | 18, 0 | 120, 0 | 0.3013 | 0, 20 | 40, 0 | 120, 0 | 0.2939 | 0, 70 | 0, 0 | 120, 0 | 0.299 |
| 1.4 | 20, 20 | 60, 20 | 0, 20 | 0, 0 | 102, 0 | 0.3013 | 0, 20 | 0, 0 | 80, 0 | 0.2939 | 0, 70 | 0, 0 | 120, 0 | 0.299 |
| 1.5 | 60, 80 | 100, 80 | 0, 80 | 18, 0 | 120, 0 | 0.3013 | 0, 80 | 40, 0 | 120, 0 | 0.2939 | 0, 70 | 0, 0 | 120, 0 | 0.299 |
| 1.6 | 20, 60 | 20, 80 | 0, 70 | 0, 0 | 47, 0 | 0.3078 | 0, 70 | 0, 0 | 40, 0 | 0.3040 | 0, 70 | 0, 0 | 120, 0 | 0.268 |
| 1.7 | 80, 60 | 80, 80 | 0, 70 | 36, 0 | 120, 0 | 0.3241 | 0, 70 | 40, 0 | 120, 0 | 0.3238 | 0, 70 | 0, 0 | 120, 0 | 0.311 |
| 1.8 | 20, 20 | 20, 40 | 0, 30 | 0, 0 | 47, 0 | 0.3078 | 0, 30 | 0, 0 | 40, 0 | 0.3040 | 0, 70 | 0, 0 | 120, 0 | 0.268 |
| 1.9 | 80, 20 | 80, 40 | 0, 30 | 36, 0 | 120, 0 | 0.3241 | 0, 30 | 40, 0 | 120, 0 | 0.3238 | 0, 70 | 0, 0 | 120, 0 | 0.311 |
| | | | | | mean | 0.3092 | | | mean | 0.3050 | | | mean | 0.2961 |
| | | | | | std. deviation | 0.0100 | | | std. deviation | 0.0130 | | | std. deviation | 0.0171 |

In a previous work we demonstrated that, to minimize the position error of a drilled hole resulting from inaccuracies on locators positions, the 2-locators on the primary datum should be positioned symmetric with respect to the X-coordinate of the hole nominal position, while the locator on the secondary datum should be positioned along the Y-coordinate of the hole nominal position [Moroni et al., 2003]. Farther and farther from the nominal centre of the hole the 2-locators on the primary datum are, preserving the symmetry, more and more stable is the optimal solution, since smaller is the gradient of the probability surface to maximise.

The optimal positions of the locators obtained for a plate with a single hole or with 2-holes pattern carried us to consider the positioning of the 2-locators on the primary datum symmetrically with respect to the X-coordinate of the pattern barycentre and as far as possible, while placing the locator on the secondary datum coincident with the Y-coordinate of the pattern barycentre; we call this rule barycentre method. In this case the average value of probability the pattern of holes falls inside the tolerance zone reduces of about 0.42%, while the standard deviation increases of 0.31% with respect to the optimal solution (see Table 1).

A further rule taken by common sense involves to position the 2-locators on the primary datum as far as possible and, therefore, on the vertices of the primary datum; we call this rule maximum distance method. In this case the average value of probability reduces of about 1.31%, while the standard deviation increases of 0.71% with respect to the optimal solution. Moreover, we observed that the distance between the centres of the 2 holes along the $X$ axis is at most equal to half of the distance between the 2 locators on the primary datum placed according to the barycentre method for all considered cases.

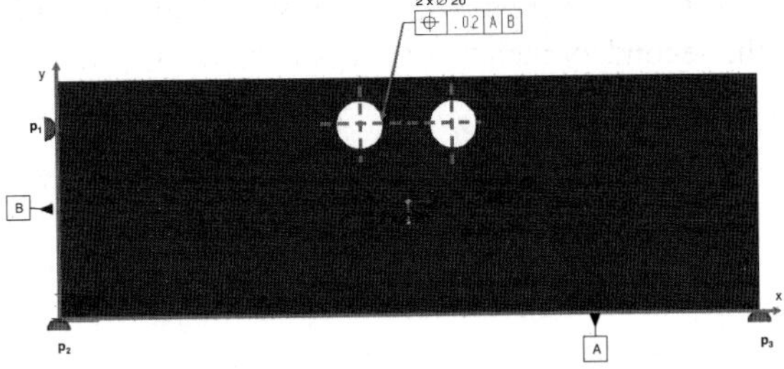


*Figure 2. Second case study*

### 3.2 Two holes pattern: case 2

A second plate (300 mm x 100 mm) was considered with a 2-holes pattern located symmetrically as regards to the median plane along the $X$ axis, as shown in Figure 2. The optimal positions of the $p_2$ and $p_3$ locators correspond to the vertices of the plate along the $X$ axis. The 2-holes pattern was moved along the $X$ axis by keeping unchanged the distance between the centres of the two holes to 40 mm. Half plate was considered to move the pattern, since the plate is symmetric as regards to the median plane of the $X$ axis. We obtained 12 different cases (see Table 2).

*Table 2. Twelve cases obtained by moving the 2-holes pattern along the X axis*

| | | | Proposed approach | | | | Barycentre method | | | | Maximum distance method | | | |
|---|---|---|---|---|---|---|---|---|---|---|---|---|---|---|
| Case | hole 1 | hole 2 | $p_1$ | $p_2$ | $p_3$ | Probability | $p_1$ | $p_2$ | $p_3$ | Probability | $p_1$ | $p_2$ | $p_3$ | Probability |
| 2.1 | 10, 80 | 50, 80 | 0, 80 | 0, 0 | 89, 0 | 0.288 | 0, 80 | 0, 0 | 60, 0 | 0.270 | 0, 80 | 0, 0 | 120, 0 | 0.248 |
| 2.2 | 20, 80 | 60, 80 | 0, 80 | 0, 0 | 102, 0 | 0.301 | 0, 80 | 0, 0 | 80, 0 | 0.294 | 0, 80 | 0, 0 | 120, 0 | 0.258 |
| 2.3 | 30, 80 | 70, 80 | 0, 80 | 0, 0 | 117, 0 | 0.310 | 0, 80 | 0, 0 | 100, 0 | 0.306 | 0, 80 | 0, 0 | 120, 0 | 0.268 |
| 2.4 | 40, 80 | 80, 80 | 0, 80 | 0, 0 | 134, 0 | 0.315 | 0, 80 | 0, 0 | 120, 0 | 0.314 | 0, 80 | 0, 0 | 120, 0 | 0.278 |
| 2.5 | 50, 80 | 90, 80 | 0, 80 | 0, 0 | 152, 0 | 0.319 | 0, 80 | 0, 0 | 140, 0 | 0.318 | 0, 80 | 0, 0 | 120, 0 | 0.287 |
| 2.6 | 60, 80 | 100, 80 | 0, 80 | 0, 0 | 170, 0 | 0.322 | 0, 80 | 0, 0 | 160, 0 | 0.321 | 0, 80 | 0, 0 | 120, 0 | 0.296 |
| 2.7 | 70, 80 | 110, 80 | 0, 80 | 0, 0 | 189, 0 | 0.323 | 0, 80 | 0, 0 | 180, 0 | 0.323 | 0, 80 | 0, 0 | 120, 0 | 0.304 |
| 2.8 | 80, 80 | 120, 80 | 0, 80 | 0, 0 | 208, 0 | 0.325 | 0, 80 | 0, 0 | 200, 0 | 0.325 | 0, 80 | 0, 0 | 120, 0 | 0.311 |
| 2.9 | 90, 80 | 130, 80 | 0, 80 | 0, 0 | 227, 0 | 0.326 | 0, 80 | 0, 0 | 220, 0 | 0.326 | 0, 80 | 0, 0 | 120, 0 | 0.317 |
| 2.10 | 100, 80 | 140, 80 | 0, 80 | 0, 0 | 247, 0 | 0.327 | 0, 80 | 0, 0 | 240, 0 | 0.326 | 0, 80 | 0, 0 | 120, 0 | 0.322 |
| 2.11 | 110, 80 | 150, 80 | 0, 80 | 0, 0 | 266, 0 | 0.327 | 0, 80 | 0, 0 | 260, 0 | 0.327 | 0, 80 | 0, 0 | 120, 0 | 0.325 |
| 2.12 | 120, 80 | 160, 80 | 0, 80 | 0, 0 | 286, 0 | 0.328 | 0, 80 | 0, 0 | 280, 0 | 0.328 | 0, 80 | 0, 0 | 120, 0 | 0.327 |
| 2.13 | 130, 80 | 170, 80 | 0, 80 | 0, 0 | 300, 0 | 0.328 | 0, 80 | 0, 0 | 300, 0 | 0.328 | 0, 80 | 0, 0 | 120, 0 | 0.328 |
| | | | | | mean | 0.3184 | | | mean | 0.3158 | | | mean | 0.2978 |
| | | | | | std. deviation | 0.0121 | | | std. deviation | 0.0171 | | | std. deviation | 0.0276 |

Table 2 presents the optimal position of the $p_2$ and $p_3$ locators along the $X$ axis given by the analytic approach. A locator is always placed on the origin of the $X$ axis, while the other moves away from the origin with the increase of the distance of the barycentre of the 2-holes from the origin of the plate. If we use the barycentre method, we obtain a reduction of the average value of probability, with respect to the optimal solution, of 0.26% and an increase of 0.52% of standard deviation (see Table 2). If we use the maximum distance method, we obtain a reduction of the average value of probability of 2.06% and an increase of standard deviation of 1.68%.

### 3.3 Two holes pattern: case 3

We considered the plate shown in Figure 2 and kept unchanged the 2-holes pattern barycentre on the $X$ axis (x = 210 mm), but we changed the holes distance. We obtained 16 different configurations, that are shown in Table 3 together with the optimal position of the $\mathbf{p_2}$ and $\mathbf{p_3}$ along the $X$ axis. A locator is always placed on the end of the plate along the $X$ axis, while the other moves near the origin of the plate with the increase of the distance between the 2 holes. If we use the barycentre method, we obtain a reduction of the average value of probability of about 0.19% and an increase of the standard deviation of about 0.26%, when the distance between the holes is lower than half of the distance between 2 locators on the X axis, as shown in Table 3. In this case if we apply the maximum distance method, we have a reduction of the probability of about 1.70% and an increase of the standard deviation of about 0.67%.

*Table 3. Sixteen cases obtained by changing the distance between the 2 holes*

| Case | hole 1 | hole 2 | Proposed approach | | | | Barycentre method | | | | Maximum distance method | | | |
|---|---|---|---|---|---|---|---|---|---|---|---|---|---|---|
| | | | $p_1$ | $p_2$ | $p_3$ | Probability | $p_1$ | $p_2$ | $p_3$ | Probability | $p_1$ | $p_2$ | $p_3$ | Probability |
| 3.1 | 209, 80 | 211, 80 | 0, 80 | 120, 0 | 300, 0 | 0.331 | 0, 80 | 120, 0 | 300, 0 | 0.331 | 0, 80 | 0, 0 | 300, 0 | 0.306 |
| 3.2 | 205, 80 | 215, 80 | 0, 80 | 119, 0 | 300, 0 | 0.330 | 0, 80 | 120, 0 | 300, 0 | 0.330 | 0, 80 | 0, 0 | 300, 0 | 0.306 |
| 3.3 | 200, 80 | 220, 80 | 0, 80 | 118, 0 | 300, 0 | 0.329 | 0, 80 | 120, 0 | 300, 0 | 0.329 | 0, 80 | 0, 0 | 300, 0 | 0.306 |
| 3.4 | 195, 80 | 225, 80 | 0, 80 | 115, 0 | 300, 0 | 0.327 | 0, 80 | 120, 0 | 300, 0 | 0.326 | 0, 80 | 0, 0 | 300, 0 | 0.305 |
| 3.5 | 190, 80 | 230, 80 | 0, 80 | 111, 0 | 300, 0 | 0.323 | 0, 80 | 120, 0 | 300, 0 | 0.323 | 0, 80 | 0, 0 | 300, 0 | 0.304 |
| 3.6 | 185, 80 | 235, 80 | 0, 80 | 106, 0 | 300, 0 | 0.320 | 0, 80 | 120, 0 | 300, 0 | 0.319 | 0, 80 | 0, 0 | 300, 0 | 0.303 |
| 3.7 | 180, 80 | 240, 80 | 0, 80 | 99, 0 | 300, 0 | 0.315 | 0, 80 | 120, 0 | 300, 0 | 0.314 | 0, 80 | 0, 0 | 300, 0 | 0.301 |
| 3.8 | 175, 80 | 245, 80 | 0, 80 | 91, 0 | 300, 0 | 0.311 | 0, 80 | 120, 0 | 300, 0 | 0.308 | 0, 80 | 0, 0 | 300, 0 | 0.300 |
| 3.9 | 170, 80 | 250, 80 | 0, 80 | 82, 0 | 300, 0 | 0.306 | 0, 80 | 120, 0 | 300, 0 | 0.301 | 0, 80 | 0, 0 | 300, 0 | 0.297 |
| 3.10 | 165, 80 | 255, 80 | 0, 80 | 72, 0 | 300, 0 | 0.301 | 0, 80 | 120, 0 | 300, 0 | 0.294 | 0, 80 | 0, 0 | 300, 0 | 0.295 |
| | | | | | mean | 0.3194 | | | mean | 0.3175 | | | mean | 0.3024 |
| | | | | | std. deviation | 0.0106 | | | std. deviation | 0.0130 | | | std. deviation | 0.0039 |
| 3.11 | 160, 80 | 260, 80 | 0, 80 | 60, 0 | 300, 0 | 0.297 | 0, 80 | 120, 0 | 300, 0 | 0.286 | 0, 80 | 0, 0 | 300, 0 | 0.306 |
| 3.12 | 155, 80 | 265, 80 | 0, 80 | 47, 0 | 300, 0 | 0.292 | 0, 80 | 120, 0 | 300, 0 | 0.278 | 0, 80 | 0, 0 | 300, 0 | 0.293 |
| 3.13 | 150, 80 | 270, 80 | 0, 80 | 34, 0 | 300, 0 | 0.288 | 0, 80 | 120, 0 | 300, 0 | 0.270 | 0, 80 | 0, 0 | 300, 0 | 0.290 |
| 3.14 | 145, 80 | 275, 80 | 0, 80 | 20, 0 | 300, 0 | 0.284 | 0, 80 | 120, 0 | 300, 0 | 0.261 | 0, 80 | 0, 0 | 300, 0 | 0.287 |
| 3.15 | 140, 80 | 280, 80 | 0, 80 | 5, 0 | 300, 0 | 0.280 | 0, 80 | 120, 0 | 300, 0 | 0.252 | 0, 80 | 0, 0 | 300, 0 | 0.284 |
| 3.16 | 135, 80 | 285, 80 | 0, 80 | 0, 0 | 300, 0 | 0.277 | 0, 80 | 120, 0 | 300, 0 | 0.243 | 0, 80 | 0, 0 | 300, 0 | 0.280 |
| | | | | | mean | 0.2864 | | | mean | 0.2651 | | | mean | 0.2851 |
| | | | | | std. deviation | 0.0074 | | | std. deviation | 0.0160 | | | std. deviation | 0.0059 |

When the distance between the holes is higher than half of the distance between 2 locators on the X axis, if we use the barycentre method the average value of probability decreases of 2.13% and the standard deviation increases of 0.86%. In this case if we apply the maximum distance method, we obtain a reduction of the probability of about 0.13, while the standard deviation increases of 0.16%.
barycentre.

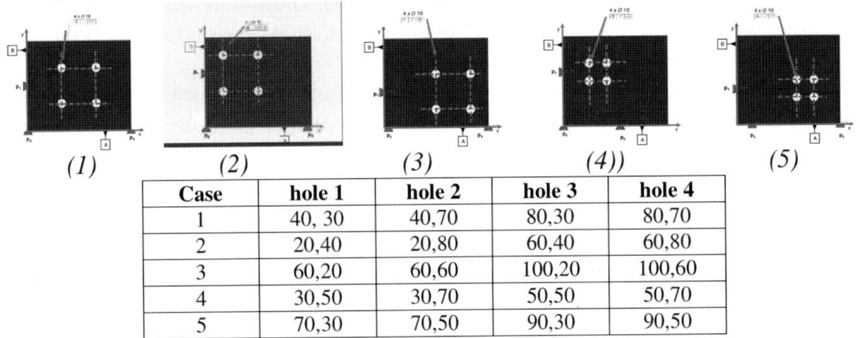

*(1)*          *(2)*          *(3)*          *(4))*          *(5)*

| Case | hole 1 | hole 2 | hole 3 | hole 4 |
|------|--------|--------|--------|--------|
| 1 | 40, 30 | 40,70 | 80,30 | 80,70 |
| 2 | 20,40 | 20,80 | 60,40 | 60,80 |
| 3 | 60,20 | 60,60 | 100,20 | 100,60 |
| 4 | 30,50 | 30,70 | 50,50 | 50,70 |
| 5 | 70,30 | 70,50 | 90,30 | 90,50 |

*Figure 3. 4-hole plates*

### 3.4 Four-holes pattern: case 4

Three 4-holes pattern were considered, as shown in Figure 3. The optimal position of the locators is obtained by following the same considerations found for 2-holes patterns. The position of the locator $p_1$ on the secondary datum is coincident with the Y-coordinate of the pattern barycentre. The locators $p_2$ and $p_3$ should be located in a quasi-symmetric configuration with respect to the X-coordinate of the pattern

If we apply the barycentre method, we obtain that the probability the pattern of holes falls inside the tolerance zone reduces of about 0.9% when the distance between the holes is lower than half of the distance between 2 locators on the $X$ axis and of about 0.09% when the distance is higher, as shown in Table 4. We obtain a reduction of 0.008% in the first case and of 0.56% in the second case, if we apply the maximum distance method (see Table 4).

*Table 4. Optimal position of 2-1 pins of the considered 4-hole plates*

| Case | Proposed approach | | | | Barycentre method | | | | Maximum distance | | | |
|------|------|------|------|------------|------|------|------|------------|------|------|------|------------|
| | $p_1$ | $p_2$ | $p_3$ | Probability | $p_1$ | $p_2$ | $p_3$ | Probability | $p_1$ | $p_2$ | $p_3$ | Probability |
| 1 | 0, 50 | 0, 0 | 120, 0 | 0.0914 | 0, 50 | 0, 0 | 120, 0 | 0.0914 | 0, 50 | 0, 0 | 120, 0 | 0.0914 |
| 2 | 0, 60 | 0, 0 | 115, 0 | 0.0833 | 0, 60 | 0, 0 | 80, 0 | 0.0742 | 0, 50 | 0, 0 | 120, 0 | 0.0832 |
| 3 | 0, 40 | 5, 0 | 120, 0 | 0.0833 | 0, 40 | 40, 0 | 120, 0 | 0.0742 | 0, 50 | 0, 0 | 120, 0 | 0.0832 |
| 4 | 0, 60 | 0, 0 | 89, 0 | 0.0998 | 0, 60 | 0, 0 | 80, 0 | 0.0988 | 0, 50 | 0, 0 | 120, 0 | 0.0942 |
| 5 | 0, 40 | 31, 0 | 120, 0 | 0.0998 | 0, 40 | 40, 0 | 120, 0 | 0.0988 | 0, 50 | 0, 0 | 120, 0 | 0.0942 |
| | | | mean | 0.09152 | | | mean | 0.08748 | | | mean | 0.08924 |
| | | | std. deviation | 0.00825 | | | std. deviation | 0.01295 | | | std. deviation | 0.00563 |

## 4. CONCLUSION

The present minimization of the position error of holes pattern resulting from inaccuracies on locators positions has shown that the optimal locators' positions involve the position of the locator on the secondary datum on the Y-coordinate of the pattern barycentre. The two locators on the primary datum should be located in a quasi-symmetric configuration with respect to the X-coordinate of the pattern barycentre, but the solution changes with the considered situation.

However, it is possible to adopt simple locator positiong rules, such as the barycentre method or the maximum distance method. The first, consisting in placing the 2-locators on the primary datum symmetrically with respect to the X-coordinate of the pattern barycentre and as far as possible, is suitable when the distance between the holes is lower than half of the distance between the 2-locators on the primary datum. The second, consisting in placing the 2-locators on the primary datum on the vertices of the plate along the $X$ axis, is suitable when the distance of the holes is higher than half of the distance between the 2-locators on the primary datum placed according to the barycentre method.

By combining these two rules we move from the optimal solution, but the differences are very small and the method is very simple.

## ACNOWLEDGEMENTS

The work has funded partially by the Italian M.I.U.R. (Italian Ministry of University and Research).

## REFERENCES

**[Armillotta et al., 1996]** Armillotta, A.; Moroni, G.; Negrini, L.; Semeraro, Q.; "Analysis of deterministic positioning on workholding fixtures", In: *Proceedings of International Conference on Flexible Automation and Intelligent Manufacturing*, pp. 274-284; Atlanta 1996.

**[Armillotta et al., 1999]** Armillotta, A.; Bigioggero, G.F.; Moroni, G.; Negrini, L.; Semeraro, Q.; "Tolerance control in workpiece fixturing", In: *Proceedings of the ASME 4th Design for Manufacturing Conference*; Las Vegas, 1999.

**[Armillotta et al., 2003]** Armillotta A.; Carrino L.; Moroni G.; Polini W.; Semeraro Q.; "An analytical approach to machining deviation due to fixturing", In: *Geometric Product Specification and Verification: Integration of Functionality*, Eds. Bourdet P. and Mathieu L., Kluwer Academic Publishers, pp. 175-184; 2003.

**[Bourdet et al., 1974]** Bourdet, P.; Clement, A.; "Optimalisation des Montages d'Usinage"; In: *L'Ingenieur et le Techniciien de l'Enseignement Technique*, pp. 8-74; 1974.

**[Bourdet et al., 1988]** Bourdet, P.; Clement, A.; "A study of optimal-criteria identification based on the small displacement screw model"; In: *Annals of the CIRP*, Vol. 37/1, pp. 503-506; 1988.

[Cai et al. 1997] Cai, W.; Jack Hu, S.; Yuan, J.X.; "A variational method of robust fixture configuration design for 3-D workpieces"; In: *Journal of Manufacturing Science and Engineering*, Vol. 119, pp. 593-602; 1997.

[Choudhuri et al., 1999] Choudhuri, S.A.; De Meter, E.C.; "Tolerance analysis of machining fixture locators"; In: *Journal of Manufacturing Science and Engineering. Transactions of ASME*, Vol. 121, pp. 273-281; 1999.

[Chu et al., 1998] Chu, Y.X.; Gou, J.B.; Wu, H.; Li, Z.X.; "Localization algorithms: performance evaluation and reliability analysis"; In: *Proceedings of the IEEE International Conference on Robotics & Automation*, pp. 3652-3657; Leuven, 1998.

[Desrochers et al., 1997] Desrochers, A.; Delbart, O.; "Determination of part positioning uncertainty within mechanical assembly using screw parameters"; In: *Proceedings of $5^{th}$ CIRP International Seminar on Computer Aided Tolerancing*, pp. 185-196; 1995; ISBN 0-412-72740-4.

[Giusarma et al., 2004] Giusarma S.; Moroni G.; Polini W.; "Inaccuracy prediction due to six-point locating principle", In: *Proceedings of $4^{th}$ CIRP International Seminar on Intelligent Computation in Manufacturing Engineering*, ICME '04, pp. 213-218; Sorrento, Italy, 2004.

[Moroni et al., 2003] Moroni G.; Polini W.; Rasella M.; "Minimal hole-drilling deviation due to six-point location principle", In: *Proceedings of the 8th CIRP International Seminar on Computer Aided Tolerancing*, pp. 321-330; Charlotte, North Carolina, USA, 2003.

[Ohwovoriole et al., 1981] Ohwovoriole, M.S.; Roth, B.; "An extension of screw theory", Transaction of ASME: Journal of Mechanical Design, vol. 103, pp. 725-735;1981.

[Weill et al., 1991] Weill, R.; Darel, I.; Laloum, M.; "The influence of fixture positioning errors on the geometric accuracy of mechanical parts"; In: *Proceedings of CIRP Conference on PE&MS*, pp. 215-225; 1991

# Tolerance Assignment Using Genetic Algorithm for Production Planning

**H. Song, Y. D. Yang, Y. Zhou, Y. K. Rong**
*100 Institute Rd, Worcester Polytechnic Institute, Worcester, MA 01609*
*hsong@wpi.edu*

**Abstract:** In production planning, series of operations are designed to control the geometry, size, and location of workpiece features. The tolerance assignment is to determine a set of manufacturing tolerances in each operation that can ensure the final product meet the design tolerance requirements. In this paper, the tolerance assignment for production planning with multi-setups is investigated. An optimal tolerance assignment strategy is developed and implemented. The optimization criteria are to minimize the manufacturing cost and cycle time while maintaining product quality. The cost model considers effective factors at machine level, part level, and feature level. Optimization of tolerance assignment plan with genetic algorithm is formulated. The Monte Carlo simulation based tolerance stack up analysis is employed to determine the satisfaction of design tolerance requirements. A case study with real product and process data shows that this approach is reliable and efficient for tolerance assignment. The developed system can be easily integrated into production planning.
**Keywords**: tolerance assignment, genetic algorithm, production planning

## 1. INTRODUCTION

Tolerancing is one of the most important engineering processes in a product development cycle. The design tolerances of workpiece, however, normally do not provide enough information for determination of process tolerances and production plan. The process planner usually has to go through time-consuming and costly try-and-error process based on experience, best guess, or available production plan information on similar product.

As the global competition driving industrial companies to pursue higher quality and lower cost, it is desired that even products developed in small volume can be manufactured with economic mass product mode, i.e., mass customization. This requires optimal production plan to be made rapidly according to the available manufacturing resources. In order to implement mass customization, industries demand more systematic and cost effective approach to assign process tolerances with known design tolerances and manufacturing resources including machines, tools, fixturing plan, and production operations. In this paper, a complete methodology is developed and implemented to assign process tolerances and locator tolerances to satisfy design tolerance requirements and optimize the assignment with genetic algorithm to reduce

213

*J.K. Davidson (ed.), Models for Computer Aided Tolerancing in Design and Manufacturing, 213–224.*
© 2007 *Springer.*

cost and/or cycle time without sacrificing product quality. This work is part of a comprehensive computer aided tolerance analysis (CATA) system developed by CAM Lab, WPI (Monte Carlo simulation based tolerance stack up analysis and quality control plan are discussed in separated papers).

## 2. LITERATURE REVIEW

Tolerance assignment is an important area where the product designer and process planner often need to work closely together. Despite the intensive studies in tolerancing, this area has been neglected by most researchers. This section reviews papers on some closely related issues, e.g. the tolerance synthesis/allocation, manufacturing cost models, and application of genetic algorithm in tolerancing.

### 2.1. Tolerance synthesis/allocation

Most of the established tolerance synthesis methods are focusing on assembly processes, allocating the assembly functional tolerance to the individual workpiece tolerance to ensure that all assembly requirements are met [Ngoi *et al.*, 1998]. No existing technique has been found by the authors that generates process and locator tolerance requirements for production plan.

A variety of techniques have been employed to allocate tolerance. Among them, integer programming for tolerance-cost optimization [Ostwald *et al.*, 1977; Sunn *et al.*, 1988], rule-based approach [Tang *et al.*, 1988; Kaushal *et al.*, 1992], feature-based approach [Kalajdzic *et al.*, 1992], knowledge-based approach [Manivannan *et al.*, 1989], and genetic algorithm [Ji *et al.*, 2000; Shan *et al.*, 2003], artificial intelligence [Lu *et al.*, 1989] has been used to optimize tolerance allocation.

In this paper, the process tolerance assignment is optimized with assist of genetic algorithm.

### 2.2. Manufacturing cost models

One of the ultimate goals of an enterprise is to make profit. Hence, every company has been struggling to reduce cost, which can be done more effectively at the design and planning stage rather than manufacturing stage. It has been shown that about 70% of production cost is determined at the early design phase [Ouyang *et al.*, 1997].

Manufacturing cost modeling at design stage has been investigated for many years and used as one of the major criteria, if not the only, for optimization of production planning. There are numerous facets in cost models. One way is to interpret the manufacturing cost as summation of processing cost, inspection cost, rework/scrap cost, and external failure cost [Mayer *et al.*, 2001]. The processing cost can then be decomposed into machine cost, tool cost, material cost, setup cost, overhead cost, energy cost, etc [Esawi *et al.*, 2003]. All terms can be further formulated if adequate information on process characteristics is known. This method gives detail analysis on each factor that contributes to final cost. However, each term normally involves assumption-orientated undetermined terms, empirical/semi-empirical formulation,

```
Feature 1: plane                    IT: 9  Ra: 125  face mill      D:1.2    L:3
PROCESS8: Tool axis direction:+Y;
Feature 4: plane                    IT: 9  Ra: 125  face mill      D:1.2    L:3

SETUP 4— locating surface: 6,3,4
PROCESS9: Tool axis direction:+X;
Feature 7,9: plane                  IT: 7  Ra: 63  slot mill       D:0.3    L:3
PROCESS10: Tool axis direction:+X;
Feature 8: plane                    IT: 7  Ra: 63  end mill        D:0.4    L:3
```

*Table III; Setup planning for the example workpiece.*

| Error source / Feature tolerance | | | Locator errors | Setup 1 | | Setup 2 | | | | Setup 3 | | Setup 4 | |
|---|---|---|---|---|---|---|---|---|---|---|---|---|---|
| | | | | P1 | P2 | P3 | P4 | P5 | P6 | P7 | P8 | P9 | P10 |
| T | Type | D | 0.08 | 0.04 | 0.03 | 0 | 0 | 0.17 | 0.16 | 0.01 | 0.04 | 0.04 | 0.04 |
| 8 | parallelism | 11 | | | | | | | | | | | |

*T: target feature; D: datum feature; P1-P10: process 1-process 10*

*Table IV; Sensitivity analysis results for selected toleranced feature.*

Guided by the simulation results and sensitivity analysis, the production plan has been revised as follows.

- Select fixture components with locator errors in the range of [-0.005mm, 0.005mm].
- Assign IT grade 6 to processes 5 and 6 in setup 2 and process 10 in setup 4.
- Assign IT grade 5 to process 2 in setup 1.

The simulation results show that the new plan is capable to produce finish products that satisfy the design tolerance requirements. This proves that the sensitivity analysis can assist tolerance assignment to generate a feasible assignment plan effectively.


## 5.  TOLERANCE ASSIGNMENT BASED ON GENETIC ALGORITHM

The sensitivity analysis based assignment can generate a feasible tolerance assignment plan. However, it may not be optimal since the cost is not considered. In this paper, genetic algorithm is adopted to for tolerance-cost optimization.

Suppose there are $n$ features, $q$ parts, and $r$ machines are involved in a production plan, the cost ($C$) is formulated as function of assigned tolerance IT grades ($IT_i$, $i = 1,2,...,n$) and other known constants such as parameters of features ($F_i$, $i = 1,2,...,n$), parts ($P_j$, $j = 1,2,...,q$), machines ($M_k$, $k = 1,2,...,r$), etc. The goal of tolerance assignment optimization is to minimize cost subject to process capability constraints and satisfaction of design requirements, i.e.

$$\text{Minimize } C = f\left(IT_i, F_i, P_j, M_k\right),\ i = 1,2,...,n;\ j = 1,2,...,q;\ k = 1,2,...,r$$

$$\text{Subject to : } \min\left[IT(F_i)\right] \le IT_i \le \max\left[IT(F_i)\right] \text{ and}$$

$$Tol_{SIM} \le Tol_{REQ} \text{ for all design tolerances}$$

where $IT(F_i)$ is feasible $IT$ for the $i$th feature; $Tol_{SIM}$ is the stack up simulation results; $Tol_{REQ}$ is the design tolerance requirement.

## 5.1. Cost model

The cost model in this study consists of machine level, part level, and feature level formulations. At the feature level, the cost depends on material machinability, feature type, size, and IT grade. With material, feature type and size as known factors retrieved from design information, the IT grade is the only variable at this level.

$$C(F_i) = \alpha \cdot \beta \cdot f_1(F_i) \cdot V(F_i) \cdot \exp(a \cdot IT_i) \qquad i = 1,2,...,n$$

where $C(F_i)$ is the manufacturing cost of the $i$th feature; $\alpha$ is the material machinability factor; $\beta$ is feature complexity factor; $f_1(F_i)$ is the cost factor associated with type of the $i$th feature; $V(F_i)$ is the volume of material to be removed in order to produce the $i$th feature; $b$ is a constant to be determined [Yang, 2005]. The cost factors for different feature types can be estimated according to previous manufacturing practice or existing cost data. Table V shows some cost factor examples. The feature complexity factor is introduced because the same type of features may result in different manufacturing cost due to different complexity. For example, a long, narrow hole is more costly compare with a short, broad hole even they have same volume of unwanted material and assigned IT grades. With this formulation, the manufacturing cost of any known single feature can be determined.

| Feature type | Cost factor | Feature type | Cost factor |
|---|---|---|---|
| Flat surface | 1 | External thread | 1.75 |
| hole | 1 | T slot | 2 |
| block slot | 1 | Internal spline | 2 |
| chamfer | 1 | Y slot | 2.25 |
| radial groove | 1.25 | External spline | 2.25 |
| Keyway | 1.5 | Internal thread | 2.25 |
| V slot | 1.5 | face groove | 2.5 |

*Table V; Manufacturing cost factors for different feature type.*

The part level cost formulation considers feature groups rather than individual. Generally, a part may go through multiple setups and a single setup may machine multiple parts. Hence, both inter-setup and intra-setup feature relationships affect the manufacturing cost. IT grades relationships are also considered in the cost model at part level. For example, it is cost effective to have different IT grades assigned to same type of feature produced by a single process. These relationships and their effects are evaluated by a relationship factor. Table VI shows some relationship factor examples used in this study. After applying relationship factors to feature costs, the part level cost are formulated.

$$C(P_j) = \Sigma\left[C(F_i) \cdot \Pi c_i\right] \qquad j = 1,2,...,q$$

where $C(P_j)$ is the manufacturing cost of the $j$th part; $C(P_j)$ is the manufacturing cost of the $i$th feature; $c_i$ represent all relationship factors applicable to the $i$th of the $j$th part.

At the machine level, the major considerations are machine type and conditions, represented by machine cost factor.

$$C(M_k) = \Sigma\left[f_2(M_k) \cdot C(F_{ik}) \cdot \Pi c_{ik}\right] \qquad\qquad k = 1,2,...,r$$

where $C(M_k)$ is the manufacturing cost at the $k$th machine/workstation; $f_2(M_k)$ is the cost factor of the $k$th machine; $C(F_{ik})$ is the cost to produce the $i$th feature on the $k$th machine; $c_{ik}$ is the relationship factor applicable to the $i$th on the $k$th machine. The overall cost is estimated by summation of costs at each machine. The focus of this modeling process is not to determine the numeric value of the cost but to provide a consistent measure to compare costs with difference tolerance assignment plan. Hence, the factors do not change with tolerance assignment have been simplified.

| Feature relationship | | | | Relationship factor |
|---|---|---|---|---|
| | Multiple parts, part number=n | | | fn |
| | Same feature | same orientation | Same IT | 0.65-0.75 |
| | | | Different IT | 0.7-0.8 |
| | | different orientation (90°) | Same IT | 0.75-0.85 |
| | | | Different IT | 0.80-0.90 |
| | | different orientation | Same IT | 0.95-1.05 |
| | | | Different IT | 1.0-1.1 |
| Intra-setup | Same feature type, different size | same orientation | Same IT | 0.75-0.85 |
| | | | Different IT | 0.8-0.9 |
| | | different orientation (90°) | Same IT | 0.85-0.95 |
| | | | Different IT | 0.90-1.00 |
| | | different orientation | Same IT | 1.05-1.15 |
| | | | Different IT | 1.1-1.2 |
| | Different feature type | same orientation | | 0.75-0.85 |
| | | different orientation | | 0.85-0.95 |
| Inter-setup | Same machine station | same feature type | | 0.85-0.95 |
| | | different feature type | | 0.95-1.05 |
| | Different machine station | | | 1.05-1.15 |

*Table VI; Complexity factors for feature geometric relationship.*

### 5.2. GA technique

Genetic algorithms have important characteristics, causing them to behave differently from traditional methods, and making them robust and computationally simple and powerful. There are five key issues affect the construction of a genetic algorithm: encoding, crossover, mutation, selection, and dealing with constraints.

The first step is to code the variable involved in the optimization, in this case, the IT grades. A chromosome vector, vi = [IT1, IT2,. . . , ITn] or a binary string can be used as codes. Then, crossover is allowed between IT grades or segments of binary string for same process in different genes at certain probability. Random mutation produces spontaneous random changes in various chromosomes by changing some binary bits from '0' to '1' or reversely. This allows the algorithm to overcome local maxima. Selection based on cost function should direct the search toward promising regions. The fitness for every chromosome is evaluated with the assumption that the lower the manufacturing cost, the higher the fitness. Penalty functions are applied to any infeasible assignment plans. Figure 2 depicts this entire process.

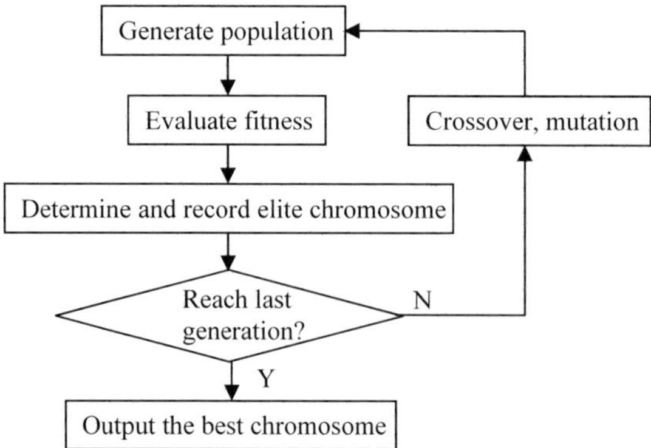


*Figure 2; Flowchart of Genetic Algorithm..*

### 5.3. Implementation and case study

Genetic algorithm parameters such as number of generations, number of populations in each generation, crossover rate, mutation rate, fitness functions, penalty functions have significant impact on it performance. Determination of those numbers could be a optimization problem. In this research, a set of parameter have been chosen as follows.

Generation =20; population =50; crossover rate=0.6; mutation rate=0.1; linear function for fitness evaluation; use maximum possible machining cost as penalty function, which is determined by assigning lowest IT grades to all features. Parameters for Monte Carlo simulation in stack up analysis module are the same as described in [Song *et al.*, 2005]. The same prismatic workpiece as in section 4 is employed. Table VII lists the optimal result along with the sensitivity analysis based assignment plan.

| Approach | Setup 1 | | Setup 2 | | | | Setup 3 | | Setup 4 | |
|---|---|---|---|---|---|---|---|---|---|---|
| | P1 | P2 | P3 | P4 | P5 | P6 | P7 | P8 | P9 | P10 |
| Sensitivity based | 9 | 5 | 9 | 9 | 6 | 6 | 9 | 9 | 7 | 6 |
| GA based | 9 | 6 | 9 | 8 | 7/8 | 6 | 9 | 9 | 8/9 | 5 |

*Table VII; Comparison of assignment results.*

It can be concluded that without performing sensitivity analysis and critical process identification, the genetic algorithm based approach can generate more cost-saving tolerance assignment plan within comparable computation time. Figure 3 shows the evolution of performance of the best gene in each generation. How to optimize the parameters of genetic algorithm and enhance its performance

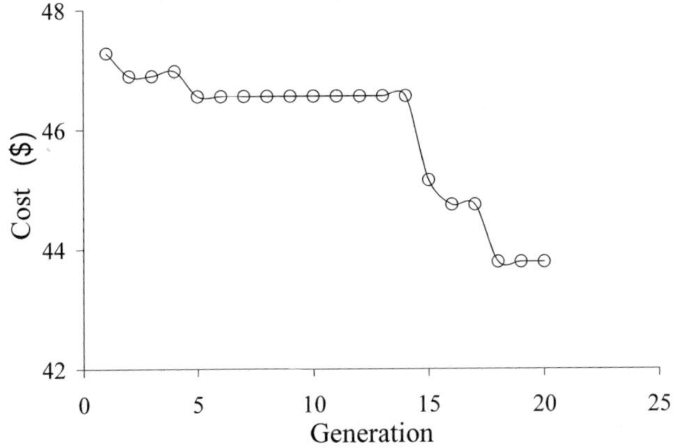

*Figure 3; Cost of the best gene improves with increase of generation.*

## 6.  SUMMARY

Tolerance assignment is an major component of tolerance analysis. It is one of the most important tasks when generating production plan from product design. This paper, both sensitivity analysis approach and genetic algorithm have been investigated to assign process tolerances upon given workpiece design tolerances. IT grade is introduced to conform to industrial standard. A three-level feature-based cost model has been developed to formulate the relationship between manufacturing cost and process tolerances. This study shows both sensitivity and GA approaches are capable of generate tolerance assignment plan when associated with well developed tolerance stack up analysis module. The sensitivity analysis provides more information on critical processes while GA method yields more optimal plan with better cost saving. How to optimize the parameters of genetic algorithm and enhance its performance need to be further studied. This study is also followed by quality control planning as the third part of the CATA system.

## REFERENCES

**[Esawi et al., 2003]** Esawi, A. M. K.; Ashby, M. F.; "Cost estimates to guide pre-selection of processes"; In: *Materials and Design*, pp. 605-616; 2003

**[Feng et al., 1996]** Fengi, J. C.-X.; Kusiak, A.; Huang, C.-C.; "Cost evaluation in design with form features"; In: *Computer-Aided Design*, pp. 879-885; 1996

**[Ji et al., 2000]** Ji, S.; Li, X.; Ma, Y.; Cai, H.; "Optimal Tolerance allocation based on fuzzy comprehensive evaluation and genetic algorithm"; In: *The International Journal of Advanced Manufacturing Technology*, pp. 461-468; 2000

**[Kalajdzic et al., 1992]** Kalajdzic, M.; Domazet, D. S.; Lu, S. C.-Y.; "Current design and process planning of rotational parts"; In: *Annals CIRP*, pp. 181-184; 1992

**[Kaushal et al., 1992]** Kaushal, P.; Raman, S.; Pulat, P. S.; "Computer-aided tolerance assignment procedure (CATAP) for design deminsioning"; In: *The International Journal of Production Research*, pp. 599-610; 1992

**[Lu et al., 1989]** Lu, S. C.-Y.; Wilhelm, R. G.; "Automating tolerance synthesis: a framework and tools"; In: *Journal of Manufacturing Systems*, pp. 279-296; 1989

**[Manivannan et al., 1989]** Manivannan, S.; Lehtihet, A.; Egbelu, P. J., "A knowledge based system for the specification of manufacturing tolerances"; In: *Journal of Manufacturing Systems*, pp. 153-160; 1989

**[Mayer et al., 2001]** Mayer, M.; Nusswald, M.; "Improving manufacturing costs and lead times with quality-oriented operating curves"; In: *Journal of Materials Processing Technology*, pp. 83-89; 2001

**[Ngoi et al., 1998]** Ngoi, B. K. A.; Ong, C.T.; "Product and process dimensioning and tolerancing techniques. A state-of-the-art review"; In: *The International Journal of Advanced Manufacturing Technology*, pp. 910-917; 1998

**[Ostwald et al., 1977]** Ostwald, P. F.; Huang, J.; "A method for optimal tolerance selection"; In: *Journal of Engineering for Industry*, pp. 558-564; 1977

**[Ouyang et al., 1997]** Ouyang, C.; Lin, T. S.; "Developing an integrated framework for feature-based early manufacturing cost estimation"; In: *The International Journal of Advanced Manufacturing Technology*, pp. 618-629; 1997

**[Shan et al., 2003]** Shan, A.; Roth, R. N.; "Genetic algorithms in statistical tolerancing "; In: *Mathematical and Computer Modeling*, 38, pp. 1427-1436; 2003

**[Shehab et al., 2001]** Shehab, E. M.; Abdalla, H. S.; "Manufacturing cost modeling for concurrent product development"; In: *Robotics and Computer Integrated Manufacturing*, pp. 341-353; 2001

**[Song et al., 2005]** Song, H.; Yang, D. Y.; Rong, K. Y.; "Monte Carlo Simulation-based Tolerance Stack-up Analysis for Production Planning."; In press

**[Sunn et al., 1988]** Sunn, H. K.; Knott, K.; "A pseudo-boolean approach to determining least cost tolerances"; In: *The International Journal of Production Research*, pp. 157-167; 1988

**[Tang et al., 1988]** Tang, X. Q.; Davies, B. J.; "Computer aided dimensional planning"; In: *The International Journal of Production Research*, pp. 283-297; 1988

**[Weustink et al., 2000]** Weustink, I. F.; Brinke, E. T.; Streppel, A. H.; Kals, H. J. J.; "A generic framework for cost estimation and cost control in product design"; In: *Journal of Materials Processing Technology*, pp. 141-148; 2000

**[Yang, 2005]** Yang, D. Y.; "Integrated quality control planning in CAMP system"; Ph.D. Dissertation, Worcester Polytechnic Institute, 2005

# Impact of Geometric Uncertainties Onto the Operating Performance of a Mechanical System

*J. M. Linares *, J. M. Sprauel *, S. Aranda*, P. Bourdet***

** EA(MS)$^2$, I.U.T., Avenue Gaston Berger - F13625 Aix en Provence cedex 1*
*** LURPA, ENS Cachan, Avenue du président Wilson - F94230 Cachan*
*linares@iut.univ-aix.fr*


**Abstract:** This paper shows that small variations of the dimensions and quality of manufactured parts may lead to great changes of the size of the contact zones and the bearing ratio curves thus increasing the stress applied to the material. In consequence, the Hertz's critical tension is modified and the ultimate stress or fatigue limit of the material may be exceeded. An industrial case is used to illustrate this assertion.


**Keywords:** Specification uncertainties, Inspection uncertainties, Operating performance, Stress, Hertz contact


## 1. INTRODUCTION

In the Eighties, the improvement of the quality of manufactured products has been the goal of a great number of research teams working on various complementary topics. The metallurgists have therefore developed enhanced methods and processes guaranteeing the properties of the manufactured alloys. This advanced metallurgy permitted controlling the behavior of the material under loading. New dimensioning methods have also been proposed by the mechanical research teams. A better characterization of the stress state of each part and of its damage mechanisms was thus obtained. The geometrical precision of the parts has been improved too, introducing numerical control technologies in the manufacturing processes. A substantial progress of the quality of the manufactured products has thus been achieved by the contribution of all these research works (figure 1).
Today, the quality of mechanical systems presents a weaker dispersion than in the Eighties. The mean quality of production also progressed in a score of years. The percentage of defects appearing during operation was strongly reduced. However each time the mechanism presents a failure, the malfunction appears at the first stage of work and leads to sudden fracture or blocking of the system. The failure modes are similar to the ones existing in electronic devices. On the previous years, the mechanisms were degraded in a gradual way whereas today the damage is quasi instantaneous. Breakdowns occur during the first cycles of loading or at low operation time. The failure mode of high quality mechanical systems is thus completely changed.

225

*J.K. Davidson (ed.), Models for Computer Aided Tolerancing in Design and Manufacturing,* 225–234.
© 2007 *Springer.*

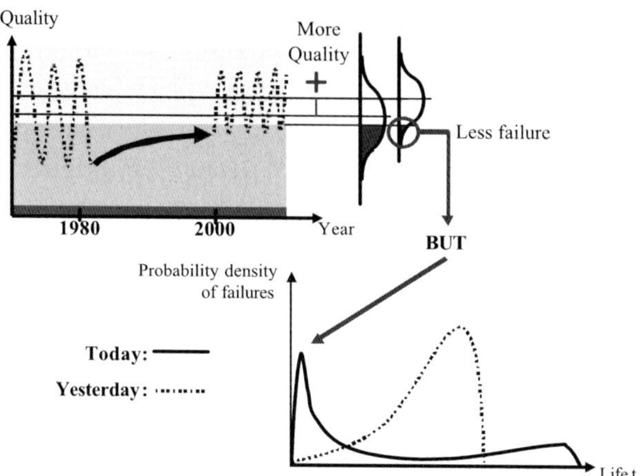


*Figure 1; Change of the failure mode*

The aim of this paper is to study the effect of geometrical errors on the operation performance of a mechanism. In a first part, the relation between the geometrical uncertainties and the new failure mode will be discussed. The second section of the document will then present the model used to describe the variations of the geometry of each part. The industrial example of a fuel injection pump will finally be used to illustrate our subject.

## 2. LINK BETWEEN GEOMETRICAL VARIATIONS AND MECHANICAL OPERATING PERFORMANCES

Power transmission in mechanical systems is made possible, thanks to two fundamental phenomena:

- macro geometrical compensations,
- micro geometrical accommodations.

### 2.1. Macro geometrical compensation

The improvement in manufacturing quality may lead to reduction of the clearances between the parts. This decrease modifies the macro geometrical compensation (MGC) of the geometrical defects in the mechanism. The MGC is characterized by the balancing of the dimensional and geometrical errors. It is made possible, but limited, by the clearances between the parts. In the mechanism of figure 2, the MGC is represented by the resulting geometrical defect $\Delta_{geo}$ between the two main components of the system: the cylinder liner and the piston. It depends on the geometrical defects ($\Delta_{geo\ i}$) of the four parts (i=1 to 4) and on the clearances existing at the different connections. In fact, in the functional cycle of the mechanical system, the geometrical errors are to be balanced by clearances; otherwise the mechanism will not run.

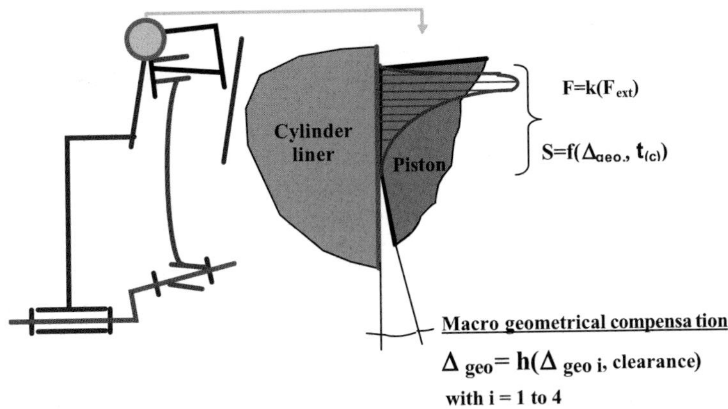


*Figure 2; Macro geometrical compensation*

The contact pressure (p) between two parts can be defined as the ratio of the transmitted force (F) and the surface of the contact zone (S).

$$p = F\!\!\Big/\!\!S$$

The in-depth stress profiles of the parts depend on this pressure. However, the size and shape of the contact area are influenced by the macro geometrical compensations ($\Delta_{geo}$). Consequently, small variations of the real geometry of the parts may induce great changes of the contact stresses. The maximum equivalent stress in the material may thus exceed the ultimate stress or fatigue limit of the material, leading to early fracture of the mechanism.

## 2.2. Micro geometrical accommodation

The micro geometrical accommodation of each surface under contact (MGA) can be characterized by the evolution of its bearing ratio curves ($t_{(c)}$ in figure 3) during operation. The bearing capacity depends on the quality of the surfaces (R roughness, W wave and form defect), the hardness of the materials and the running time ($N_{cycles}$). It evolves during the life of the part, due to local plastic deformations or abrasions. Such micro geometrical accommodations modify the distribution of the matter around the mean surface profile, increasing the number of points under contact (figure 3).

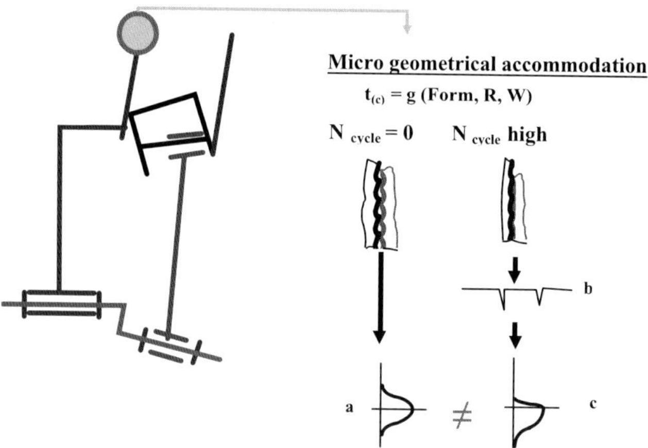


*Figure 3; Micro geometrical accommodation*

The ideal surface profile for a mechanical contact and the related bearing ratio curve are represented in part b and c of figure 3. Such surface consists of large plateaus which will carry load during the practical lifetime of the part and valleys which will retain lubricant. The topography of the surfaces is modified during operation. In fact, the peaks of the roughness profile will typically be worn off or down during the run-in period, while most valleys remain unchanged. The surface comes thus closer to the ideal profile. However, most actual manufacturing processes (hard turning, grinding...) lead to surfaces of very high quality. For such feature, the points around the mean profile are distributed according to a Gaussian probability density, containing only few peaks and valleys (curve a, of figure 3). Moreover, the use of very hard enhanced materials does no longer permit any plastic deformation or abrasion of the roughness peaks. The initial surface profiles remain thus practically unchanged during first operation of the parts. This phenomenon has been observed on most industrial parts we had to expertise (like axels of suspension damper of automotives, bearings, gears, ...). The MGA between the functional surfaces is no longer possible. Consequently, excess loads induced by macro-geometrical errors will no longer be lowered by local accommodations of the contact surfaces. Any small variation of the initial geometry of the part will thus result in an early failure of the mechanical system.

## 3. BASIS OF THE GEOMETRICAL UNCERTAINTY MODEL

Any variation of the geometry will influence the operating performance of the mechanical system. This variation can be defined by an uncertainty. It has been divided in three types by the ISO TC/213 standard [Srinivasan, 2001]: correlation uncertainties, specification uncertainties and measurement uncertainties. A model has already been proposed in previous papers to define the two last uncertainties [Sprauel *et al.*, 2001], [Linares *et al.*, 2003]. In this model, each surface is not only described by its mean parameters, but its variations are defined by a random vector $\hat{a}(a_i)$.

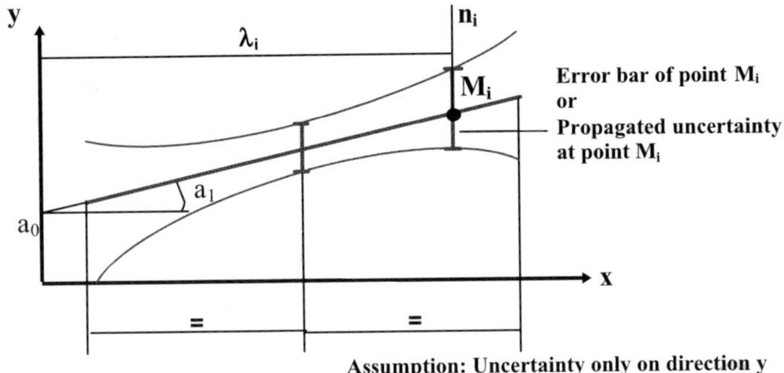


*Figure 4; Principle of geometric uncertainty model*

The points of the surface are assumed to be randomly distributed around the mean feature. As a first approximation, a Gaussian distribution was assumed. For such distribution the geometrical errors can be characterized by their variance $\sigma^2_r$. The variance-covariance matrix $Cov(\hat{a})$ of the parameters can also be calculated. Next, this data can be propagated to each point $M_i$ of any feature. This allows computing the variance in a given direction ($n_i$):

$$Cov(M_i) = J(M_i).Cov(\hat{a}).J^t(M_i)$$

$$var(M_i / n_i) = n_i.Cov(M_i).n_i{}^t$$

In the case of a line of a 2D plot (figure 4) the following relations are obtained:

$$Cov(a_0, a_1) = \begin{pmatrix} var\,a_0 & cov_{(a_0, a_1)} \\ cov_{(a_0, a_1)} & var\,a_1 \end{pmatrix}$$

$$var(M_i / n_i) = var(a_0) + 2.\lambda_i.cov(a_0, a_1) + \lambda_i^2.var(a_1)$$

$$u(Mi/n_i) = \sqrt{var(Mi/n_i)}$$

The error bar of the propagated uncertainty at point $M_i$ can finally be estimated in the direction $n_i$ for any level of risk $\alpha$: $U_{(Mi/ni)}=k_{(\alpha)}.u_{(Mi/ni)}$. The localization of the mean surface with this risk $\alpha$ is thus known. This determination has been extended to most classic surfaces. Figure 5 presents the uncertainty zones obtained for a cylinder and a sphere. In the next section, this uncertainty model is used to demonstrate the impact of the geometrical variations on the failure of a mechanism.

J. M. Linares *et al.*

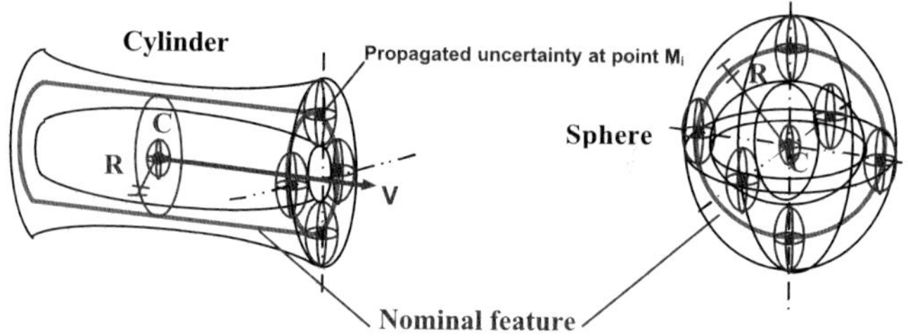

*Figure 5; Case of a cylinder and a sphere*

## 4. INDUSTRIAL CASE STUDY

The industrial example of a fuel pump will be used to illustrate the impact of geometrical variations on the operation performance of mechanical systems. In the last years, the injection pressure in diesel engines has greatly increased. Consequently, the internal mechanical loads have become stronger and failures of several injection pumps appeared.

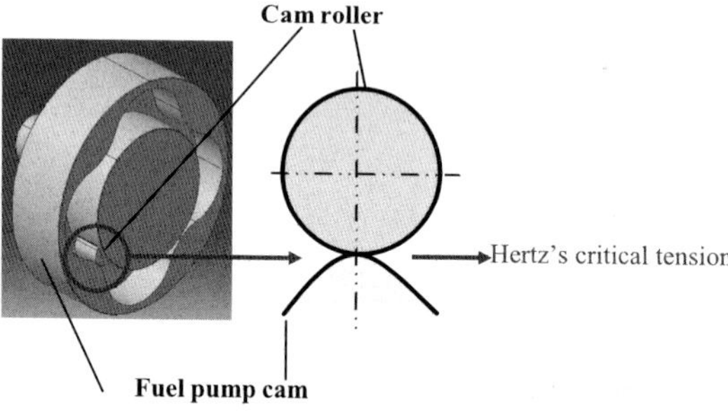


*Figure 6; Industrial case*

These failures are quasi instantaneous. They occur during the first running cycles or at low operation time. The most critical point of the pump is located in the contact zone between the cam and the cam roller. The aim of this section is to demonstrate the impact of the geometrical uncertainties on the stresses applied to the system.

### 4.1. Adjustment of the contact pressure

The maximum normal load applied to the cam roller is about 10000 N. A perfect cylindrical geometry would lead to a distribution of the contact pressure with strong edge effects (figure 7). To avoid this impediment, a curvature C of 5 µm has been imposed in the design of the cam roller. This allows adjusting the shape and size of the contact zone in order to obtain an elliptic contact area which leads to an ellipsoidal distribution of pressure.

The design of the cam roller is presented in the right hand side of figure 8. It has the external shape of a torus. According to the general description of such surface (left hand side of figure 8), it is defined by the direction vector of its axis, and two radii (R and r).

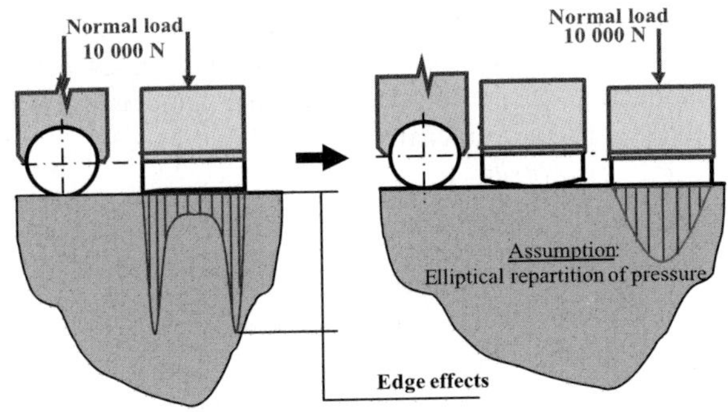



*Figure 7; Adjustment of the contact pressure*

In local coordinates its equations are the following:

$$\begin{cases} X = (R - r.\cos\theta).\cos\varphi \\ Y = (R - r.\cos\theta).\sin\varphi \quad (1) \\ \quad Z = r.\sin\theta \end{cases}$$

$$r = \frac{1}{2}.\left(\frac{L^2}{4.C} + C\right) \quad (2)$$

$$R = r - \frac{D_1}{2} - C \quad (3)$$

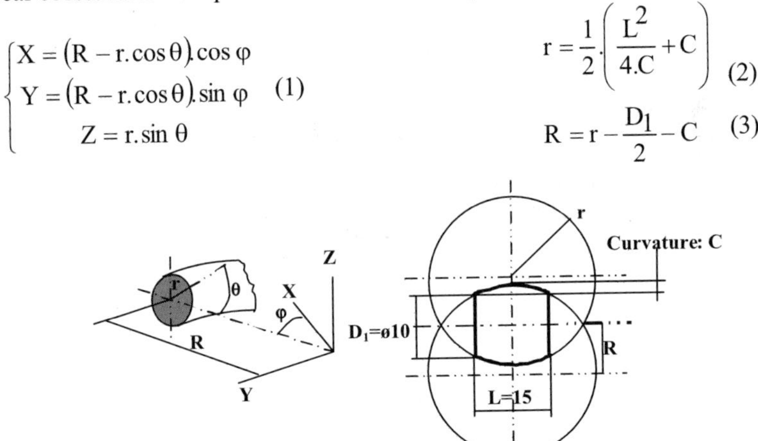



*Figure 8; Description of the cam roller design by a torus.*

The nominal value for the curvature of the cam roller has been specified to 5 μm. The two radii R and r of the torus are then easily deduced from equations 2 and 3. Their values are respectively 5625 and 5630 mm.

### 4.2. Simulation of local geometrical defects

Now, the nominal parameters of the torus are completely defined in a local coordinate system (coordinates of the center of its axis (0,0,0), cosines of its direction vector (0,0,1), radii R and r). The real part is however not perfect. In order to reproduce the geometrical defects found in a measurement of the roller, a Monte Carlo simulation method was used to build a set of 120 acquisition points, evenly distributed on the surface. In this procedure, the matter is assumed to be randomly distributed around the nominal feature, following a Gaussian distribution with a standard deviation of 2 μm.

### 4.3. Estimation of the parameters of the torus

The geometrical uncertainties model presented in section 3 has been applied to the torus. It allowed estimating the uncertainties of the simulated measurement for a confidence level of 99.7%. These results are presented in figure 9.

|        | Mean value | Uncertainty |
|--------|------------|-------------|
| Point  | -0.000627  | 0.000728    |
|        | 0.000199   | 0.000723    |
|        | 0.160817   | 0.860496    |
| Vector | -0.000065  | 0.000169    |
|        | 0.000019   | 0.000000    |
|        | 1.000000   | 0.000000    |
| R      | -7155.503  | 3253.940    |
| r      | 7160.503   | 3253.939    |

*Figure 9; Estimation of the parameters of the torus*

The deviations of the points around the nominal surface (standard deviation = 2 $\mu$m), the low width (L=15 mm) and the size of the nominal torus lead to high uncertainties of the two radii, but the whole surface remains in an envelop of about 4 $\mu$m.

### 4.4. Estimation of the variation of the curvature of the cam roller

The curvature is linked to the second radius r of the torus. Its value C is derived from equation 3:

$$C = r - \sqrt{r^2 - \left(\frac{L}{2}\right)^2}$$

The maximal and the minimal values of this curvature are determined considering the maximal and minimal values of r ($r_{max.}$ ~ 10414 mm and $r_{min.}$ ~ 3907 mm):

$$C_{max.} = 7.2\ \mu m,\ C_{min.} = 2.7\ \mu m$$

| Curvature $\mu$m | a mm | b mm | Contact area mm$^2$ |
|-------------|-------|-------|--------------|
| 2.5         | 14.43 | 0.86  | 9.75         |
| 5           | 10.95 | 0.99  | 8.51         |
| 7.5         | 9.29  | 1.07  | 7.81         |

*Figure 10; Contact area*

### 4.5. Estimation of the equivalent stress

The equivalent stress can be calculated by a Hertz analytical approach [Hills et al., 1993], the finite element method (FEM) or the boundary element method (BEM) [Guyot et al., 2000], [Moro et al., 2002], [Aliabadi et al., 2000], [Banerjee, 1994] using Von Mises criterion.

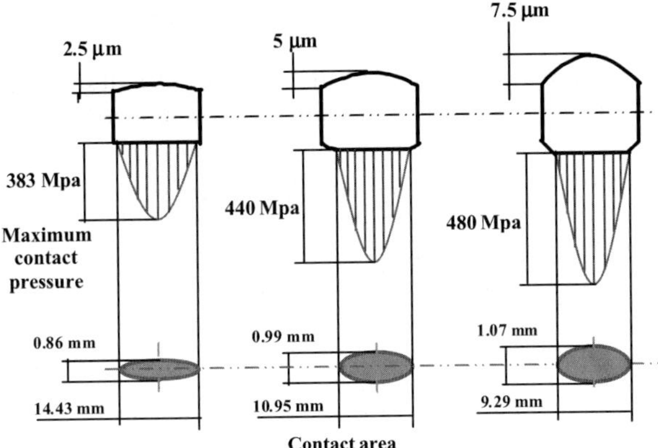


*Figure 11; Effect of the geometrical uncertainty on the contact area and the contact pressure*

The generalized Hertz model has been used to determine the in-depth stress profiles of the pieces under contact. The model will be not described in this paper but a good presentation of the method is proposed in the book written by Hills et al. [Hills et al., 1993]. A normal load of 10000 N has been considered for that purpose. The calculations have been performed for the nominal geometry of the cam roller (curvature C = 5 µm) and the extreme cases (these values were rounded to 2.5 and 7.5 µm). The maximum contact pressure and the axis a and b of the contact area are first determined.

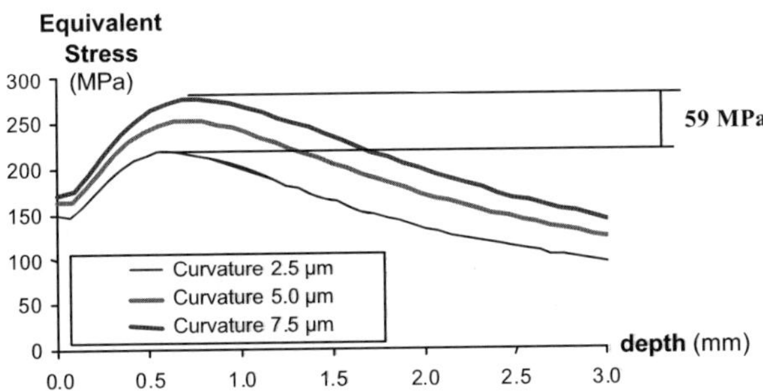


*Figure 12; Effect of the geometrical uncertainty on the equivalent stress*

The size of the contact area decreases when the roller curvature increases (figures 10 and 11). In consequence, the maximum value of the ellipsoidal contact pressure rises (figure 11). The stress field is then calculated using previous results. The equivalent stress profile is finally determined using Von Mises criterion (figure 12). The extreme values of the equivalent stress are 218 MPa for a roller of 2.5 µm curvature and 277 MPa for 7.5 µm. This represents a variation of 23%.

CONCLUSION

The constant improvement of the manufacturing processes allowed an increase in the geometrical quality of the mechanical systems leading to a reduction of the clearances between the parts. This paper brings to the fore the change of failure modes of such high quality mechanisms where malfunctions appear during the first running cycles or at low operation time. In fact the high quality of the surfaces and the increase of the hardness of the materials, modify the macro-compensation of the geometrical defects and does no longer permit any micro-accommodation. In this paper, an industrial example has been used to illustrate the impact of the geometric variations on the failure of the mechanical systems. Using a geometrical uncertainty model and a generalized Hertz contact calculation method, a great impact of geometrical variations on the stress profiles has been shown. In consequence, the ultimate stress or fatigue limit of the material may be exceeded and the operation performance of the mechanical system is deteriorated.

REFERENCES

**[Srinivasan, 2001]** V. Srinivasan, "An integrated view of Geometrical Product Specification and verification", *In Proceedings of the 7th CIRP International Seminar on Computer Aided Tolerancing*, pp.7-17, 2001.

**[Sprauel et al., 2001]** J.M. Sprauel, J.M. Linares, P. Bourdet, "Contribution of the non linear optimization to the determination of measurement uncertainties", *In Proceedings of the 7th CIRP International Seminar on Computer Aided Tolerancing*, pp.285-295, 2001.

**[Linares et al., 2003]** J.M. Linares, J. Bachmann, J.M. Sprauel and P. Bourdet, "Propagation of specification uncertainties in tolerancing", *In Proceedings of the 8th CIRP Seminar on Computer Aided Tolerancing*, pp.301-310, 2003.

**[Hills et al., 1993]** D.A. Hills, D. Nowell, A. Sackfield, Mechanics of Elastic Contacts, *Butterworth - Heinemann Ltd, Oxford*, 1993.

**[Guyot et al., 2000]** N. Guyot, F. Kosior, G. Maurice, "Coupling of finite elements and boundary elements methods for study of the frictional contact problem", *Computer Methods in Applied Mechanics and Engineering*, Vol.181, Issues 1-3, pp.147-159, 2000.

**[Moro et al., 2002]** T. Moro, A. El Hami, A. El Moudni, "Reliability analysis of a mechanical contact between deformable solids", *Probabilistic Engineering Mechanics*, Vol.17, Issue 3, pp.227-232, 2002.

**[Aliabadi et al., 2000]** M. H. Aliabadi, D. Martin, "Boundary element hyper-singular formulation for elastoplastic, contact problems", *International Journal for Numerical Methods in Engineering,* Vol.48, pp.995–1014, 2000.

**[Banerjee, 1994]** P. K. Banerjee, The Boundary Element Method in Engineering (2nd. edition), *McGraw Hill, London*, 1994.

# Influence of the Standard Components Integration on the Tolerancing Activity

**J. Dufaure*, D. Teissandier**, G. Debarbouille***
**Open CASCADE SA, Domaine Technologique de Saclay, Immeuble Ariane,*
*4 rue Rene Razel, 91400 Saclay France*
***Université Bordeaux 1, Laboratoire de Mécanique Physique UMR 5469 CNRS,*
*351 Cours de la Libération, 33405 Talence Cedex France*
*j-dufaure@opencascade.com, d.teissandier@lmp.u-bordeaux1.fr,*
*gilles.debarbouille@opencascade.com*


**Abstract:** Nowadays, designers use more and more standard components in the design process of mechanical products (aeronautics and automobile industries, etc.). Numerous products are made, for the most part, with standard components. By standard components we understand screw, nut, washer, transmission by belt or gear, ball bearing, etc. With current CAD modellers, the geometrical description of the standard components is easily integrated in the product description by the way of the standard geometric exchange formats (dxf, STEP) or the native formats (CATIA, Pro-Engineer). In consequence, the geometric description of standard components comes close to perfection. Nevertheless, what is the added value for the product? In terms of visualization the advantage is important but in terms of functional description of the product, nothing is added. The answer is the use of product models which allow to describe both functional and geometrical description. By the use of this approach, we try to demonstrate how the integration of standard components influences the tolerancing schema of the product. The functional requirements induced by the integration of the standard components have to be transferred on the parts of the mechanism according the ISO geometric specifications. Moreover this approach allows to distinguish the geometric specifications linked with the standard components from the others specifications.

**Keywords**: tolerance and functionality, tolerancing and life cycle issues, functional requirement


## 1. INTRODUCTION

In the tolerancing domain, lot of works have been done on tolerancing tools to describe geometric specifications and on 3D dimension chain computation. About the description of the geometric specifications we can cite the variationnal approach ([Gossard *et al*, 1988], [Gaunet, 1994] where the real geometry is described by variation of the nominal geometry and the tolerance zone approach ([Requicha, 1983]) where a tolerance is obtained by offset of the nominal geometry. Numerous computation tools have been developed to calculate 3D dimension chains. In [Teissandier *et al*, 1999], [Giordano *et*



*J.K. Davidson (ed.), Models for Computer Aided Tolerancing in Design and Manufacturing*, 235–244.
© 2007 Springer.

*al*, 1993], [Fleming, 1987], [Roy *et al*, 1999] the computation is realized using operations on polyhedrons which represent deviation and clearance domains. An other approach [Ballu *et al*, 1999] uses surface graph to choose type of the geometric specifications. In this paper we do not present a new computation or description tool of the geometric specification. We present a framework to allow designers to represent functional and geometric requirements at any stages of the design process. Moreover we focus on the influence of the standard component integration on the tolerancing schema of the product. The integration of standard components introduces new geometric requirements on the product. These geometric requirements represent the respect of the mounting conditions and the intrinsic characteristics (i.e. internal clearances of a ball bearing) of the standard component. With the presented approach we are able to distinguish the geometric specifications induce by the standard component integration from the geometric specifications corresponding to the functional requirements on the product.

## 2.  IPPOP PRODUCT MODEL

The presented product model is the result of the IPPOP project. The IPPOP project is interested in the Integration of Product, Process, Organization for Performance enhancement. IPPOP is a project labelled by the French Ministry of Economy, Finances and Industry (http://www.opencascade.org/IPPOP/).

To ease the work around the product in a collaborative context, we purpose to describe a product with three main concepts and three main links between these concepts as presented in [Dufaure *et al*, 2003], [Dufaure *et al*, 2004] and [Noel *et al*, 2004]. A reminder of the concepts of the product model is shown on Figure 1. A complete description of these concepts is made in the next paragraph.

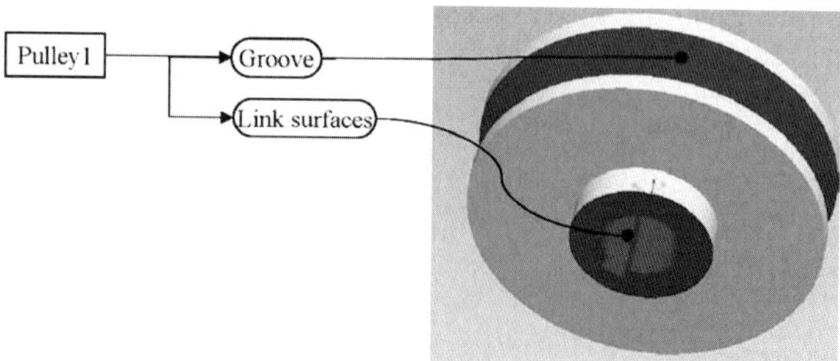


*Figure 1 ; Description of the component and interface concepts*

In consequence, a product is described as a set of components, interfaces and functions which are linked together. At the beginning of the design process, the minimal description of the product is represented by one component (the mechanism), one interface (the handle of the component) and one function (the main function of the mechanism) as follow:

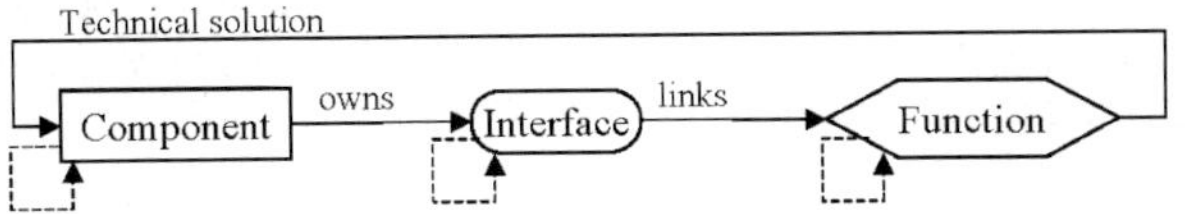


*Figure 2 ; Product description at the beginning of the design process*

This schema represents that the mechanism is a technical solution of the function and the component owns interfaces to link it with the external medium.

## 2.1. Three main concepts

The three objects presented in this paper are component, interface and function. These objects allow to represent both geometric and functional descriptions of the product. In this paper the semantic of these concepts is given with a tolerancing point of view.

The geometric description of the product is represented using components and interfaces. A component describes the structural decomposition of the product. A component can be an assembly, a sub-assembly, a part and generally any partition of the product. To describe the full product structure, each component can be decomposed into several ones. For example a mixing machine (Figure 3) is decomposed into five components that represent the motor, the transmission, the reducer, the bowl and the frame.

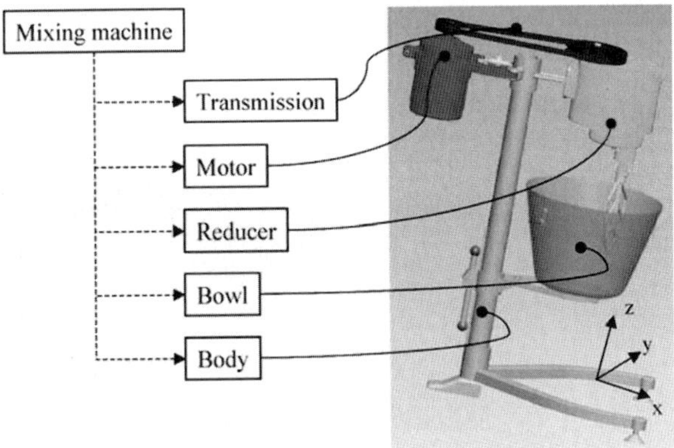


*Figure 3 ; Structural decomposition*

The interface object allows to describe the geometric elements of a component which are in relation with the external medium. In the tolerancing activity, the interfaces can be a surface, a line or a point. According to this definition we illustrate the concept of interface on the reducer component (Figure 4).

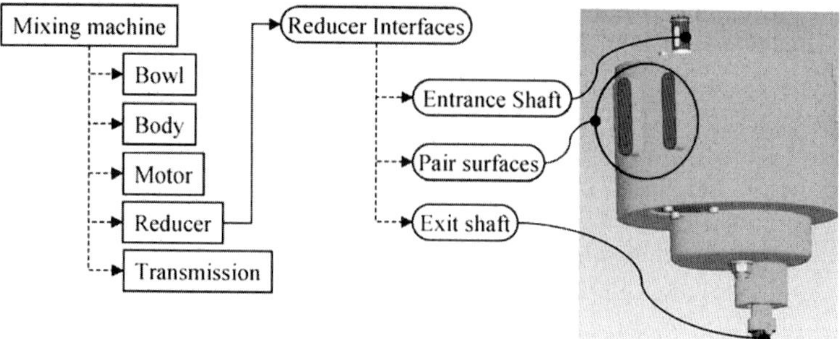


*Figure 4 ; Description of the interfaces of the reducer*

The third object is function. In tolerancing activity, a function can represent a functional requirement, a kinematic joint or a geometric specification. The specifications supported by the product model are geometric specifications by tolerance zone, dimensional specifications (linear and angular) and roughness specifications. We propose to group together these requirements under the term of function to ensure the traceability of the geometric specifications from the conceptual to the detail design. On the example of a mixing machine, we are able to express the functional requirement (transmit a rotational movement between the motor and the reducer). This function is described as a decomposition of the root function of the product as shown in Figure 5.

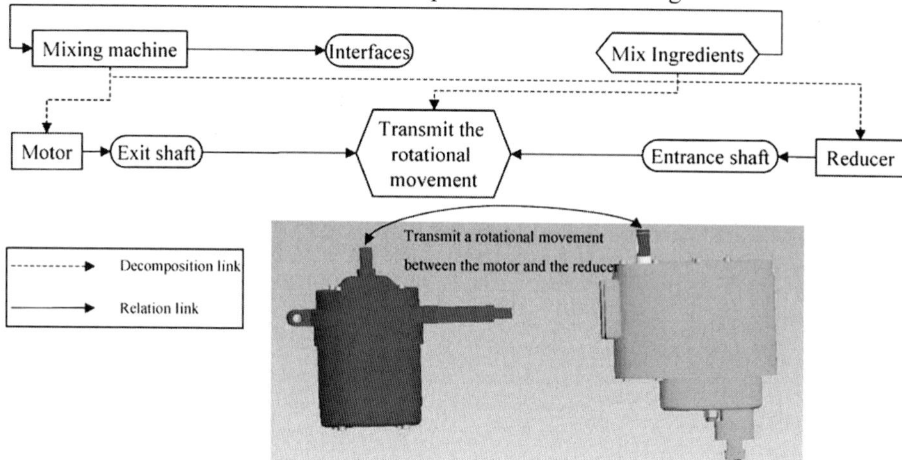


*Figure 5 ; Description of a functional requirement*

We have proved in this paragraph that we are able to represent both structural and functional decomposition of a product. We obtain three trees corresponding to component decomposition, interface decomposition and function decomposition.

## 2.2. Three main links between component, interface and function

As mentioned in [Summers *et al*, 2001], we have to link the structural and the functional description of a product to ensure the consistency of the product.
In the purposed product model three types of link are available:

The semantic of the link between component and interface is that the interfaces belong to one component. The link between function and interface corresponds to the definition that the function links two or more interfaces. The third link between function and component corresponds to the component is a technical solution of the function. These links are instantiate as shown in Figure 2.

## 3. STANDARD COMPONENT DESCRIPTION

We have presented how to describe a product at the first stages of the design process. In the mechanical industries, designers use numerous standard components to ensure the functions of mechanisms. Designers do not design the standard components, they only know their intrinsic characteristics (i.e. thread of a screw, clearances of a ball bearing) and the surfaces, which are used to mount the standard component (helix of a screw, cylindrical surface of the inner ring of a ball bearing). The mounting conditions of a standard component have to be described in the product because these requirements have an influence on the product tolerancing schema. A standard component is a set of components, interfaces and functions. We focus on the function "transmit the rotational movement" and this solution is described in the product model as follows:

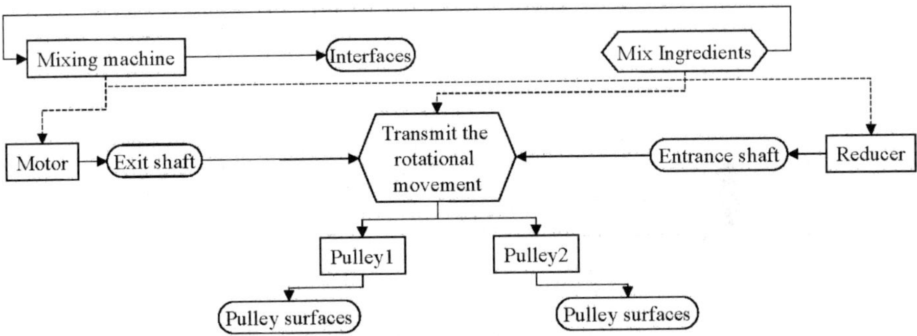


*Figure 6 ; Standard component description*

When the standard component is integrated in the product model we have to describe the links between the interfaces of the standard component and the other parts of the mechanism (Figure 7).

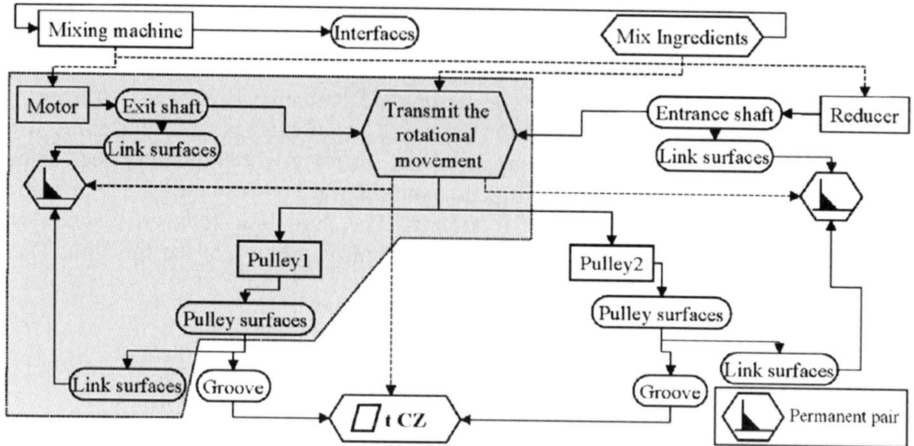


*Figure 7 ; Complete integration of the standard components*

### 3.1. Transfer of the mounting condition of a standard component

When the structure of the standard component is described in the product model, we have to transfer the links between the interfaces of the transmission, the reducer, the motor and the function on the interfaces of the pulleys 1 and 2. We only describe the assembly between the pulley1 and the exit shaft of the motor (Figure 7). To transfer the assembly condition we have to take into account the geometric specifications on the pulleys. The specifications on the pulley1 impose the dimensional specification on the shaft of the motor. As shown in Figure 8, the cylindrical surface of the motor is specified to ensure that the pair between the pulley1 and the motor is a cylindrical pair. The valuation of this specification is deduced from the dimensional specifications on the cylinders of the pulleys.

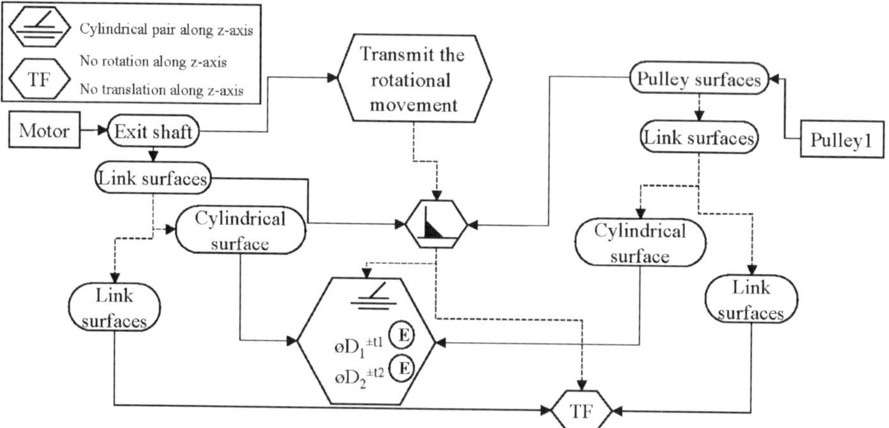


*Figure 8 ; Transfer of the mounting conditions into geometric specifications*

### 3.2. Transfer of the geometric specifications induced by the choice of the standard components

To ensure the function "transmit a rotational movement" using a transmission by belt, we have to specify the position and the orientation between the two median planes of the pulley grooves. In ISO language, this geometric requirement can be expressed with a flatness specification in common zone. This specification has to be integrated in the tolerancing schema of the product and transferred on the parts of the mechanism (Figure 9).

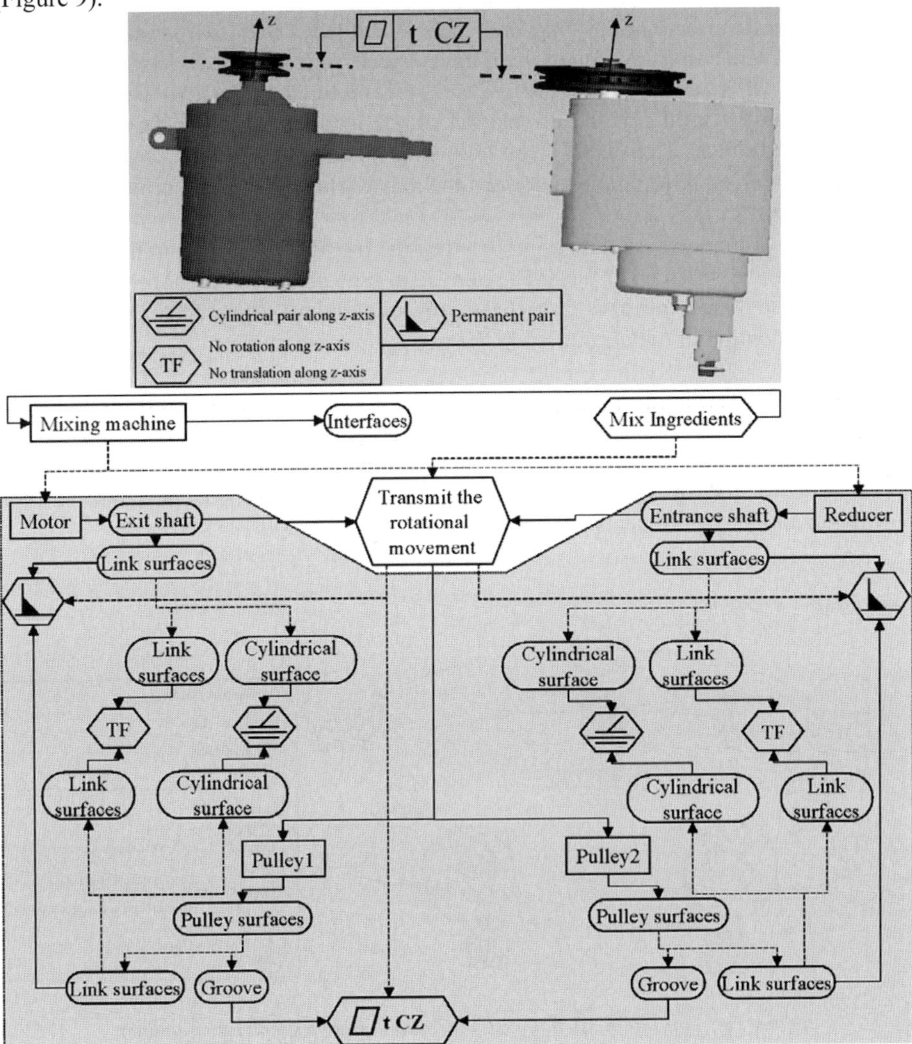


*Figure 9 ; Description of the flatness specification in the product model*

In the presented mechanism, the position between the motor and the reducer is adjustable along the z-axis. With the presented product model we are able to describe a surface graph. In the surface graph, an interface is represented using a circle named with a small letter. In opposition a component is described by a circle around interfaces. A functional condition is described by the symbol of the cinematic link or the symbol of the ISO specifications. This representation is often used in the tolerancing activity and has to be taken into account (Figure 10). A qualitative transfer of the flatness specification is done using this representation.

In the presented surface graph, only one cycle is influent on the flatness specification. The influent cycle corresponding to the flatness specification is the cycle which contains the surfaces called a and b for the reducer, the motor and the two pulleys. These surfaces (a and b) influence the orientation of the median planes of the pulleys. The transfer of the flatness is ensured by the following specification:

- a parallelism specification between the axis of the cylindrical surfaces (a) of the reducer and the motor,
- a perpendicularity specification between the median plane (b) and the axis of the cylindrical surface (b) for each pulley. This specification has not to be valuated because it is an intrinsic characteristic of the standard component (pulley),
- the valuation of the clearances in the two cylindrical pairs.

The result of this transfer can be stored in the presented product model as shown in Figure 11. The geometric specifications (parallelism and perpendicularity) which ensure the flatness specification in common zone are described as sub-functions of the function corresponding to the flatness. By this way, we keep the link between the functional requirements and the geometric specifications.

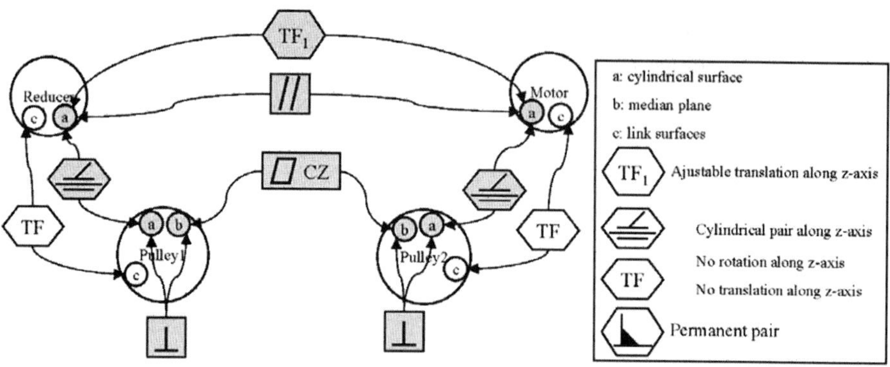


*Figure 10 ; Transfer of the flatness specification using a graph representation*

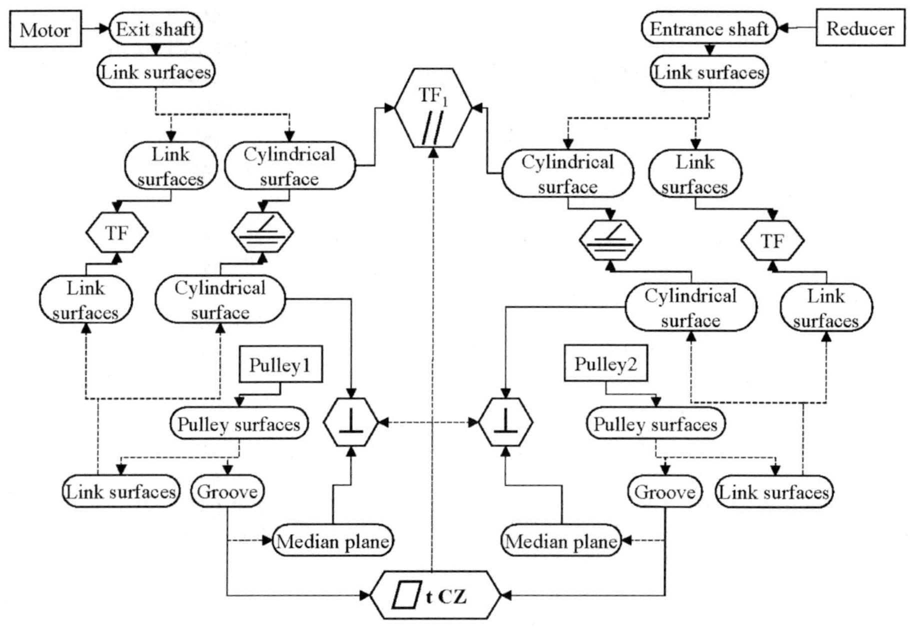


*Figure 11 ; Description of the flatness specification transfer in the product model*

## 4. CONCLUSION

This paper presents the concepts and the use of a framework in the tolerancing activity. This framework allows to describe both functional and geometric descriptions of a product and enhances the traceability of the geometric specifications at any stages of the design process. Moreover we present in this paper the influence of standard component integration on the tolerancing of a product. We think it is important to distinguish the geometric specification induce by the standard component integration from the geometric specifications corresponding to the respect of the functional requirements. With this approach if a standard component changes during the design process, we can easily update the tolerancing schema of the product and compute only the 3D dimensions chains which evolve. Future works will allow designers to describe design process alternatives (structural and/or functional). By this way, in one product model we will be able to represent different tolerancing schema corresponding to the integration of numerous standard components which answer to a same function.

REFERENCES

[**Ballu et al, 1999**]  A. Ballu et L. Mathieu, "Choice of functional specifications using graphs within the framework of education". CIRP CAT, Enschede (NL), 1999.

[**Dufaure et al, 2003**]  J. Dufaure et D. Teissandier, "Geometric tolerancing from conceptual to detail design". CIRP, Charlotte (north carolina, USA), 2003.

[**Dufaure et al, 2004**]  J. Dufaure, D. Teissandier et G. Debarbouille, "Product model dedicated to collaborative design: A geometric tolerancing point of view". IDMME, Bath (UK), 2004.

[**Fleming, 1987**]  A. Fleming. "Analysis of uncertainties and geometric tolerances in assemblies of parts", PhD thesis, 1987.

[**Gaunet, 1994**]  D. Gaunet. "Modèle formel de tolerancement de position. Contributions à l'aide au tolerancement des mécanismes en CFAO.", PhD thesis, 1994.

[**Giordano et al, 1993**]  M. Giordano et D. Duret, "Clearence Space and Deviation Space. Espace Jeu, Espace Ecart." CIRP CAT, 1993.

[**Gossard et al, 1988**]  D. C. Gossard, R. P. Zuffante et H. Sakuria, "Representing dimensions, tolerances, and features in MCAE Systems." IEEE 1: 51-59,1988.

[**Noel et al, 2004**]  F. Noel, L. Roucoules et D. Teissandier, "Specification of product modelling concepts dedicated to information sharing in a collaborative design context". IDMME, Bath (UK), 2004.

[**Requicha, 1983**]  A. A. G. Requicha, "Toward a theory of geometric tolerancing." The International journal of Robotics Research **2**: 45-60 n4,1983.

[**Roy et al, 1999**]  U. Roy et L. Bing, "3D variational polyhedral assembly configuration". CIRP CAT, Enschede (NL), 1999.

[**Summers et al, 2001**]  J. D. Summers, N. Vargas-Hernandez, Z. Zhao, J. J. Shah et Z. Lacroix, "Comparative study of representation structures for modeling function and behavior of mechanical devices". DETC Computers in Engineering, Pittsburgh, 2001.

[**Teissandier et al, 1999**]  D. Teissandier, V. Delos et Y. Couetard, "Operations on polytopes: application to tolerance anlysis". 6th CIRP CAT, Enschede (The Netherlands), 1999.

# Surfaces Seam Analysis

***J.-P. Petit*, S. Samper*, I. Perpoli****
** Laboratoire de Mécanique Appliquée,*
*LMécA/ESIA, BP 806,*
*74016 ANNECY Cedex, France*
*jean-philippe.petit@univ-savoie.fr*

**Abstract:** we present an analysis of gap and flush defects for automotive exterior body panels. We will consider position deviations of parts and those deviations are traduced through the concept of small displacements torsor. Thus it is possible to use the model of clearances and deviations domains. This method allows modeling geometric specifications on parts (deviation domains) and clearances in joints (clearance domains). The seam analysis is made in order to verify if the tolerances are compatible with specifications. By our solving, we obtain the set of relative displacements between two parts. This set of displacements is a 6D domain (3 translations plus 3 rotations) and is not visible for the designer, thus we compute a corresponding zone on parts in order to see the results. This post processor zone is shown in a CAD environment. It is then possible to the designer to see consequences of its tolerancing choices.
**Keywords**: tolerancing analysis, surfaces seam, deviation domains, CAD.

## 1. INTRODUCTION

A mechanism is composed of manufactured parts which present deviations. To answer to functional requirements (assemblability, accuracy, non interference…), the designer has to define limits of those geometry deviations through a tolerancing. Once this work is done, it is possible to check if the chosen tolerancing satisfy requirements thanks to an operation called tolerance analysis. We distinguish two kind of tolerance analysis: the statistical analysis (not detailed in this paper) and the worst case analysis. We propose in the following a worst case method of tolerance analysis based on the model of clearances and deviations domains. Several works deal with domains (volumes) to translate geometrical specifications or to propose analysis methods. For example, Shah and Davidson use the hypothetical volume called Tolerance Map (T-map) in their approach [Davidson *et al.*, 2002]. For its part, Teissandier has developed a tolerancing model leading on a tool named Proportioned Assembly Clearance Volume (PACV) [Teissandier *et al.*, 1999]. In our point of view, those methods are limited and can deal only with simply cases of assemblies. We propose through our approach to treat a concrete industrial example of tolerance analysis. This example is the analysis of gap and flush defects for automotive exterior body panels. It consists in verifying if the chosen tolerances on functional contact surfaces of an assembly (mechanism in open-



*J.K. Davidson (ed.), Models for Computer Aided Tolerancing in Design and Manufacturing,* 245–254.
© 2007 *Springer.*

chain constituted of four parts) allow a functional requirement between two of these parts. A representation of the consequences of the designer's choices on the geometry variations of the assembly will be given in a CAD environment.


## 2.    MODEL OF CLEARANCES AND DEVIATIONS DOMAINS

International standards [ISO 1101] allow to represent every geometric specification by a tolerance zone. This zone is built on the nominal geometry of the toleranced feature. The tolerance will be validate if the real feature (the theoretical geometric element actually) associated to the nominal feature lies inside the tolerance zone.

In our model, a general datum frame is built for the mechanism. A datum frame is attached to each functional associated feature and another one is attached to each nominal feature. The displacements of the associated feature inside the tolerance zone are supposed small enough. It is then possible to express, at the centre of the general frame $O$, the positions of the associated frame regarding the nominal frame in the shape of a small displacements torsor [Bourdet *et al.*, 1995]. Six components (3 translations plus 3 rotations) characterize this torsor called deviation torsor [Giordano *et al.*, 1993]. The general form of a deviation torsor is given below:

$$E = \left\{ \begin{matrix} Tx & Rx \\ Ty & Ry \\ Tz & Rz \end{matrix} \right\}_O \tag{1}$$

The set of the values of all deviation torsors defines a domain in the 6D configuration space called the deviation domain noted *[E]* (Cf. Figure 1).

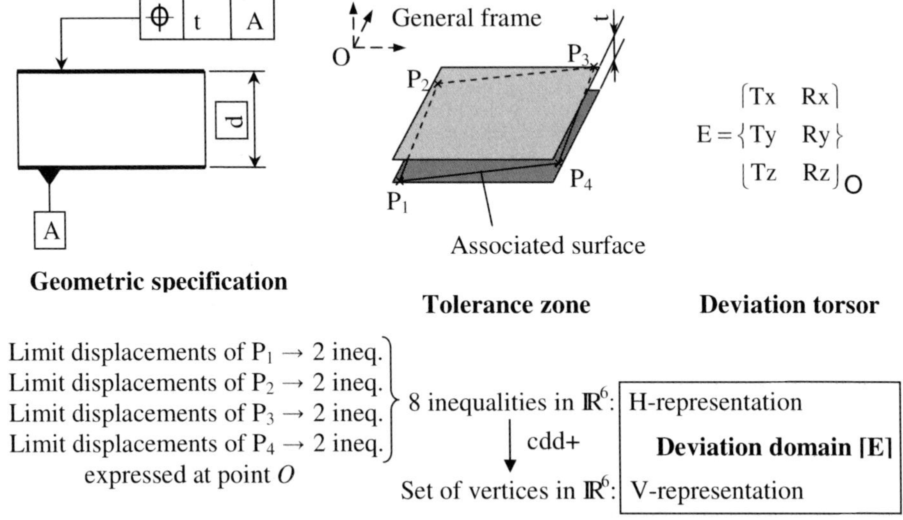


*Figure 1; Deviation domain associated to a tolerance zone*

Associated feature is built from the polygonal convex hull containing the outline of the nominal feature. Defining the deviation domain is reduced to consider each maximum displacement of characteristic vertices of the associated feature inside the tolerance zone. Those displacements are translated by an inequalities system expressed at a fixed point ($O$ in our example).

With a polyhedral computation code *cdd+* developed by K. Fukuda, all vertices of the convex 6-polytope are generated in $\mathbb{R}^6$ from the set of inequalities [Fukuda *et al.*, 1996]. This double definition (*Vertex-representation* and *Halfspace-representation*) is necessary for several geometric operations on different domains defined in our model.

A clearance domain is built with the same method. First a clerance torsor defined. A clearance zone is translated in a set of inequalities. This one gives the corresponding domain in a 6D space.

## 3.  STUDIED SEAM

Definition: a seam is a neutral zone between two fixed parts and/or opening allowing the assembly of the various elements of a body car and thus avoiding any interference of opening parts with other elements.

There are many functions to a seam, one can gather them in three principal categories: technical functions (to satisfy assemblability, to improve aerodynamism of the car...), aesthetic functions (lights on and between various surfaces...) and ergonomic functions (to allow the disassembling of the elements...).

From the functional surfaces geometry resulting from CAD model and annotations stipulating the functional requirements from the assembly specification, we will check if the tolerancing associated to each part intervening in the positioning of the headlight checks the functional requirements imposed by the customer.

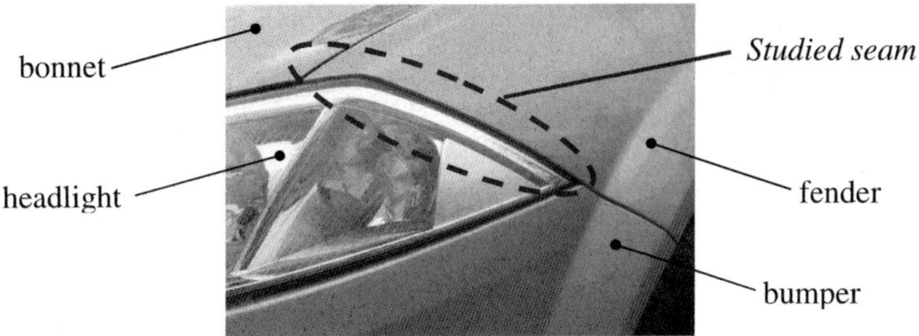


*Figure 2; Car parts and studied seam.*

The functional requirement considered in this application concerns the seam line between the fender and the headlight. This requirement breaks up into a condition of evolutionary flush and a condition of minimum clearance. Four control points (*a, b, c* and *d*) are defined on the studied curve. With each one of these points a tolerance on the clearance and a position tolerance are associated limiting the evolutionary flush. A measurement of flush is done according to the normal on the datum surface: the fender in our case. One thus defines a normal $\vec{n}_x$ associated with each control point (Figure 3). The clearance is measured perpendicularly to the seam line and following the direction defined on the definition drawing. For each control point defined by his coordinates in the general frame, one builds a normal $\vec{j}_x$ perpendicular to the seam line and normal to the datum surface.

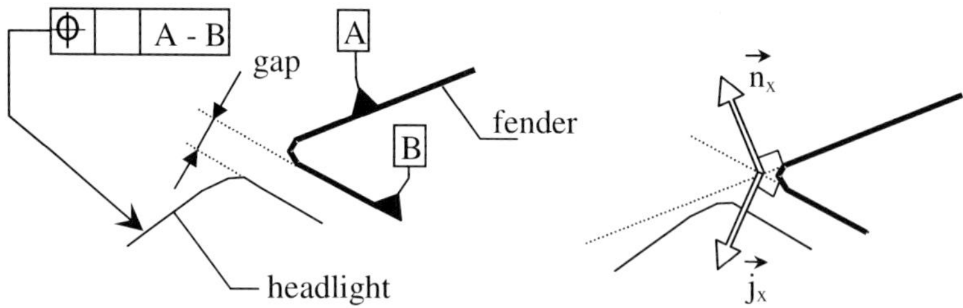



*Figure 3 ; Specification on assembly.*

Various values of the tolerances allow to translate the functional requirement studied by four acceptable zones built on control points. These zones are represented on the right figure 4. Here, the fender is the datum. Clearances and cumulated deviations on parts (intervening in the chain of the headlight positioning) must allow maxima displacements of headlight control points inside their associated zones.

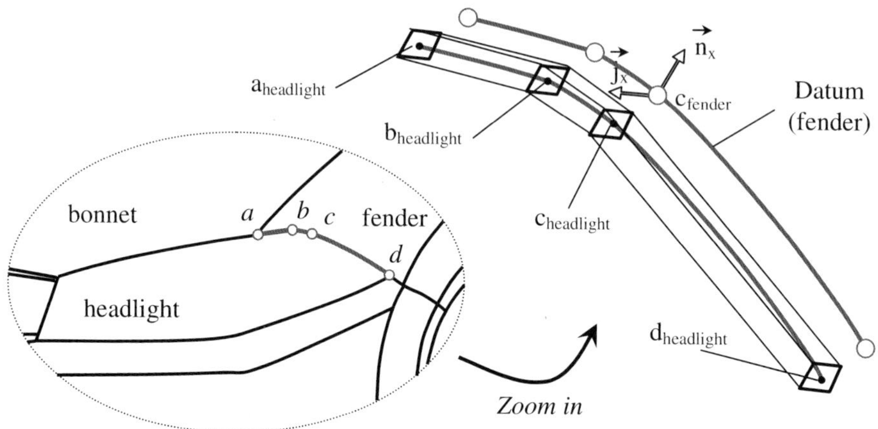



*Figure 4; Specified zone on headlight curve.*

## 4.  TOLERANCING ANALYSIS

The positioning of the headlight regarding the fender is carried out by a sub-assembly composed of four parts forming an open chain: the headlight, the technical front face (F.A.T.), the case and the fender (Figure 5). The functional requirement study between the fender and the headlight requires to consider the joint between these two parts and thus to close the chain by transforming the analysis graph into a single loop graph. The three full contacts of the assembly will be modelled by isostatic positioning broken up each one into three elementary joints: a planar joint (three point slider joints) blocking three degrees of freedom, an edge slider joint (two point slider joints) blocking two degrees and a point slider joint blocking the last degree of freedom of the link between the two parts. These contact points are obtained from CAD model, their coordinates are expressed in the global frame of the mechanism.

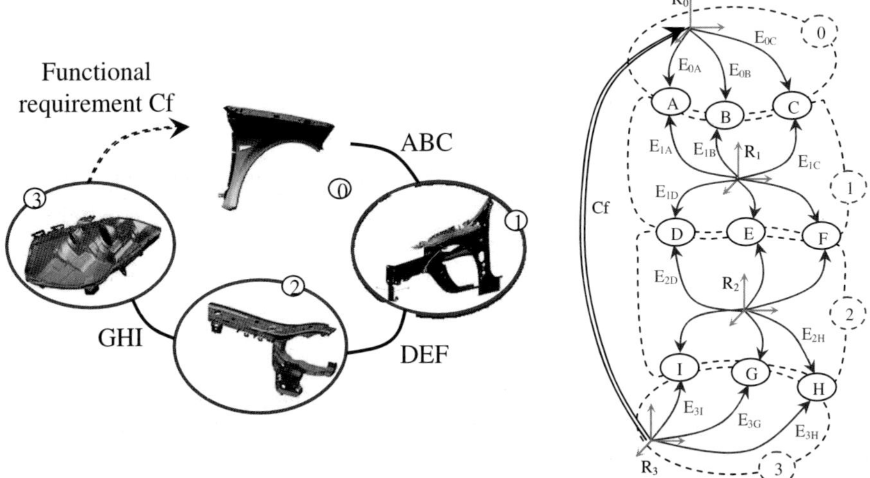



*Figure 5; Contact graph and analysis graph*

Contact surfaces of these nine elementary joints (three planar joints, three edge slider joints and three point slider joints) are known as functional features and limits of their geometrical deviations are fixed by tolerancing. For confidentiality reasons, we will not give in this paper the definition drawings provided by the customer. A definition by tolerance zone is given for each type of contact surfaces. The tolerance for a surface of a planar joint is 0.3 mm, of an edge slider joint 0.2 mm and of a point slider joint 0.1mm. Each of the three tolerance zone is modelled by a deviation domain expressed at the point $O$ representing the centre of the general frame. This operation is made for all the specifications in order to obtain the corresponding deviation domain.

This fact, the relative position of the part *(0)* compared to the part *(1)* depends on the deviations on contact surfaces $A_1$, $B_1$ and $C_1$ belonging to part *(1)* and of the deviations on surfaces $A_0$, $B_0$ and $C_0$ belonging to part *(0)*. Finally, the whole of the positions of the the frame $R_1$ associated to the part *(1)* regarding the frame $R_0$ associated to the part *(0)* is defined by the resulting deviation domain *[R$_{01}$]* calculated in the following way:

$$[R_{01}] = [[E_{0A}] \oplus [E_{A1}]] \cup [[E_{0B}] \oplus [E_{B1}]] \cup [[E_{0C}] \oplus [E_{C1}]] \qquad (2)$$

Remarks:

- *[E$_{iX}$]* is the deviation domain of surface *Xi* expressed at point *O*.
- $\oplus$ is the operator of the Minkowski addition [Fukuda, 2003].
- $\cup$ is the intersection operator.

  Each deviation domain is a 6-polytope characterized by its H-representation and its V-representation. It is the same for *[R$_{01}$]*.

The four parts being in series, one obtains with final the whole of the possible and reachable positions of the $R_3$ (thus of the headlight) compared to $R_0$ (attached to the fender) defined by the domain *[R$_{03}$]* calculated in the following way:

$$[R_{03}] = [R_{01}] \oplus [R_{12}] \oplus [R_{23}] \qquad (3)$$

The relative position of part *(3)* compared to part *(0)* is thus modelled by a composition of the contact surfaces deviation ensuring the assembly.

The studied functional requirement relates to the seam line between the headlight (part *(3)*) and the fender (part *(0)*). The analysis of tolerance thus passes by the definition of the whole of the positions of one of these two parts regarding the other. This definition is given by the resulting deviation domain *[R$_{03}$]*, the analysis thus passes by the calculation of this field in space 6D of small displacements.

The domain *[R$_{03}$]* models the whole of the positions of the headlight compared to the fender expressed at the point *O* according to the deviations of contacts surfaces forming the chain between these two parts. The functional specification fixed by the customer enabled us to calculate the tolerance zones of the control points of the headlight (*a*, *b*, *c* and *d*) when the fender is the datum.

Each vertex of the V-representation characterizing the domain *[R$_{03}$]* corresponds to a small displacements torsor. This torsor corresponds to a relative position of the part *(3)* compared to the part *(0)*. It is thus possible to calculate the components in translation of each one of these torsors (set of the vertices of the 6-polytope) in any point of space [Samper *et al.*, 2004]. While transporting the domain *[R$_{03}$]* of each four control points, we can then determine their 3D projections in space (Tx, Ty, Tz) modelling maxima displacements of these points in Euclidean space. These calculated zones are represented in figure 6.

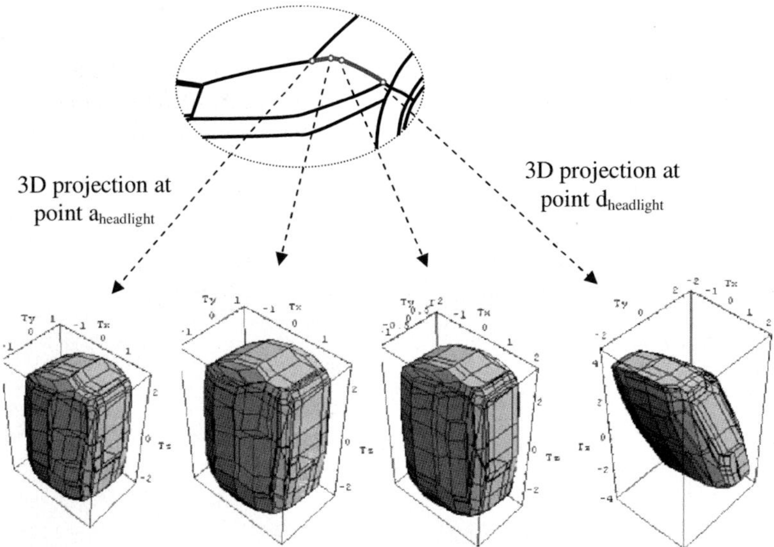



*Figure 6; 3D projections at control points.*

A post-processing procedure was developed [Petit, 2004]. It allows to change a 6D object belonging to the mathematical model (which is the domain) in a 3D calculated zone which is then injected into CAD model. The calculated zone can then, at each of the four control points, be numerically and graphically compared with the specified zone translating the functional condition in order to validate or not the designer's tolerancing.

## 5.   REPRESENTATIONS IN A CAD ENVIRONMENT

Figure 7 gives the representation in a CAD software of the nominal geometry of the fender (on the right) taken in datum as well as the 3D calculated and specified zones for the points *a* and *b* belonging to the headlight. A smoothing of the 2D specified zones allows to visualize in which volume the toleranced curve must be contained so that the functional requirement is checked and this, whatever the selected tolerance. A smoothing on the calculated zones can be carried out to represent the possible variations of geometry of this curve according to the tolerances.

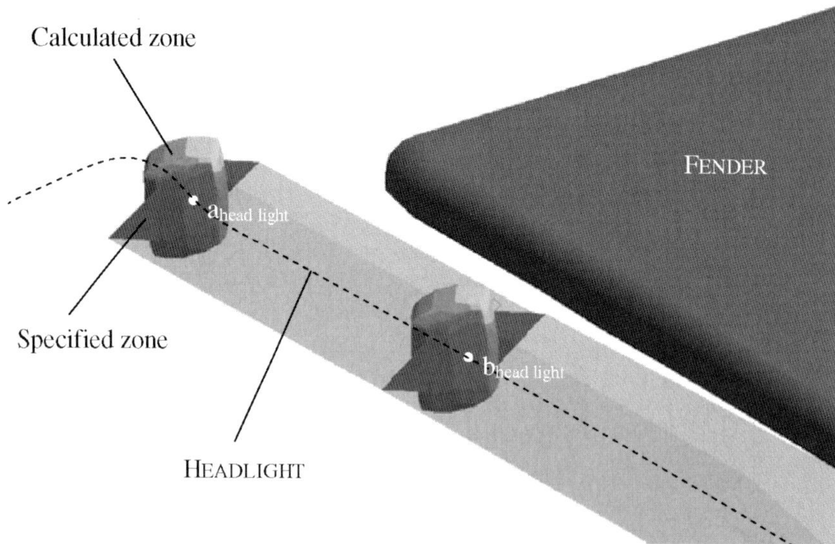



*Figure 7; CAD representation.*

In a more rigorous way, the functional condition checking carries on four 2D specified zones. Knowing topologies of these specified zones at each control point, 2D sections of the calculated zones according to the plans containing the tolerances zones allow to check graphically the functional condition at each point. The figure below represents this comparison at point *a*.

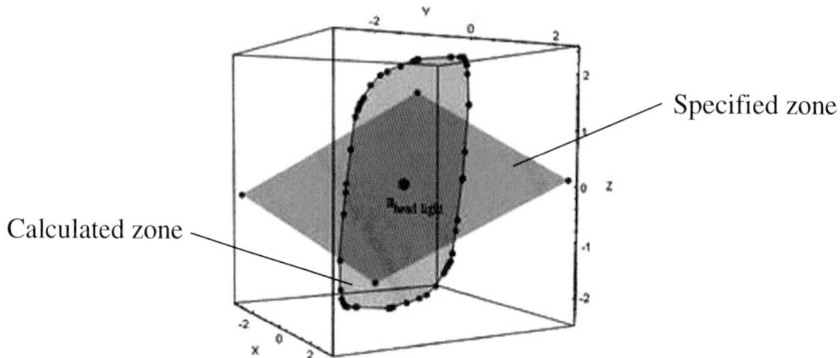



*Figure 8; 2D sections of zones.*

The inclusion of one polygon inside another is verified numerically from the vertices coordinates of the 2D section of the calculated zone and the inequalities defining the specified zone.

    Remark: one can note that the functional condition is not verified here in all the cases (vertices outside the specified zone). The vertices of the calculated zone represent the extreme cases of possible displacements for the various control points. Let us note however that the calculated zone is very close to the specified zone. This result shows that the tolerancing is accurate and the designer tolerancing methods are good.


## 6.  CONCLUSION

The method of tolerance analysis of surfaces seam proposed in this paper allows the designer to visualize within its product designing tool (CAD software) the consequences of its tolerancing choices on the variations of the assembly geometry. Through this representation, all worst cases are considered. In the example treated here and in particular thanks to the representation given figure 7, he will be able to check that its tolerancing will authorize in worst case a minimum clearance between the fender and the headlight but that the flush condition will not be always checked. An iterative procedure relating to the modification of the tolerances (from a quantitative point of view) can then be imagined to shrink the calculated zones until they are included in the specified zones.

In addition, a study relating to the fender flexibility was treated to integrate the influence of this phenomenon into the analysis method and to obtain from this fact a finer modeling [Perpoli, 2004]. We can quote other works going in this direction as a study where both variations of geometry coming from tolerancing, and displacements coming from car door deformations are taken into account to provide the global variations of the final assembly of this part [Cid *et al.*, 2004].


## REFERENCES

**[Bourdet et al., 1995]** Bourdet, P.; Mathieu, L.; Lartigue, C.; Ballu, A.; "The concept of the small displacement torsor in metrology"; *International Euroconference, Advanced Mathematical tool in Metrology*, Oxford 1995

**[Cid et al., 2004]** Cid, G.;Thiébault, F.; Bourdet, P.; "Taking the deformation into account for components' tolerancing"; *Proceedings of the 5<sup>th</sup> International Conference on Integrating Design and Manufacturing in Mechanical Engineering*, Bath 2004

**[Davidson et al., 2002]** Davidson, J.K.; Shah, J.J.; "Geometric Tolerances: A New Application for Line Geometry and Screws"; *IMechE Journal of Mechanical Engineering Science, Vol. 216 Part C*, pp. 95-104; 2002

**[Fukuda et al., 1996]** Fukuda, K.; Prodon, A.; "Double description method revisited"; *Combinatorics and Computer Science, Volume 1120 of Lecture Notes in Computer Science*, pp. 91-111; 1996

**[Fukuda, 2003]** Fukuda, K.; "From the zonotope construction to the Minkowski addition of convex polytopes"; *Journal of Symbolic Computation*, 2003

**[Giordano et al., 1993]** Giordano, M.; Duret, D.; "Clearance space and deviation space. Application to three dimensional chain of dimensions and positions"; *3$^{rd}$ CIRP Seminar on Computer-Aided Tolerancing*, pp. 179-196; Cachan 1993

**[ISO 1101]** "Technical drawings. Geometrical tolerancing. Tolerancing of form, orientation, location and run-out. Generalities, definitions, symbols, indications on drawings", 1983

**[Perpoli, 2004]** Perpoli, I.; "Tolérancement des mécanismes flexibles - Application à une aile de carrosserie"; *Mémoire CNAM*, Université de Savoie 2004

**[Petit, 2004]** Petit, J-Ph.; "Spécification Géométrique des Produits : Méthode d'analyse de tolérances. Application en Conception Assistée par Ordinateur.", *PhD Thesis*, Université de Savoie 2004

**[Samper et al., 2004]** Samper, S.; Petit, J-Ph.; "Computer aided tolerancing - solver and post processor analysis"; *Proceedings of the 14$^{th}$ International CIRP Design Seminar*, Le Caire 2004

**[Teissandier et al., 1999]** Teissandier, D.; Delos, V.; Couetard, Y.; "Operation on polytopes: application to tolerance analysis"; *Proceedings of the 6$^{th}$ CIRP International Seminar on Computer-Aided Tolerancing*, pp. 425-434; Enschede 1999

# Statistical Tolerance Analysis of Gears by Tooth Contact Analysis

*J. Bruyere, J.-Y. Dantan, R. Bigot, P. Martin*

*Laboratoire de Génie Industriel et de Production Mécanique E.A. 3096,*

*E.N.S.A.M. de Metz, 4 rue A. Fresnel, 57070 METZ Cedex, France*

*jerome.bruyere@metz.ensam.fr*


**Abstract:** To analyze the influence of geometrical variations of parts on functional characteristics, a powerful way consist to simulate its geometrical behaviour. For gears, a specific method based on Tooth Contact Analysis is developed to evaluate influence of intrinsic and situation deviations of pinion and wheel on the assembly. Moreover, a statistical analysis allows to determine probability distribution of the functional characteristic - kinematic error for a list of random geometrical deviations.

**Keywords**: tolerance analysis, statistical simulation, bevel gear, Tooth Contact Analysis


## 1. INTRODUCTION

Tolerance analysis involves evaluating the effect of geometrical variations of individual part or subassembly dimensions on designated functional characteristics or assembly characteristics of the resulting assembly.

To do so, the best way to analyze the tolerances is to simulate the influences of parts deviations on the geometrical requirements. Usually, these geometrical requirements limit the gaps or the displacements between product surfaces. For this type of geometrical requirements, the influences of parts deviations can be analyzed by different approaches as Variational geometry [Roy et al.,1999], Geometrical behavior law [Ballot et al.,1997], Clearance space and deviation space [Giordano et al.,1993], [Teissandier,1999], Gap space [Zou et al.,2004], quantifier approach [Dantan et al.,2005], kinematic models [Sacks et al.,1997], …

In the case of gears, their geometrical deviations impact the kinematic error, the tooth contact position, meshing interference, gap … To ensure a quality level, designers limit these parameters by requirements. To simulate the influences of gears deviations on these requirements, the classical approaches can not possibly be employed because the tolerance analysis of gears includes a kinematic aspect and a determination of contact position. There are important questions that would need to be looked upon:

255

*J.K. Davidson (ed.), Models for Computer Aided Tolerancing in Design and Manufacturing, 255–265.*

256                                J. Bruyere *et al.*

- How to simulate the influences of gears deviations?
- How to modelize the gears tolerances?
- And how to analyze these tolerances with a simulation tool?

In this paper, we focus on the first and the third questions. We propose a geometrical deviation model (a parameterization of deviations from theoretic geometry, the real geometry of parts is apprehended by a variation of the nominal geometry. The substitute surfaces model these real surfaces) and the Tooth Contact Analysis to describe the geometrical behavior. This aspect is detailed in section 2.

For the third question, we propose to use a statistical simulation (Monte Carlo Simulation) of the gears deviations. The different models and analysis are presented in section 3.

## 2. HOW TO SIMULATE THE INFLUENCES OF GEARS DEVIATIONS?

The best way to determine the optimal tolerances is to simulate the influences of deviations on the geometrical behavior of the mechanism. Usually, for mathematical formulation of tolerance analysis, the geometrical behavior is described using different concepts. We principally need a detailed description of each variation to characterize the geometrical behavior [Ballot et al.,1997], [Dantan et al.,2005].The approach used in this paper is the parameterization of deviations (section 2.1) and the Tooth Contact Analysis (TCA) for geometrical behavior model (section 2.3). The principle idea of TCA is based on simulation of tangency of tooth surfaces being in mesh. The TCA program provides the information about the function of transmission errors [Litvin, 2004].

### 2.1. Geometrical description of variations of bevel gears

Usually, geometrical behavior description needs to be aware of the surface deviations of each part (situation deviations and intrinsic deviations) and relative displacements between parts according to gap (gaps and functional characteristics). Compared with the nominal model, each substitute surface has position variations, orientation variations and intrinsic variations [Ballot et al.,1997], [Dantan et al.,2005].

In the case of gears, the geometrical behaviour description principally needs an exact analytical definition of tooth surface. A tooth surface $\Sigma$ is described by expressing its cartesian coordinates x, y, z in the local coordinate system ($S_7$ substitute surface gear 1 or $S_8$ substitute surface gear 2) as functions of two parameters ($\varphi$, $\theta$) or ($\alpha$, $\gamma$) in a certain closed interval. In this paper, only the mathematical models of the active tooth surfaces of the pinion and the gear are expressed and discussed. The position vectors and unit normal vectors of tooth surface of pinion and gear are denoted as $\mathbf{r}_k^{(i)}$ and $\mathbf{n}_k^{(i)}$ for part (i) in the coordinate system "k".

The model of the gear box with geometrical variations can be simulated by changing the settings and orientations of the coordinate systems. The odd index are use for the wheel 2 and even one is used for pinion 1. The coordinate systems of part 1 are (Figure 1):

- S7 is rigidly attached to the first substitute surface of pinion 1,
- S5 is rigidly attached to one of nominal active surfaces of part 1,

-    S3 is rigidly attached to the hole,
-    S1 is rigidly attached the frame. The first axis of S1 is rotational axes of pinion 1,
-    Sf is rigidly connected with the frame. It's common coordinate system.

In the same way, the coordinate systems of bevel wheel 2 S8, S6, S4 & S2 are defined.

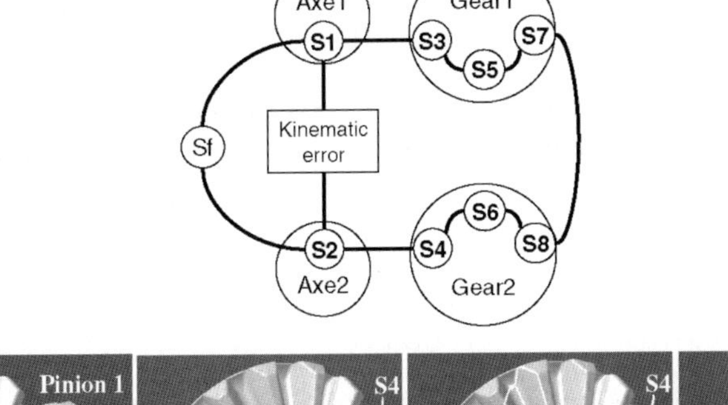


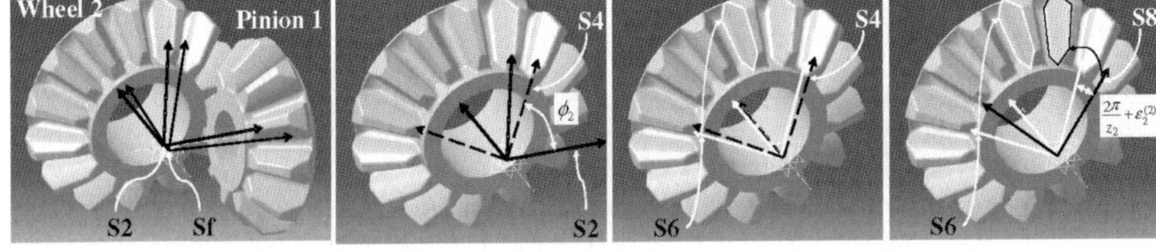


*Figure 1: Coordinate systems of gear.*

From S5 to S7, the nominal transformation is a rotation ($2n\pi/z1$ with n an integer). Some deviations can be introduced in this transformation: an error on cumulative angular pitch.

From S3 to S5, some deviations can be introduced: position and orientation deviations between the axis of the teeth and the hole axis.

From S1 to S3, between fixed coordinate system and rotational coordinate system, rotational parameter of pinion ($\phi_1$) is introduced.

From Sf to S1, some misalignments can be introduced.

The matrix $M_{ij}$ is the homogenous coordinate transform matrix from Sj to Si. In the case of pinion 1, the matrixes are:

$$M57=\begin{bmatrix} 1 & 0 & 0 & 0 \\ 0 & \cos(\sigma^{(1)}) & \sin(\sigma^{(1)}) & 0 \\ 0 & -\sin(\sigma^{(1)}) & \cos(\sigma^{(1)}) & 0 \\ 0 & 0 & 0 & 1 \end{bmatrix}; \quad M35=\begin{bmatrix} \cos(e2) & 0 & -\sin(e2) & TR1 \\ 0 & 1 & 0 & TR2 \\ \sin(e2) & 0 & \cos(e2) & TR3 \\ 0 & 0 & 0 & 1 \end{bmatrix}$$

$$M13(\phi_1)=\begin{bmatrix} 1 & 0 & 0 & 0 \\ 0 & \cos(\phi_1) & \sin(\phi_1) & 0 \\ 0 & -\sin(\phi_1) & \cos(\phi_1) & 0 \\ 0 & 0 & 0 & 1 \end{bmatrix}; \quad Mf1=\begin{bmatrix} \cos(\Delta) & -\sin(\Delta) & 0 & 0 \\ \sin(\Delta) & \cos(\Delta) & 0 & 0 \\ 0 & 0 & 1 & 0 \\ 0 & 0 & 0 & 1 \end{bmatrix} \quad (1)$$

Angle $\sigma^{(1)}$ is equal to $(n-1)\dfrac{2\pi}{z_1}+\varepsilon_n^{(1)}$ with n, the order of tooth and $\varepsilon^{(1)}_n$ the cumulative pitch step error form the first tooth. TR1, TR2, TR3 and e2 are coaxial errors between gear teeth and hole. $\Delta$ is real angle between revolution axes of wheels; it equal to sum of nominal angle and an error $\varepsilon_\Delta$. $\phi_1$ is the rotation angle on pinion.
In the same way, the matrixes M68, M46, M24, Mf2 are defined.

Therefore, in the frame, the equations of $\Sigma1$ and $\Sigma2$ are:

$$\begin{cases} r_f^{(1)}(\varphi,\theta,\phi_1) = M_{f1}.M_{13}(\phi_1).M_{35}.M_{57}.r_7^{(1)}(\varphi,\theta) \\ r_f^{(2)}(\alpha,\gamma,\phi_2) = M_{f2}.M_{24}(\phi_2).M_{46}.M_{68}.r_8^{(2)}(\alpha,\gamma) \end{cases} \quad (3)$$

The nominal active surfaces are conic surfaces with involute to a sphere as lead line:

$$X^{(i)}(\alpha,R) = \begin{cases} R.b^{(i)}.\cos(a^{(i)}\alpha) \\ R.(a^{(i)}.\cos(\alpha).\sin(a^{(i)}.\alpha)+\cos(a^{(i)}).\sin(a^{(i)}.\alpha)) \\ R.(a^{(i)}.\sin(a^{(i)}).\cos(a^{(i)}.\alpha)-\cos(a^{(i)}).\cos(a^{(i)}.\alpha)) \end{cases} \quad (4)$$

With $\delta^{(i)}_b$ the half top base cone angle ; $a^{(i)} = \sin(\delta^{(i)}_b)$ ; $b^{(i)} = \cos(\delta^{(i)}_b)$ ; R is the sphere radius and $\alpha$ is the parameter of lead line.

Therefore, form deviations (waves : $w^{(2)}(\alpha,\gamma)$) are added on the active surfaces of wheel. A form deviations (waves : $w^{(1)}(\varphi,\theta)$) and a longitudinal crowning ($\eta(\theta)$) are added on active surfaces of pinion. Indeed, formulas of finale surfaces are:

$$\begin{cases} \mathbf{r}_7^{(1)}(\varphi,\theta) = \mathbf{X}^{(1)}(\varphi,\theta) + \left[w^{(1)}(\varphi,\theta)+\eta(\theta)\right]\mathbf{n}_7^{(1)} \\ \mathbf{r}_8^{(2)}(\alpha,\gamma) = \mathbf{X}^{(2)}(\alpha,\gamma) + w^{(2)}(\alpha,\gamma)\mathbf{n}_8^{(2)} \end{cases} \quad (5)$$

For example : $w^{(1)}(\varphi,\theta) = A^{(i)}\cos((\varphi+B^{(1)}\theta)C^{(1)})$ and $\eta(\theta) = D+E*\theta+F*\theta^2$

### 2.2. Geometrical behavior description

The mathematical formulation of tolerance analysis is based on the expression of the geometrical behavior of the mechanism; various equations and inequations modeling the geometrical behavior of the mechanism are defined:
-    the equations between displacements of surfaces of parts,
Composition relations of displacements in the various topological loops express the geometrical behavior of the mechanism. The composition relations define compatibility equations between deviation, gaps, …

- the inequations and equations of contacts between parts surfaces nominally in contact,

Interface constraints limit the geometrical behavior of the mechanism and characterize non-interference or association between substitute surfaces, which are nominally in contact [Giordano et al.,1993], [Teissandier et al.,1999], [Dantan et al.,2005].

In the case of gears, the interface constraints are defined by the Tooth Contact determination.

### 2.3. Tooth Contact Analysis (TCA)

The aim of TCA is to obtain the real gear ratio at the mean contact point during the meshing, contact path and orientation and size of contact ellipse [Litvin, 2004]. Only first one is evaluate in this study.

If the teeth surfaces and the relative positions are perfect, the instantaneous gear ratio would be constant. Due to misalignment and parts deviations, this instantaneous kinematic relationship is changing. The relative variations of real gear ratio are minor but accelerations induced are not negligible. Indeed, jump of angular velocity must be avoided in order to reduce noise level and vibrations.

The most difficult in TCA is to solve the system of non-linear equations that traduce contact between the two surfaces. When the position of contact point(s) is (are) known, it's easy to determinate the real gear ratio and transmission error, the contact paths on the gear tooth surfaces, lengths and orientation of contact ellipses.

During the meshing, surfaces $\Sigma 1$ and $\Sigma 2$ are tangential and it is well known the necessary and sufficient conditions for this situation are:

$$\begin{cases} \mathbf{r}_f^{(1)}(\varphi,\theta,\phi_1) = \mathbf{r}_f^{(2)}(\alpha,\gamma,\phi_2) \\ \mathbf{n}_f^{(1)}(\varphi,\theta,\phi_1) = \mathbf{n}_f^{(2)}(\alpha,\lambda,\phi_2) \\ (\varphi,\theta) \in E^{(1)} \\ (\alpha,\gamma) \in E^{(2)} \\ (\phi_1,\phi_2) \in E^{(3)} \end{cases} \quad (6)$$

$$\begin{cases} \mathbf{r}_f^{(1)}(\varphi,\theta,\phi_1) = \mathbf{r}_f^{(2)}(\alpha,\gamma,\phi_2) \\ \mathbf{N}_f^{(1)}(\varphi,\theta,\phi_1) = c.\mathbf{N}_f^{(2)}(\alpha,\lambda,\phi_2) \\ (\varphi,\theta) \in E^{(1)} \\ (\alpha,\gamma) \in E^{(2)} \\ (\phi_1,\phi_2) \in E^{(3)} \\ c \in \text{Real} \end{cases} \quad (7)$$

$E^{(i)}$ is the domain of admissible values.

$$\text{with } \mathbf{n}_f^{(1)} = \frac{\dfrac{\partial \mathbf{r}_f^{(1)}}{\partial u} \times \dfrac{\partial \mathbf{r}_f^{(1)}}{\partial \theta}}{\left\| \dfrac{\partial \mathbf{r}_f^{(1)}}{\partial u} \times \dfrac{\partial \mathbf{r}_f^{(1)}}{\partial \theta} \right\|} \text{ and } \mathbf{n}_f^{(2)} = \frac{\dfrac{\partial \mathbf{r}_f^{(2)}}{\partial v} \times \dfrac{\partial \mathbf{r}_f^{(2)}}{\partial \lambda}}{\left\| \dfrac{\partial \mathbf{r}_f^{(2)}}{\partial v} \times \dfrac{\partial \mathbf{r}_f^{(2)}}{\partial \lambda} \right\|}, \text{ the surfaces unit normal vector.}$$

Like normal vectors are unit length, the second vector equation of (6) leads 2 independent scalar equations. So, the numerical resolution of this equations system is difficult. To solve the system easier, it can be replaced by the system (7), with $\mathbf{N}$ the surface normal vector. The second equation expresses the colinearity between normal vectors. This system has 6 equations with 6 unknowns.

In some cases, the contact point can reach the boundary of one surface. The meshing goes on with the contact point stay on a surface boundary. In this case, the surface normal is not defined and the system (6) or (7) is not valid. It is necessary to replace the normal equation by another. Assume contact point is on line $L\varphi^{(1)}$ defined in S7:

$$\mathbf{L}\varphi_7^{(1)}(\theta) = \mathbf{r}_7^{(1)}(\varphi_0, \theta) \; ; \quad \theta \in D \subset E^{(1)}, \; \varphi_0 \text{ fixed} \tag{8}$$

So, when the contact occurs on $\mathbf{L}\varphi_f^{(1)}$, the tangent of $\mathbf{L}\varphi_f^{(1)}$ at this point is perpendicular to $\mathbf{N}_f^{(2)}$. Then, the new equation is:

$$\frac{\partial \mathbf{r}_f^{(1)}(\varphi_0, \theta, \phi_1)}{\partial \theta} \cdot \mathbf{n}_f^{(2)}(\alpha, \gamma, \phi_2) = 0 \; \text{ or } \; \frac{\partial \mathbf{r}_f^{(1)}(\varphi_0, \theta, \phi_1)}{\partial \theta} \cdot \mathbf{N}_f^{(2)}(\alpha, \gamma, \phi_2) = 0 \tag{9}$$

The new equations of meshing are in this case:

$$\begin{cases} \mathbf{r}_f^{(1)}(\varphi_0, \theta, \phi_1) = \mathbf{r}_f^{(2)}(\alpha, \lambda, \phi_2) \\ \dfrac{\partial \mathbf{r}_f^{(1)}(\varphi_0, \theta, \phi_1)}{\partial \theta} \cdot \mathbf{n}_f^{(2)}(\alpha, \gamma, \phi_2) = 0 \\ \theta \in D \subset E^{(1)} \\ (\alpha, \gamma) \in E^{(2)} \\ (\phi_1, \phi_2) \in E^{(3)} \\ \varphi_0 \text{ fixed} \end{cases} \tag{10} \quad \text{or} \quad \begin{cases} \mathbf{r}_f^{(1)}(\varphi_0, \theta, \phi_1) = \mathbf{r}_f^{(2)}(\alpha, \lambda, \phi_2) \\ \dfrac{\partial \mathbf{r}_f^{(1)}(\varphi_0, \theta, \phi_1)}{\partial \theta} \cdot \mathbf{N}_f^{(2)}(\alpha, \gamma, \phi_2) = 0 \\ \theta \in D \subset E^{(1)} \\ (\alpha, \gamma) \in E^{(2)} \\ (\phi_1, \phi_2) \in E^{(3)} \\ \varphi_0 \text{ fixed} \end{cases} \tag{11}$$

System of equations for meshing is known {(7) or (11)} and the Tooth Contact Analysis is make up to determine $\phi_2$, $\varphi$, $\theta$, $\alpha$ and $\gamma$ in function of $\phi_1$. This mathematical problem has not an explicit solution in general case. We may only have an approximate numerical solution. In this aim, the following method is used:

1. To choose a series of values for $\phi 1$.
2. For each value of $\phi 1$, to solve the system of equation (7) (6 scalar equations and 6 unknowns).
3. If one of the parameters $\{\varphi, \theta, \alpha, \gamma\}$ is not in his validity domain, a new equations system like (11) (4 scalar equations and 4 unknowns) must be solved.
4. To analyze the instantaneous kinematics error, paths of contact …

To solve the non-linear system of equations, Cloutier & Tordion [Cloutier et al., 1967] used a Newtow-Raphson method. It exist new faster method, based on Newton-Raphson method, as Aitken or Steffenson method. But this methods are fastidious to program, they need to compute the Jacobien matrix of $\mathbf{r}_f^{(1)}$, $\mathbf{r}_f^{(2)}$, $\mathbf{N}_f^{(1)}$ and $\mathbf{N}_f^{(2)}$. To avoid this, resolution and analysis are programmed on Matlab®. Levenberg-Marquardt algorithm [Moré, 2005] is employed.

## 3. HOW TO ANALYZE GEAR TOLERANCES WITH A SIMULATION TOOL?

Tolerance analysis involves evaluating the effect of geometrical variations of individual part (gear) on functional characteristics (kinematic error, …). It can be either worst-case or statistical.

In worst-case tolerance analysis (also called deterministic or high-low tolerance analysis), the analysis considers the worst possible combinations of individual tolerances and examines the functional characteristic. In the case of tolerance analysis of gears, we are not sure that the worse kinematic error corresponds to the worse possible configurations of tolerances.

Statistical tolerancing is a more practical and economical way of looking at tolerances and works on setting the tolerances so as to assure a desired yield.

### 3.1. Statistical tolerances analysis

Usually, statistical tolerance analysis uses a relationship of the form:

$$Y = f(X_1, X_2, \ldots X_n) \tag{12}$$

where $Y$ is the response (characteristic such as gap or functional characteristics) of the assembly and $X = \{X_1, X_2, \ldots X_n\}$ are the values of some characteristics (such as situation deviations or/and intrinsic deviations) of the individual parts or subassemblies making up the assembly.

The function f is the assembly response function. The relationship can exist in any form for which it is possible to compute a value for $Y$ given values of $X = \{X_1, X_2, \ldots X_n\}$. It could be an explicit analytic expression or an implicit analytic expression, or could involve complex engineering calculations or conducting experiments or running simulations. The input variables $X = \{X_1, X_2, \ldots X_n\}$ are continuous random variables. In general, they could be mutually dependent.

There are a variety of methods and techniques available for the above computational problem. Essentially, the methods can be categorized into four classes [Nigam,1995]:

·     Linear Propagation (Root Sum of Squares)
·     Non-linear propagation (Extended Taylor series)
·     Numerical integration (Quadrature technique)
·     Monte Carlo Simulation

The linear propagation can possibly be employed if the assembly response function is a linear analytic function. If the assembly response function is non-linear, application of linear propagation could lead to serious errors. In such a case, an extended Taylor series approximation for the relationship f can possibly be employed. For this, f needs to be available in analytic form.

If the function f is not available in analytic form, the quadrature technique and the Monte Carlo simulation can be employed. The quadrature technique need the definition of an approximation of the function f. Monte Carlo technique is easily the most popular tool used in tolerancing problems. Indeed, the appeal of Monte Carlo lies in its applicability under very general settings and the unlimited precision that can be achieved. In particular, Monte Carlo can be used in all situations in which the above three techniques can be used and can yield more precise estimates [Skowronski et al.,1997].

### 3.2. Monte Carlo Simulation

In the case of the statistical tolerance analysis of gears, the function f is not available in analytic form, the determination of the value of Y involve the running simulation. Therefore, we use a Monte Carlo simulation. Monte Carlo simulation proceeds:
Pseudo random number generators are used to generate a sample of numbers $x_1, x_2, \ldots$ $x_n$, belonging to the random variables $X_1, X_2, \ldots X_n$, respectively. The value of Y , $y_1 =$ $f(x_1, x_2, \ldots x_n)$, corresponding to this sample is computed. This procedure is replicated a large number of times, N times. This would yield a random sample, $\{y_1, \ldots y_N\}$ for Y . Standard statistical estimation methods are then used to analyze the distribution of Y . The precision of this statistical analysis increases proportional to $\sqrt{N}$ and therefore unlimited precision can be achieved through large number of replications.

### 3.3. Application

We use the statistical tool to analyze effect of variations of geometrical parameters defining wheels and the assembly on the kinematic error without torque. We have fixed an interval for variation of each parameter
For this simulation, gear parameters are given in following tabular:

Number of teeth for pinion :                 $z_1 = 11$

Number of teeth for wheel :                  $z_2 = 15$

Nominal shaft angle :                        $\Delta = \pi/2$

Sphere radius at small tip :                 $Rp = 50$ mm

Sphere radius at big tip :                   $Rg = 100$ mm

Maximum of crowning :                        $B = 1$ mm

Tooth height :                               $h = 0.1613*$(sphere radius)

In first time, we have run simulations for each kind of failing then with all of them. All simulations have been done with form deviations due to the manufacturing process. Only some of these results are shown here.

Figure 2A presents effect of the misalignment between rotation axis on the kinematic error. $\varepsilon$ is the angle deviation and {A1, A2, A3} is the position vector from top cone of wheel 2 to top cone of pinion 1. As attempted, profiles of actives surfaces being spherical involute profile, kinematic error is independent form $\varepsilon_\Delta$, for $\varepsilon_\Delta \in [-0.02 , 0.02]$. Each value of A1, A2 and A3 is chosen with uniform random manner. For 20 shots, results are shown in figure A. Kinematic error is on the inner interval $[-7.10^{-4} , 5.5.10^{-3}]$.

The program is replicated 1010 times and the kinematic error amplitude is stored. We count them number in each class of $10^{-4}$ radian. The result is the probability function of the kinematic error amplitude (figure 2B).

During manufacturing, misalignments could occur between active surfaces and rotation axis. This default is simulated by an angle between axis and a vector of translation. This parameters can take any value with a uniform distribution, in interval [-

0.02, 0.02] for angle and [-0.02, 0.02] for each component of translation vectors. The result is on figure 2C.

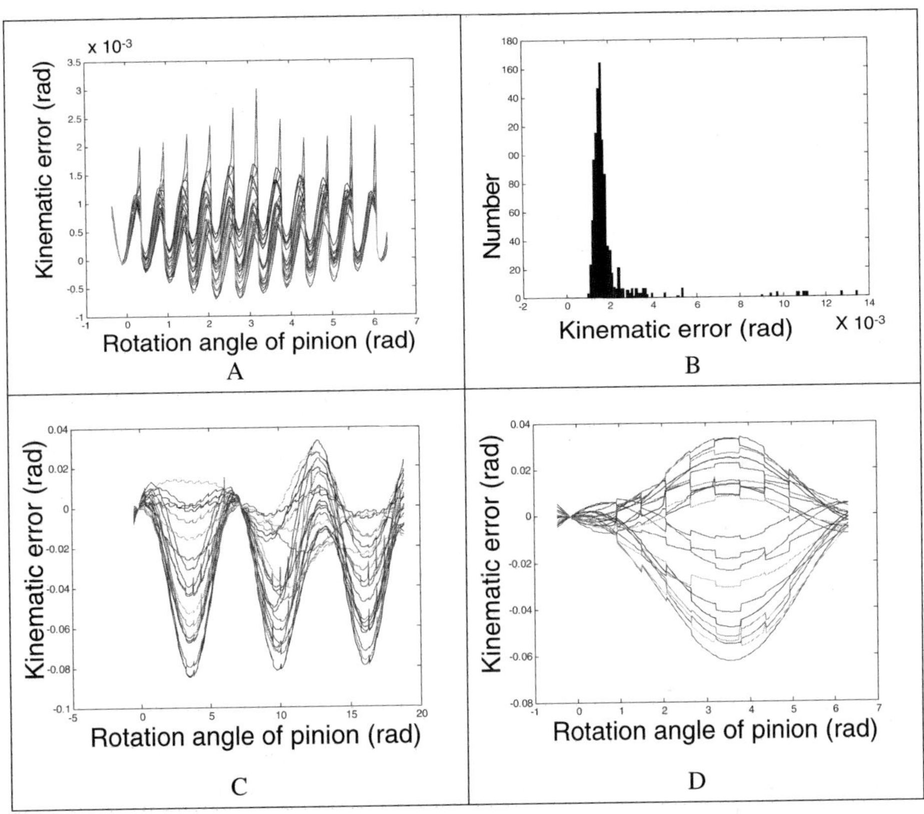

*Figure 2: Result of statistical analysis.*

More simulations of individual default have been done but not presented here. Figure 2D shows 20 simulations results with all random defaults (waves on actives surfaces, misalignment and displacement between pitch cones, errors on cumulative circular pitch, misalignment between actives surfaces and rotation axis).

## 4. CONCLUSION

Statistical tolerance analysis seems to be the best tool to analyze the behavior of a gear box. We have applied this method to analyze the kinematic error in a bevel gear. We modelize the geometrical deviation and the geometrical surfaces; we use Tooth Contact Analyze tool to compute kinematic error at each position; and we analyse the impacts of deviations on the kinematic error by Monte Carlo simulation. Therefore, numerous calculi allow to extract some pertinent information about the geometrical behaviour of gear.

To increase relevance of this work, we will take account the teeth distortions, ie analyse the transmission errors under load (Load Tooth Contact Analysis).

## REFERENCES


**[Ballot et al.,1997]** Ballot E, Bourdet P. "A Computation Method for the Consequences of Geometric Errors in Mechanisms". *Proceedings of CIRP Seminar on Computer Aided Tolerancing*, Toronto, Canada, April 1997.

**[Dantan et al.,2005]** Dantan JY, Mathieu L, Ballu A, Martin P. "Tolerance synthesis : quantifier notion and virtual boundary". *J. Computer Aided Design, 2005*, Vol.37, n°2, pp.231-240.

**[Godet et al.,1967]** Godet J-C, Raffy M. "Le calcul des engrenages coniques droits à développante de cercle sphérique", *Société d'Etude de l'industrie de l'engrenage*, Volume 53, 1967

**[Giordano et al.,1997]** Giordano M, Duret D. "Clearance space and deviation space, application to three-dimensional chain of dimensions and positions". *Proceedings of CIRP Seminar on Computer Aided Tolerancing*, ENS Cachan, France, May 1993.

**[Litvin,2004]** Litvin F.L., "Gear Geometry and Applied Theory", *PTR Prentice Hall*, Englewood Cliffs, NJ, 2004.

**[Nigam et al.,1995]** Nigam SD, Turner JU. "Review of statistical approaches to tolerance analysis". *J. Computer Aided Design*, 1995, Vol.27, n°1, pp.6-15.

**[Roy et al.,1999]** Roy U, Li B. "Representation and interpretation of geometric tolerances for polyhedral objects". *J. Computer Aided Design*, 1999, Vol.31, n°4, pp.273-285.

**[Sacks et al.,1997]** Sacks E, Joskowicz L. "Parametric kinematic tolerance analysis of planar mechanisms". *J. Computer Aided Design,* 1997, Vol. 29, n°5, pp. 333-342.

**[Skowronski, et al.1997]** Skowronski VJ, Turner JU. "Using Monte Carlo variance reduction in statistical tolerance synthesis". *J. Computer-Aided Design*, 1997, Vol.29, n°1, pp.63-69.

**[Teissandier et al.,1999]** Teissandier D, Delos V, Couetard Y. "Operations on polytopes application to tolerance analysis". *Proceedings of CIRP Seminar on Computer Aided Tolerancing*, Enschede, Netherlands, March 1999.

**[Zou et al.,2004]** Zou Z, Morse EP. "A gap-based approach to capture fitting conditions for mechanical assembly". *J. Computer Aided Design*, Vol.36, pp691-700, 2004.

**[Cloutier et al.,1967]** Cloutier L-J., Tordion G-V., "Méthode générale d'analyse du contact des engrenages du type Wildhaber-Novikov aux axes quelconques", *Société d'Etude de l'industrie de l'engrenage*, Volume 51, 1967

**[Moré, 1977]** Moré, J. J., "The Levenberg-Marquardt Algorithm: Implementation and Theory," Numerical Analysis, ed. G. A. Watson, Lecture Notes in Mathematics 630, Springer Verlag, pp. 105-116, 1977.

# Tolerance Analysis and Allocation Using Tolerance-Maps for a Power Saw Assembly


*A. D. Jian, G. Ameta*, J. K. Davidson*, J. J. Shah**

**Mechanical and Aerospace Engineering Department,*
*Arizona State University, Tempe, AZ -85287, USA*
*ameta@asu.edu , j.davidson@asu.edu*



**Abstract:** This paper concerns the tolerance analysis of a swingarm power saw assembly to determine the sensitivities of the tolerances that influence both the orientation and the side-to-side position of the circular blade. The stackup conditions are developed using the new bi-level mathematical model (Tolerance-Maps) for geometric tolerances that has been under development at Arizona State University. The model is compatible with the ASME Y14.5 standards for geometric tolerances. Each Tolerance-Map$^{®1}$ (T-Map$^®$) is a hypothetical Euclidean point-space, the size and shape of which reflect all variational possibilities for a feature. Each is the range of points that result from a one-to-one mapping from all the variational possibilities of that feature. The saw assembly combines several individual features, such as an axis, a round face, and a rectangular face, in the stackup. Therefore, T-Maps for these features are added together to form an accumulation map. The paper includes an abbreviated summary of T-Maps, a description of the saw assembly, the development of the stack up equations, and an allocation scheme for tolerances. Material condition (bonus tolerance) is not considered. The source of this work is [Jian, 2001].
**Keywords:** tolerance allocation, tolerance analysis, Tolerance-Map.


## 1. TOLERANCE-MAPS FOR SELECTED FEATURES

The methods for creating Tolerance-Maps that represent tolerance-zones have been presented in [Davidson, *et al.*, 2002], [Bhide, *et al.*, 2001], [Mujezinović, *et al.*, 2004], and [Bhide, *et al.*, 2005]. Each T-Map is a convex hypothetical Euclidean point-space that represents all the variations possible for a feature, or target feature, in an assembly. The tolerances that are specified for a feature are reflected by the size and shape of the T-Map.

The 3-D Tolerance-Maps for round and rectangular faces are shown in Figs. 1(b) and (c). As one example, the end of a rectangular bar of cross-sectional dimensions $d_y$ x $d_x$ is

---


$^1$ Patent No. 69638242.





*J.K. Davidson (ed.), Models for Computer Aided Tolerancing in Design and Manufacturing, 267–276.*
© 2007 *Springer.*

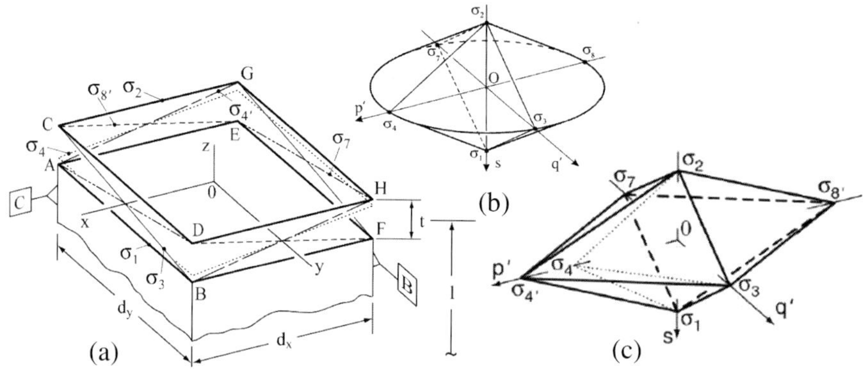

*Figure 1. (a) The end of a rectangular bar with size tolerance t; the vertical scale in the tolerance-zone is exaggerated. Drawn with $d_y > d_x$. (b) The double-cone T-Map® (three dimensional range of points), for the size tolerance t applied to a round bar; the double cone has dimension $\sigma_1\sigma_2 = t$ and rim-radius $O\sigma_1 = t$. (c) The T-Map® for the tolerance-zone on the rectangular bar shown in (a); $\sigma_3\sigma_7 = t$ and $\sigma_{4'}\sigma_{8'} = td_y/d_x$.*

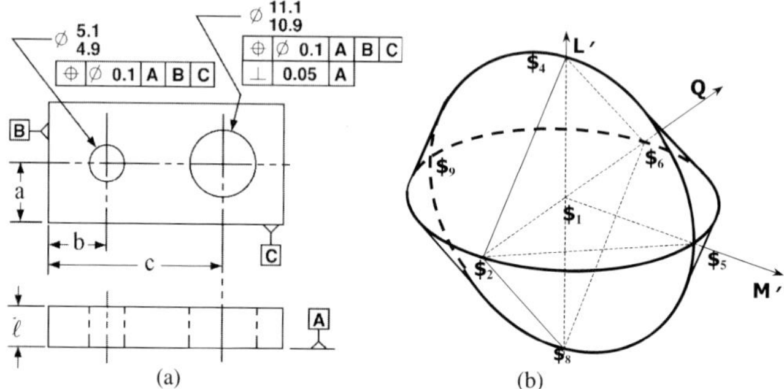

*Figure 2. a) Two holes in a plate of thickness $\ell$. Both holes are located with the tolerance t = 0.1 mm. The larger hole is to be held perpendicular to Datum A with the tolerance t″ = 0.5 mm. b) One of the 3-D hypersections (L′M′Q) of the T-Map (hypothetical 4-D point-space) that represents the range of the position variation of an axis (tolerance t″ is not applied). The only edges are the two circles shown; both have diameter t. The points $\$_i$ are points of the T-Map that correspond to lines in the tolerance-zone [Bhide, et al. 2005].*

shown in Fig. 1(a) with a highly exaggerated tolerance on its length $l$. According to the Standards [e.g. ASME, 1994], all points of the end-face must lie between the limiting planes $\sigma_1$ and $\sigma_2$, and within the rectangular limit of the face. The T-Map for this rectangular face is developed by identifying the planes $\sigma_1$, $\sigma_2$, $\sigma_3$, and $\sigma_4$ as the basis-planes in the *tolerance-zone* and then establishing corresponding basis-points $\sigma_1$, $\sigma_2$, $\sigma_3$, and $\sigma_4$ in the *hypothetical T-Map space* as shown in Fig. 1(c). To avoid confusion, the same labels are used. If the face were circular instead of rectangular, the shape of the T-Map would be as shown in Fig. 1(b). The $p'$- and $q'$-axes of both T-Maps represent the orientational variations of the plane while the $s$-axis represents the translational variations of the plane. Therefore, it is quite evident from Fig 3 that the orientational and translational variations of the plane are uncoupled. If additional orientational control for either parallelism or perpendicularity, using a tolerance $t''$, were desired, the T-Maps (Figs 1(b) and (c)) would be truncated at tolerance $t''$ along the appropriate orientational axes labeled $p'$ and $q'$.

By positioning the basis-points $\sigma_1$, $\sigma_2$, $\sigma_3$, and $\sigma_4$, as in Figs. 1(b) and (c), the dipyramidal shape in Fig. 1(b) for the rectangular face conforms with diagrams and spaces presented by others for representing size-tolerances. [Whitney, *et al.*, 1994] obtained the shape using an intuitive argument. [Roy and Li, 1999] used inequalities to establish a variation zone of acceptable ranges of the coordinates for any plane in the tolerance-zone. [Giordano, *et al.*, 1999, 2001] get a dipyramidal deviation space using the same method.

The T-Map for an axis, such as for one of the holes in the plate of Fig. 2(a), is a 4-D solid of points. Although this solid cannot be viewed directly, it can be visualized with 3-D hypersections in which one of the four coordinates, $L'$, $M'$, $P$, and $Q$, is held fixed; one representative hypersection of it is shown in Fig. 2(b). (Note: In the plan (top) view of Fig. 2(a), $L'$ and $M'$ represent tilts of the axis to the left-or-right and fore-or-aft, respectively, and $P$ and $Q$ represent translations of the axis in the same directions, respectively.) The method of [Giordano, *et al.*, 1999, 2001] also gives the shape as that in Fig. 2(b).

## 2. DESCRIPTION OF THE SAW-ARM ASSEMBLY

The overall arrangement of parts is shown in Figs. 3 and 4. The important features and dimensions can be grouped according to their appearance on three parts: the saw arm, the gear cover, and the arbor.

The saw arm has Datums A, B and C (points) for clamping the raw casting in the first machining set-up; they are considered perfect. The arm contains the important face, Datum $D_1$, which engages the gear cover. A size tolerance $t_M = 0.16$mm is specified on the distance $l_{MD} = 47.13$mm between surface $D_1$ and the reference plane M. In addition, a tolerance $t''_M = 0.10$mm controls the amount of orientational variation between faces M and $D_1$. Two short coaxial holes E and F are bored separately on the two sides of the arm-casting, and they engage the hinge-pin that is mounted to the frame. A positional tolerance $t_E = 0.6$mm locates the axis of hole E (length $h_E$) relative to datums A, B and C, and the positional tolerance $t_F = 0.07$mm locates the axis of hole F relative to axis of hole E (Datum E).

Datum Face M in the arm engages with the frame and limits axial motion in one direction along the hinge.

At the left in Fig. 4(c) coaxial hole $H_1$ is for mounting the needle bearing that supports the left end of the spindle, and coaxial hole $G_1$ provides the radial alignment of the gear cover. Hole $H_1$ is positioned relative to datums A, B and C with the tolerance $t_{H1} = 0.3$mm; it is kept perpendicular to Datum $D_1$ with the small tolerance $t''_{H1} = 0.05$mm. Hole $G_1$ has a size tolerance of 0.026mm, and it is positioned relative to the axis of $H_1$ with the tolerance $t_{G1} = 0.05$mm. At the right in Fig. 4(c) the gear cover has three important features: face $D_2$ that engages face $D_1$ on the gear casing on the saw arm, cylinder $H_2$ that maintains radial

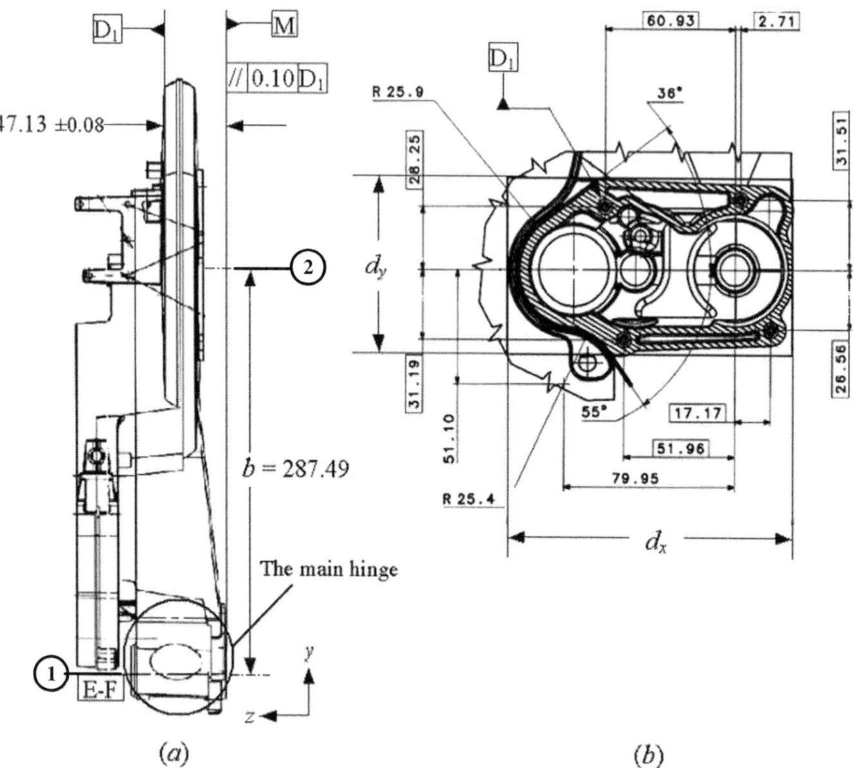

(a)        (b)

*Figure 3. Selected dimensions and tolerances for the arm and (a) Front view of the arm. (b) Partial side view at a larger scale showing details of surface $D_1$ on the arm that will be approximated by a $d_x$x$d_y$ rectangular surface).*

| Stackup frames | Axis | Plane |
|---|---|---|
| 1 | E-F | M |
| 2 | $H_1$-$H_2$ | $D_1$ |
| 3 | J-K | Q |
| 4 | J-K | $P_a$ |
| 5 | J-K | $P_b$ |

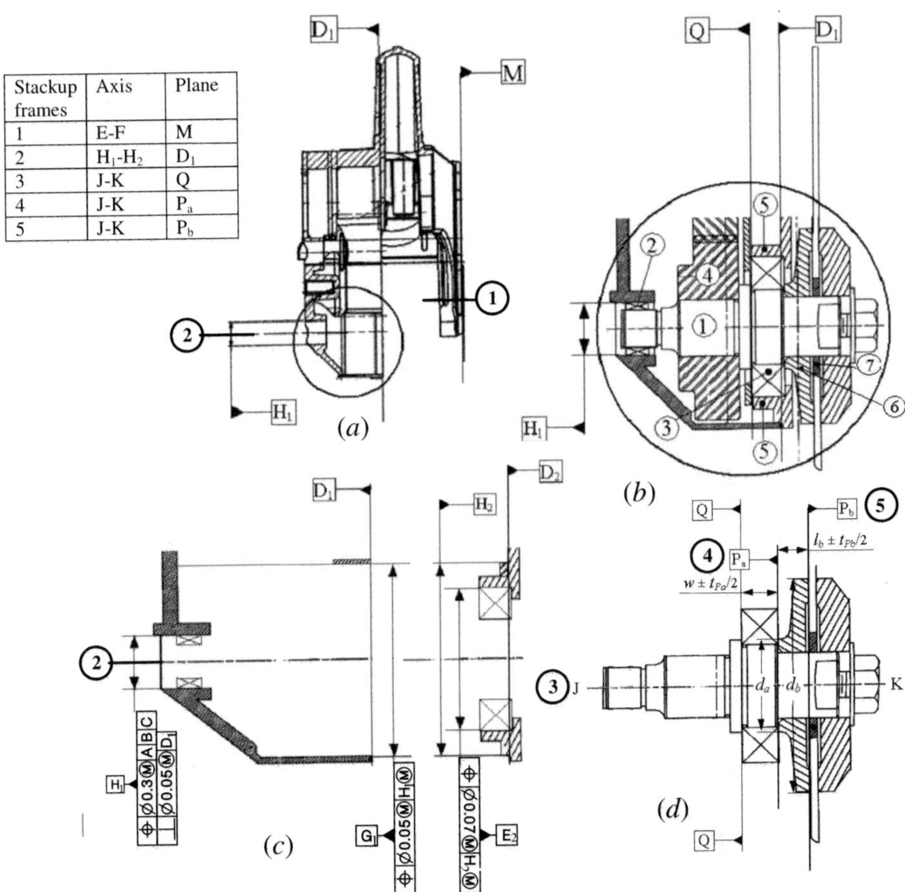



*Figure 4. The arrangement of the parts that support the spindle in the gear casing of the arm. (a) A bottom view of the gear casing and a cross-section of the arm from the main hinge. (b) A more detailed view of the spindle subassembly at a larger scale. 1) Spindle; 2) Needle bearing; 3) Ball bearing; 4) Gear; 5) Gear case cover; 6) Collar; 7) Blade. (c) The mating between the gear casing on the arm (left) and the gear case cover (right); the datums and the target features on both parts are shown. (d) The contacting plane $P_a$ is the contact between the ball bearing and the collar; its tolerance on size is controlled relative to Plane Q. The contacting plane $P_b$ is the contact between the collar and the saw blade; its tolerance on size is controlled relative to Plane $P_a$. Note that both planes $P_a$ and $P_b$ are round faces of diameters $d_a$ and $d_b$, respectively.*

alignment with the gear cover by fitting tightly into hole $G_1$ of the gear casing, and pocket $E_2$ for mounting the ball bearing that supports the right end of the spindle.  The position tolerance $t_{E2} = 0.07$mm keeps pocket $E_2$ concentric with $H_2$, and tolerance $t_{Cs}$ controls the depth of the pocket relative to $D_2$.

The spindle (Fig. 4(d)) has cylindrical surfaces J and K that engage the needle and ball bearings, respectively. The coaxiality of surface K with Datum J is controlled by a position tolerance of $t_K = 0.038$mm. Four parts are clamped together against the annular face Q on the spindle: the inner race of the ball bearing (size tolerance $t_w = t_{Pa}$ on the width of *each race*), the collar (size tolerance $t_{Pb}$), the saw blade, and the washer. Face Q is held perpendicular to the spindle axis J-K within the tolerance $t''_Q = 0.10$mm. Overall the stack up frames are as listed in Fig. 4 (numbered 1 through 5), starting with the coordinate frame formed by axis plane pair E-F & M and ending with the coordinate frame formed by axis plane pair J-K & $P_b$.


## 3. TOLERANCE ALLOCATION AND ANALYSIS

The objective of this analysis is to control the orientation and lateral position of the saw blade. The sequence of engaging parts from the frame to the saw blade are :  axis E-F (orientation of the arm) and face M (position of the arm) ; surfaces D, $G_1$, and $H_1$; seating of the ball bearing in the pocket $E_2$; support of the spindle at the two bearings; and clamping of the blade, collar, and inner race of the ball bearing against face Q on the spindle. The target is the face of diameter $d_b$ on the collar (Fig. 4(d)). Starting at the blade, the T-Map in Fig. 5(a) is for the stackup from Q (bearing inner race and the collar); its height is the sum of the last 3 terms in eqn (2), and its radius is the last 2 terms in eqn (3). It was formed using the methods in [Davidson, *et al.*, 2002]. Fig. 5(a) reflects a truncation of Fig. 1(b) by a cylinder. This shape comes from the orientational tolerances $t''_{Pa}$ and $t''_{Pb}$, respectively, that could be applied to the faces $P_a$ and $P_b$.

The T-Map in Fig. 5(b) is for the stackup to face Q from the Datum $D_2$ on the gear cover. The maximum tilt of Q comes from the orientational tolerance $t''_Q$ on Q relative to J-K and from $t_S/h_{H1\text{-}E2}$, where $t_S = t_{G1} + t_{E2} + t_{ecc1} + t_{ecc2} + t_K$ is the positional misalignment of the two axes J and K on the spindle. Values $t_{ecc1}$ and $t_{ecc2}$ are the circular runouts of the two bearings. However, the maximum translational variation of Q is the sum of the tolerances on pocket depth ($t_{Cs}$) and width of the outer bearing race ($t_w$).  Therefore, position and tilt variations are not coupled, and the T-Map is a cylinder with height $t_{Cs} + t_w$ and radius $t_S$.

The T-Map in Fig. 5(c) is for the stackup from the hinge-pin on the frame to the target $d_x$ x $d_y$ rectangular face $D_1$. Then, face M alone (tolerance $t_M$) accounts for the lateral position of the arm from its contact with the frame of the saw, but it forms only a point contact. Presume at first that offset $b$ (Fig. 3(a)) is zero.  Since tilt variations come from tilt of the axis E-F on the arm (limit $t_E + t_F$) and are uncoupled to positional variations (limit $t_M$), the T-Map would be a rhombic prism with height $t_M$ and diagonal of base $\sigma_{4'}\sigma_{8'} = t_E + t_F$ (Fig. 1(c)). When the offset $b$ is introduced, any misalignments at holes E and F produce an additional lateral displacement of the blade. This skews the prismatic T-Map vertically, as shown in Fig. 5(c) (see [Bhide, et al., 2001]). When these three T-Maps are combined with the Minkowski sum, the accumulation T-Map arises; its size and shape (Fig. 5(d)) are represented with dimensions $a$ and $c$ (eqns (1)) in Fig. 5(e). T-Maps are always convex [Davidson, et al., 2002].

The tolerances in the entire assembly should be adjusted so the accumulation T-Map (Fig. 5(d)) will just fit inside a functional T-Map that represents (i) all the variations in orientation that are acceptable to woodworkers and (ii) the unit-to-unit variations in position acceptable to the manufacturer of the power saw, as reflected in variations of target face $P_b$ on the round collar. Since the target face is circular, its functional T-Map will be a double cone that is truncated, thereby allowing for some additional orientation control. It will have the same shape as Fig. 5(a) but have a vertex-to-vertex dimension $t_f$. Following the line of thought in [Mujezinović, et al., 2004], stackup equations can be found by fitting the Minkowski sum of Figs. 5(b) and (c) within a Minkowski difference of the functional T-Map and Fig. 5(a). A cross-section of this fit is shown Fig. 5(e); the dimension $e$ of the figure, along with dimensions $a$ and $c$ are given by

$$a = \frac{d_b}{2}\left(\frac{t_F}{h_{EF}} + \frac{t_E}{h_E}\right) + \frac{d_b}{d_Q}t''_Q + \frac{1}{2}\left(\frac{d_b}{h_{H1-E2}}\right)t_S; \qquad e = t''_f - t''_{Pb} - \left(\frac{d_b}{d_a}\right)t''_{Pa};$$

$$c = \frac{d_b}{d_Q}t''_Q + \frac{1}{2}\left(\frac{d_b}{h_{H1-E2}}\right)t_S - \frac{d_b}{2}\left(\frac{t_F}{h_{EF}} + \frac{t_E}{h_E}\right)$$

$$(1)$$

The stackup equation can be written as $t_f = a + t_M + t_Q + b(t_F/h_{EF} + t_E/h_E) + t_{Pa} + t_{Pb} + (d_b - d_a)t''_{Pa}/d_a$; further, the cylinder radius of the accumulative T-Map is $t''_a = a + t''_{Pb} + d_b t''_{Pa}/d_a$. In these, the quantities $t_f$ and $t''_a$ represent accumulated positional and orientational variations (relative to the frame) for the edge of the 44mm diameter face of the collar that engages the blade. Expanding the stack up equations we get

$$t_f = t_M + \left(t_{Cs} + t_w\right) + \left(b + \frac{d_b}{2}\right)\left(\frac{t_F}{h_{EF}} + \frac{t_E}{h_E}\right) + \left(\frac{d_b}{d_Q}\right)t''_Q$$

$$+ \left(\frac{d_b}{2h_{H1-E2}}\right)\left(t_{G1} + t_{E2} + t_{ecc1} + t_{ecc2} + t_K\right) + t_{Pb} + t_{Pa} + \left(\frac{d_b}{d_a} - 1\right)t''_{Pa}$$

$$(2)$$

$$t''_a = \left(\frac{d_b}{2}\right)\left(\frac{t_F}{h_{EF}} + \frac{t_E}{h_E}\right) + \left(\frac{d_b}{d_Q}\right)t''_Q + \left(\frac{d_b}{2h_{H1-E2}}\right)\left(t_{G1} + t_{E2} + t_{ecc1} + t_{ecc2} + t_K\right)$$

$$+ t''_{Pb} + \left(\frac{d_b}{d_a}\right)t''_{Pa}$$

$$(3)$$

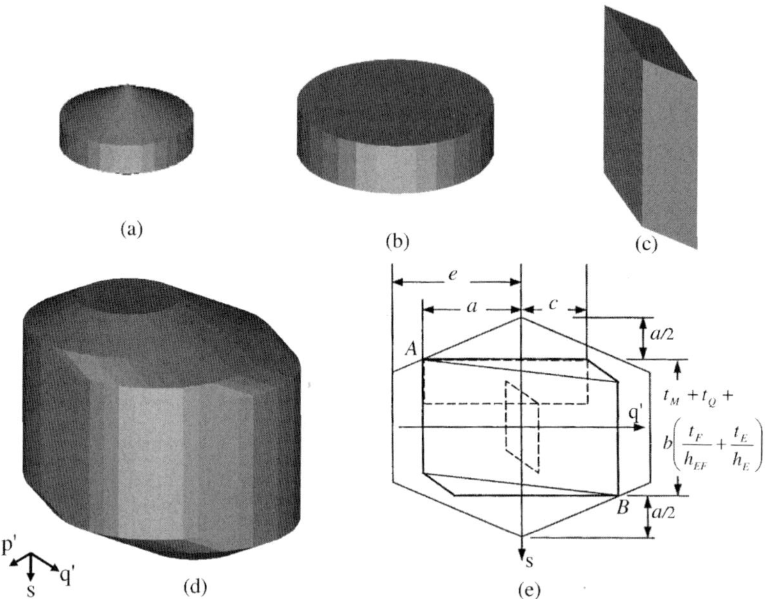



*Figure 5. Three Tolerance-Maps and their sum, all of which are conformable to target face* $P_b$ *on the collar. a) The T-Map for* $P_b$ *relative to face Q on the spindle. b) The T-Map for target face Q relative to face D. c) The T-Map for target face* $D_1$ *which accounts for the amplification of positional variations arising from the offset b between axis E-F and the center of* $D_1$*. d) The accumulative T-Map is a cylinder with four flattened sides and complex ends and results from the Minkowski sum of (a), (b), and (c). e) The q'-s section of the Minkowski sum (interior solid lines) of (b) and (c), together with the hexagonal Minkowski difference of the functional T-Map, which has the form of (a). The points of contact are labeled A and B. The dimensions a, b and c are represented in equation (1).*

*Table 1. Values of tolerances obtained from eqn (2) when each term is set to 0.0833 mm.*

| $t_M$ | $t_{Cs}$ | $t_w (= t_{Pa})$ | $t_E$ | $t_F$ | $t''_Q$ |
|---|---|---|---|---|---|
| $t_f/12$ <br> =0.0833mm | $t_f/12$ <br> =0.0833mm | $t_f/(12*2)$ <br> =0.0416mm | $t_f/(12*15.56)$ <br> =0.0053mm | $t_f/(12*4.06)$ <br> =0.016mm | $t_f/(12*2.0)$ <br> =0.041mm |
| $t_{G1}$ | $t_{E2}$ | $t_{ecc1}$ | $t_{ecc2}$ | $t_K$ | $t_{Pb}$ |
| $t_f/(12*0.51)$ <br> =0.163mm | $t_f/(12*0.51)$ <br> =0.163mm | $t_f/(12*0.51)$ <br> =0.163mm | $t_f/(12*0.51)$ <br> =0.163mm | $t_f/(12*0.51)$ <br> =0.163mm | $(1/12)t_f$ <br> =0.0833mm |

The sensitivities on the tolerances in eqn (2) can be balanced by assigning an equal value to each of the 12 terms.  Certainly $t''_{Pa}$, an orientational tolerance assigned by the bearing manufacturer to the width dimension of the bearing, may, or may not, be distinct from $t_{Pa}$, the size tolerance on this same width. For purposes of this sample computation, we set $t''_{Pa} = t_{Pa}$ and combine these terms with the one for $t_w$. Then, when the value of $t_f$, the functional tolerance, is normalized to 1mm, each term becomes $(1/12)t_f = 0.0833$mm.  The values obtained for the tolerances are shown in Table 1. Although the equality of terms in eqn (2) may or may not give the lowest manufacturing cost, nonetheless, eqn (2) shows the relative importance of all the tolerances in positioning the blade laterally.

Equation (3) represents the accumulation of angles (tilt) from the hinge supporting the arm to the saw blade. The accumulated orientational variations, $t''_a$, is calculated by substituting the allocated tolerance values from Table 1 into equation (3), which gives,

$$t''_a = (22/2)*(0.016/76.27+0.0053/12.06)+2*0.041+0.51*(5*0.163)+0.0833+2*0.00416$$

$$= 0.5964\text{mm}$$

This number represents an orientational tolerance-zone of diameter 44mm where the collar contacts the blade. From the above calculation we see that the value for $t''_a$ already gives substantial control of orientation. If additional orientational control $t''_f$ $(t''_f < t''_a)$ were required, it could be obtained by imposing tighter orientational tolerance $t''_{Pb}$ on the ends of collar.


CONCLUSION

In this paper we have demonstrated the procedure for tolerance analysis using T-Maps. This example shows how to combine the variations of an axis and a plane that contribute to the desired control of the variations at the target feature. The procedure for obtaining the stack up equation from the accumulation T-Map using the target feature was elucidated. Tolerance allocation was achieved using a simple scheme of equal contribution from the terms of the stack up equation.


ACKNOWLEDGEMENT

The authors are grateful for funding provided to this project by National Science Foundation Grants #DMI-9821008 and #DMI-0245422.

REFERENCES

[**ASME Standard, 1994**] ASME Y14.5M.; "Dimensioning and Tolerancing"; The American Society of Mechanical Engineers, NY.

[**Bhide, *et al.*, 2001**] Bhide, S., Davidson, J.K., and Shah, J.J.; "Areal Coordinates: The Basis for a Mathematical Model for Geometric Tolerances," In *Proc., 7th CIRP Int'l Seminar on Computer-Aided Tolerancing*, Ecole Norm. Superieure, Cachan, France, April 24-25, (ed. P. Bourdet and L. Mathieu), pp. 35-44. Kluwer.

[**Bhide, *et al.*, 2005**] Bhide, S., Ameta, G. Davidson, J.K., and Shah, J.J.; "Tolerance-Maps Applied to the Straightness and Orientation of an Axis," In *CD ROM Proc., 9th CIRP Int'l Seminar on Computer-Aided Tolerancing*, Arizona State University, Tempe, AZ, USA, April 10-12.

[**Davidson, *et al.*, 2002**] Davidson, J.K., Mujezinović, A., and Shah, J. J. "A New Mathematical Model for Geometric Tolerances as Applied to Round Faces", *ASME Transactions, J. of Mechanical Design*, **124**, pp. 609-622.

[**Giordano, *et al.*, 1999**] Giordano, M., Pairel, E., and Samper, S. (1999). "Mathematical representation of tolerance zones." In *Global Consistency of Tolerances*, Proc., 6th CIRP Int'l Seminar on Computer-Aided Tolerancing, Univ. of Twente, Enschede, Netherlands, March 22-24 (ed. F. vanHouten and H. Kals), pp. 177-86.

[**Giordano, *et al.*, 2001**] Giordano, M., Kataya, B., and Samper, S. "Tolerance analysis and synthesis by means of clearance and deviation spaces." In *Geometric Product Specification and Verification*, Proc., 7th CIRP Int'l Seminar on CAT, Ecole Norm. Superieure, Cachan, France, April 24-25, (eds. P. Bourdet and L. Mathieu), pp. 345-354.

[**ISO, 1983**] International Organization for Standardization ISO 1101. (1983). *Geometric tolerancing—Tolerancing of form, orientation, location, and run-out—Generalities, definitions, symbols, and indications on drawings.*

[**Jian, 2001**] Jian A.D. (2001), *The Tolerance -Map*® *and its application to one stack up in a power saw*. M.S. Thesis, Arizona State University.

[**Mujezinović, *et al.*, 2004**] Mujezinović, A., Davidson, J.K., and Shah, J. J. "A New Mathematical Model for Geometric Tolerances as Applied to Polygonal Faces", *ASME Trans., J. of Mechanical Design*, **126**, pp. 504-518.

[**Roy and Li, 1999**] Roy, U. and Li, B. (1999). "Representation and interpretation of geometric tolerances for polyhedral objects– II.: Size, orientation and position tolerances", *Computer-Aided Design*, **31**, pp. 273-285.

[**Whitney, *et al.*, 1994**] Whitney, D. E., Gilbert, O. L., and Jastrzebski, M. (1994). "Representation of geometric variations using matrix transforms for statistical tolerance analysis in assemblies", *Research in Engineering Design*, **6**, pp. 191-210.

# Error Analysis of a NanoMechanical Drill


*A. Bryan*, J. Camelio*, S. J. Hu**
**Department of Mechanical Engineering,*
*University of Michigan, Ann Arbor, MI 48108*
*jackhu@umich.edu*

*N. Joshi*, A. Malshe**
**Mechanical Engineering Department,*
*University of Arkansas, Fayetteville, AR 72701*
*apm2@engr.uark.edu*



**Abstract:** With the use of new materials and nanoprocessing techniques such as layered deposition and surface micromachining, a three dimensional nanodrill has been successfully manufactured [O' Neal *et al.*, 2002]. The nanodrill is intended for drilling holes on the order of a few hundred nanometers. Several applications can be envisioned for such a device, from uses in data storage technologies to the creation of microfluidic channels. Due to the high accuracies often required for technologies on this scale, the dimensional quality of the final hole is of interest. The error analysis performed in this paper is used to determine the final error in the size and position of the drilled hole due to static and kinematic effects. A linearized sensitivity approach is used to identify the most important factors influencing the hole's quality. The results indicate that the high tolerances in the existing drill's architecture make it impossible to obtain holes of the proposed dimensions.

**Keywords**: Nanotechnology, Error, Dimensional Quality, Kinematic, Static


## 1. INTRODUCTION

Electromechanical devices on the micro and nano scales have found a wide range of applications. Traditionally, these devices were planar and were produced with the same techniques used in the manufacture of integrated circuits. However, novel manufacturing techniques, specific to micro and nano devices, have emerged [Judy, J., 2001]. These new techniques have led to the development of 3-D mircro and nano structures [O' Neal *et al.*, 2002]. Two examples of these new 3-D structures are MIT's microengine [Khanna, R., 2003] and IBM's nanodrive [Vettiger *et al.*, 2003]. The emergence of these 3-D devices necessitates the development of novel manufacturing processes and technologies. The University of Arkansas has proposed and developed the nanodrill for the drilling of nanochannels as one such new manufacturing technique.

The nanodrill is intended for drilling holes of 100-300nm in diameter and 50-100nm deep. An actuator system connected to a drive gear rotates the load gear on




*J.K. Davidson (ed.), Models for Computer Aided Tolerancing in Design and Manufacturing, 277–287.*
© 2007 *Springer.*

which the nanodrill is mounted as shown in Figures 1 (a)-(b). The drill's system consists of several layers of surface micromachined materials as shown in the schematic in Figure 1 (c). Although the development of a new single axis machine on the macro-scale may go unnoticed, the development of such a tool on the nanoscale is a significant achievement.

However, several limitations can be envisioned when drilling holes with this proposed device. Some of the concerns that immediately arise include the capability of the drill to actually produce a hole, the potential material applications, the expected life of the drill and the achievable dimensional quality of the hole. In addition to these macro-scale considerations, micro and nano devices experience high forces of attraction which may affect the tool's functionality.

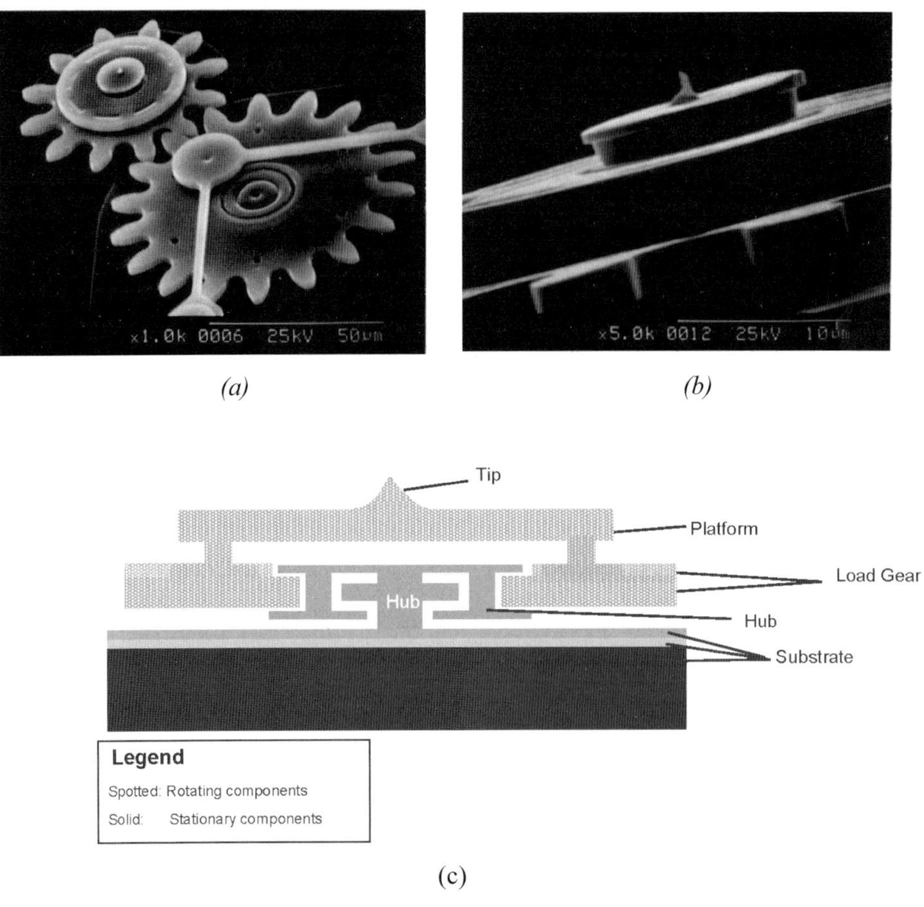


*Figure 1 (a) Nanodrill drive mechanism; (b) Nanodrill gear, platform and tip; (c) Cross-sectional view of nanodrill substrate, hub, gear, platform and tip*

The dimensional quality of the hole, defined by its geometric accuracy, is influenced by errors in the positioning of the drill tip with respect to the workpiece. Therefore, a detailed study of the geometric error developed in the drill is undertaken. The drill tip's geometric error is a function of geometric accuracies in the machine architecture, force induced errors and thermally induced errors [Okafor et al., 2000].

This paper presents an analysis of the combined effects of kinematic and static errors of the drill on errors in the hole. Tolerance analysis methods proposed by [Daniel *et al.*, 1986] and [Hu *et al.*, 1997] are adopted in this study.

## 2.  DESCRIPTION OF THE NANODRILL

The nanodrill is composed of two existing technologies: (1) A surface micromachined microengine which was developed at Sandia National Laboratories [Tanner *et al*, 1998] and (2) A standard Atomic Force Microscope (AFM) tip [O'Neal *et al.*, 2002]. The AFM tip will be referred to as "tip" in the remainder of the paper. The entire mechanism is manufactured from layers of polysilicon in a batch process. This manufacturing technique allows for the creation of linkages by the etching away of material during surface micromachining, precluding the need for traditional assembly methods.

The microengine consists of two linear comb drive actuators which are used to impart rotational motion to a drive gear (Figure 1 (a)). The drive gear in turn rotates the load gear on which the AFM tip is mounted (Figure 1 (b)). Both the drive gear and the load gear rotate about stationary hubs which are mounted on substrates. This gear train has a mechanical advantage of 1.5:1. A platform is used to stabilize the tip onto the load gear (Figure 1 (c)). Figure 1 (c) is a schematic of a cross-section of the drill architecture and is not to proportion. The tip drills holes by making initial indentations into the work piece in a manner similar to other AFM devices. Load induced errors in the tip result from deflections in the load gear as a result of interaction with the drive gear and cutting forces at the tip-workpiece interface.

## 3.  ERROR ANALYSIS

Since the required dimensions of the holes created by the nanodrill are very small, approximately 100 nm in diameter and 50nm in depth, the acceptable errors in these dimensions are also small, approximately <25% of the original dimensions [O'Neal *et al.*, 2002]. Therefore, a high precision drill is required to produce the final feature. In a laboratory setting, Sandia National Laboratories is capable of producing the layers of the nanodrill with standard deviations in thickness of approximately 1.7% of the nominal dimensions. With improvements in fabrication techniques, it is believed that tolerances of ±5% may be attainable for mass production in the near future. Limitations in the wavelength of light make the measurement of holes as small as the target hole difficult. Although  measurement  techniques   such  as  Scanning  Electron  Microscopy

(SEM) have been used to measure features on this scale, these methods are not very reliable and are incapable of measuring receded features such as the depth of a cylindrical hole. Therefore, an error analysis is very important. This type of analysis can help to verify the feasibility of the final feature, reduce manufacturing costs and validate part measurements.

In a previous work, [Joshi *et al.*, 2003] identified the key product characteristics (KPC's) as the depth, diameter and run-out of the drilled hole. Errors in the run-out and depth are affected by errors in the tip's horizontal and vertical positions respectively. These errors are in turn affected by the tip and workpiece materials, the geometry and aspect ratio of the tip, the gear to hub engagement and spacing, the frictional forces between the mating teeth of the gears, the cutting forces and the temperature. Assuming no interaction between the kinematic, static and dynamic factors and using a worst case tolerance analysis, the error in the tip ($U_{tip}$) can be obtained from the superposition of the errors caused by the above factors: ($U_{kinematic}$, $U_{static}$, $U_{dynamic}$, $U_{thermal}$), i.e.

$$U_{tip} = U_{kinematic} + U_{static} + U_{dynamic} + U_{thermal} \qquad (1)$$

The determination of the kinematic and static tip errors are presented in Sections 3.1 and 3.2. Assuming small vibrations and deflections, the dynamic factors are assumed to be negligible. Also, assuming that this analysis is performed in the cold state, thermal effects are neglected.

Although the worst case tolerance analysis method is used, $U_{tip}$ can also be found using the root sum square (RSS) method.

### 3.1. Kinematic Analysis

Positional and geometric errors in the tip, $U_{tip}$, cause dimensional errors in the drilled hole. Errors in the KPC's are influenced by the orientation and position of the centreline of the tip (Figure 2) and the thicknesses of the components of the drill's structure. The first two factors are functions of the horizontal and vertical gap between the hub and the gear, $H_{gap}$ and $V_{gap}$, which are created by removal of a sacrificial oxide layer during surface micromachining. Although translational effects of the gear in the spacing can have some influence, this error analysis assumes that the gear is in its nominal position in the hub spacing.

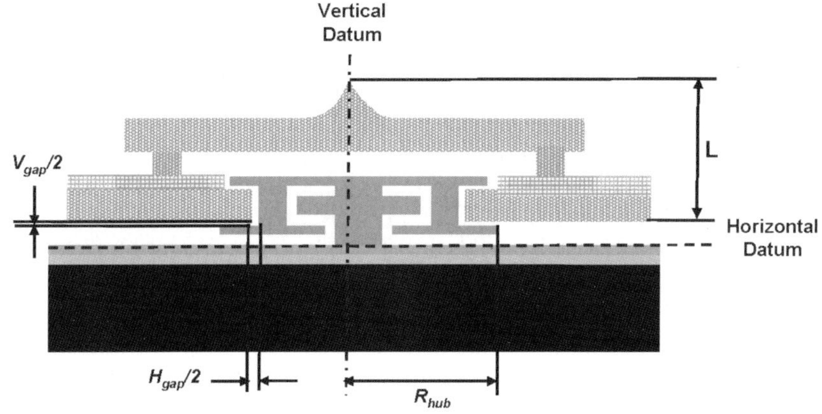


*Figure 2 Geometry of the structure of the nanodrill*

Several assumptions are made in the determination of the effect of kinematic factors on the tip's error. The top layer of the substrate is assumed to be the horizontal datum and the center of the substrate, the vertical datum (Figure 2). In its initial configuration, all layers of the nanodrill are assumed to be perfectly parallel with the reference datum. Since this particular assembly is manufactured by silicon micromachining, the structure is created by the deposition of successive layers of material. With the addition of each new layer, there is an opportunity to correct planar errors in the orientation of the previous layer. Therefore, the assumption that the layers are parallel is viable. The centreline of the hub is assumed to coincide with the vertical reference datum and the gear is assumed to be centered in the gap. This configuration represents the nominal orientation of the gear and the tip in the device. These assumptions allow for the simplification of calculations and the determination of the effect of the gap between the gear and the hub. In order to further simplify calculations, it is assumed that there are no errors in the geometry of the workpiece. Since the goal of this study is to determine the contribution of the nanodrill to errors in the drilled hole, it is logical to isolate the desired errors by assuming no errors in the workpiece.

The following equations describe the geometric relationships that lead to the rotation of the drill system

$$\theta = \tan^{-1}\left(\frac{0.5V_{gap}}{R_{hub}}\right) \approx \left(\frac{0.5V_{gap}}{R_{hub}}\right) \tag{2}$$

$$U_x^k = L\sin(\theta) \tag{3}$$

$$U_y^k = L - L\cos(\theta) \approx 0 \tag{4}$$

$U_x^k$ and $U_y^k$ are the horizontal and vertical deviations of the drill due to a rotation, $\theta$, of the drill's structure (Figure 3). $\theta$ is considered to be the rotation of the top of the gear. $H_{gap}, V_{gap}, R_{hub}$ and $L$ are illustrated in Figure 2.

     A. Bryan *et al.*

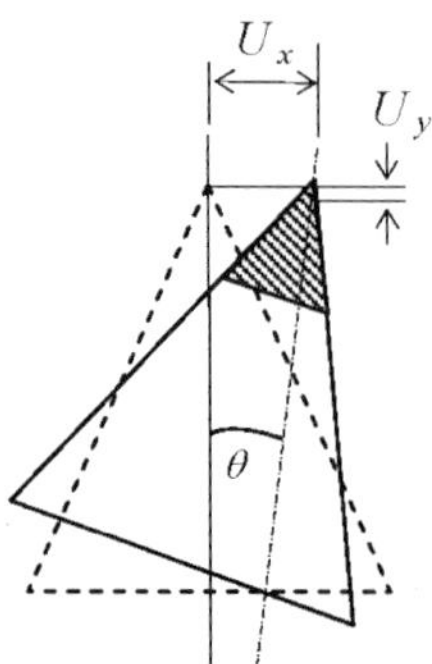

*Figure 3 Drill tip errors.*

In addition to affecting the location, $U_x^k$ also affects the run-out of the desired hole. $U_y^k$ determines the error in the drill depth of the hole. Since $\theta$ is small, $U_y^k$ can be considered to be negligible.

### 3.2. Static Analysis

The static analysis is used to determine the effect of loads on the orientation of the tip. The geometric assumptions made for the kinematic analysis are assumed to apply to the static analysis. Several additional assumptions are made in order to perform the quasi-static analysis of the dynamic drill system.

The drive gear is assumed to apply tangential and radial forces ($F_t$ and $F_r$) on the load gear. $F_t$ causes the rotation of the system and would not contribute to stresses and deflections. Since the load gear rotates about a stationary hub, the radial forces ($F_r$) are assumed to be balanced by a side load ($R_r$) acting on the base of the tip (Figure 4 (a)). This side load causes the deflection of the gear, platform and tip from nominal as drilling occurs. In order to make the system truly static, a tangential force ($R_t$) and a reaction moment ($M_z$) are also applied at the base of the tip to equilibrate the tangential force and torque respectively. The inertial forces due to the weight of the drill assembly are assumed to be negligible. The cutting forces ($F_{cutting}$) are applied at the top of the tip. Due to the small scale of the top of the tip compared with the tip's overall dimensions, the tip's geometry is altered from conical to cylindrical in the FEM analysis. This allows for easier application of the cutting force at the top of the tip. The reaction to the vertical cutting force ($R_{cutting}$) is assumed to act on the bottom of the hub, where the hub interfaces with the substrate. The free body diagram of the system is shown in Figure 4 (a).

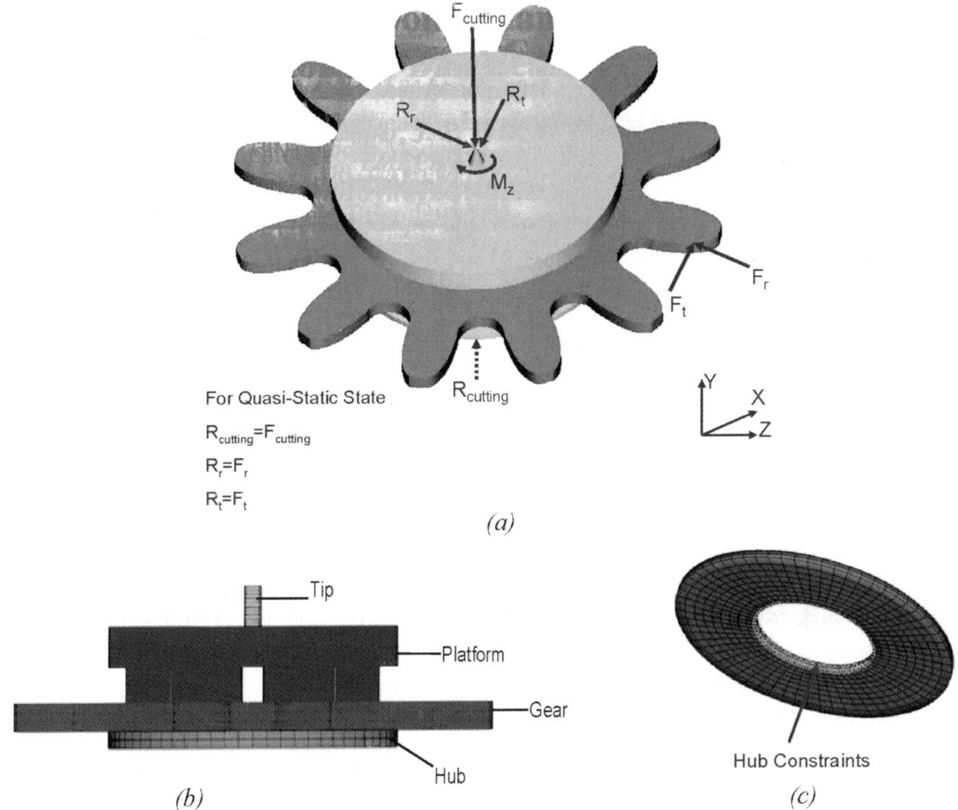


*Figure 4  (a) External loads;  (b) Finite element mesh of the structure; (c) Finite element mesh showing constraints on the hub*

Hypermesh and ANSYS are used to perform the static analysis. The meshes of the tip, and platform and platform and hub are assembled together with coupled nodes (Figure 4 b). The inner diameter of the hub and gear are fixed in the x, y and z directions as shown in Figure 4 (c).  The substrate is not included in the FEM mesh as it is not needed.  The reaction load, $R_{cutting}$, replaces the effect of the substrate.

The material properties of the polysilicon used to manufacture the layers are as follows: Young modulus (E) =161 $\mu N/\mu m^2$ and Poison ratio (v) = 0.22. The boundary conditions (Table I) applied to the structure were obtained from [Tanner *et al.*, 2000] and [O'Neal *et al.*, 2002].  As deformations are assumed to be small, a linear static analysis is performed.

*Table I  Input forces for static analysis.*

| $F_{cutting}$ | $F_r$ | $F_t$ | $M_z$ |
|---|---|---|---|
| 1.0 µN | 4 µN | 2.5 µN | 61 µN- µm |

The errors in the key characteristics due to load, are determined from the FEM analysis. Since the in-plane deviation of the tip cannot be assumed to be symmetric about the x and z axes (in plane axes), the total in-plane deviation was found from the following relation,

$$U_x^s = \sqrt{X_1^2 + Z_1^2}$$
(5)

## 4.  RESULTS

### 4.1 Kinematic Error

For the kinematic analysis, the nominal value of $\theta$, $U_x^k$ and $U_y^k$ are considered to be zero since translational effects are not being considered. The error caused by the rotation of the load gear in the gap between the hub and the gear had a dominant effect on the kinematic error of the tip. Using the values of $R_{hub}$=10µm, $V_{gap}$=0.3µm and L=11µm, the errors obtained for this particular drill architecture are reported in Table II.

*Table II  Error in the drill tip due to kinematic factors.*

| Parameter | Value |
|---|---|
| $\theta$ | $\pm 0.86°$ |
| $U_x^k$ | $\pm 0.165$µm |
| $U_y^k$ | -0.001µm |

From the results, it can be observed that the error in the tip's horizontal position (0.165nm) is more significant than the error in the vertical direction (0.001nm). This large horizontal deflection will have a significant impact on the run-out and diameter of the drilled hole.

### 4.2  Sensitivity Analysis of Forces

Tip deflection can be caused by the external forces exerted on the system, and it is important to understand how sensitive the system's KPCs are to changes in the magnitude of external forces. The cutting force ($F_{cutting}$) causes a vertical tip deflection of -0.108µm.  However, it has no influence on the horizontal deflection of the tip.  The in plane gear forces causes a nominal in plane deviation of 0.075µm.

*Table III  Sensitivity analysis for gear force.*

| $F_r$ (µN) | $F_t$ (µN) | $M_z$ (mm) | $\triangle$ | $U_x^s$ (µm) | $\dfrac{\Delta U_x^s}{U_x^s}$ | $U_y^s$ (µm) | $\dfrac{\Delta U_y^s}{U_y^s}$ |
|---|---|---|---|---|---|---|---|
| 4 | 2.5 | 61.05 | 0 | 0.075 | 0 | -0.108 | 0 |
| 4.4 | 2.75 | 67.16 | 10% | 0.083 | 10.1% | -0.108 | 0.4% |

Since the tip's deflection varies proportionally with the gear forces over a ±10% range, a linearized sensitivity analysis is used to determine the effect of the gear forces on the tip's error [Liu *et al.*, 1997].  From the sensitivity analysis, it is determined that there is negligible change in the vertical deflection of the tip due to gear forces.  However, the gear forces do have an impact on the horizontal deflection of the tip.  Therefore, the gear forces will also have a significant impact on the diameter of the drilled hole.

### 4.3 Tip Error

The tip error in the x and y directions ($U_x^{tip}$ and $U_y^{tip}$) are determined by the superposition of the mean kinematic and static errors (Table IV).  The error due to forces is considered with the forces at their nominal dimensions.

*Table IV  Tip error due to kinematic and static effects.*

| Parameter | Error |
|---|---|
| $U_x^{tip}$ | 0.240µm |
| $U_y^{tip}$ | -0.109µm |

Table IV indicates that the error in the horizontal direction is much more significant than the error in the vertical direction.  These horizontal and vertical errors in the tip due to combined kinematic and static effects exceed the desired dimensions of the hole.  Therefore, a hole of such a small size may not be obtainable from the current drill architecture.

From the previous two sections, it is observed that the tip's error due to static effects is 45% of that due to kinematic effects.  Therefore, correction of the kinematic error could lead to significant improvements in the tip's position and the final hole's geometry.  A reduction in the size of the gap between the hub and the gear and an increase in the width of the hub can significantly decrease the angle of rotation of the drill structure, thereby improving the run-out and positional accuracy of the final hole. However, even if these kinematic adjustments are made, the error that would be caused by the gear forces would still be significant.


## 5.  SUMMARY

In this paper, the determination of the horizontal and vertical errors in the position of the tip of a nanodrill is presented.  As a result of this study, it is determined that the horizontal error in the tip can be quite large, making it practically infeasible to obtain the holes with diameters on the nanoscale with the current drill architecture.  Although, both the kinematic and static errors are large, the kinematic error is found to have a dominant effect on the final error of the horizontal position of the drill tip.  Vertical errors in the drill tip are found to be negligible.  However, changes in geometry, e.g., making the hub wider, will lead to significant improvements in the error of the final drilled hole.


## 6.  ACKNOWLEDGEMENTS

The authors would like to acknowledge the financial support of the SGER grant from DMII-National Science Foundation (NSF, Grant # 0236465).  We would also like to thank Mr. Kumar Virwarni for his timely support, Dr. Springer for his advice and Mr. Ming Zhu for running the FEM analysis.


## 7.  REFERENCES

**[Joshi et al, 2003]** Joshi, N.; Malshe, A. P.; Bryan, A.; Camelio, J.; and Hu, S. J.; "Geometric Error Assessment of A Nanomechanical Drill"; *American Society of Mechanical Engineers, Micro-Electromechanical Systems Division (MEMS)*; v5; pp. 271-276.

**[Khanna, R., 2003]** Khanna, R.; "MEMS Fabrication Perspective from the MIT Microengine Project"; *Surface and Coatings Technology*; v163-164; pp 273-280

**[Vettiger et al, 2003]** Vettiger, P.; Binnig, G. ; "The Nanodrive Project"; Scientific American; v288; n1, pp 34-41

**[Judy, J., 2001]** Judy, J. W.; "Microelectromechanical Systems (MEMS): Fabrication, Design and Applications"; *Smart Materials and Structures*; v10; n6; pp1115-1134.

**[O'Neal et al., 2002]** O'Neal, C. B.; Malshe, A. P.; Virwani K. R.; Schmidt W. F.; "Design Consideration, Process and Mechanical Modeling, and Tolerance Analysis of a MEMS based Mechanical Machining System-on-a-chip for Nanomanu-facturing"; *Society of Mechanical Engineers, Electronic and Photonic Packaging, Electrical Systems and Photonics Design and Nanotechnology*; v2; pp. 529-534

**[Wittwer et al, 2002]** Wittwer, J. W.; Gomm, T.; Howell, L.; "Surface Micromachined Force Gauges: Uncertainty and Reliability"; *Journal of Micromechanics and Micro-engineering*; v12; n 1; pp. 13-20

**[Chen et al., 2001]** Chen, G.; Yuan, J.; Ni, Y.; "A Displacement Measurement Approach for Machine Geometric Error Assessment"; *International Journal of Machine Tools & Manufacture*; v41; pp. 149-161

**[Okafor et al., 2000]** Okafor A. C.; Ertekin, Y. M.; "Derivation of Machine Tool Error Models and Error Compensation Procedure for Three Axes Vertical Machining Center using Rigid Body Kinematics"; *International Journal of Machine Tools & Manufacture;* v40; pp. 1199–1213

**[Tanner et al, 2000]** Tanner D. M.; Smith, N. F.; "MEMS Reliability, Infrastructure, Test Structure, Experiments and Failure Modes"; *Sandia National Laboratories*; Report SAND2000-0091

**[Tanner et al, 1998]** Tanner D. M.; Miller, W. M.; Eaton, W. P.; "The Effect of Frequency on the Lifetime of a Surface Micromachined Microengine Driving a Load"; *IEEE International Reliability Physics Symposium Proceedings*; pp 26-35

**[Hu, 1997]** Hu, S. J.; "Stream of Variation Theory for Automotive Body Assembly"; *Annals of the CIRP*; v 46; n1; pp.1-6

**[Liu, et al, 1997]** Liu, S. C. ; Hu, S. J.; "Variation Simulation for Deformable Sheet Metal Assemblies Using Finite Element Methods "; *Journal of Manufacturing Science and Engineering, Transactions of the ASME*; v119; n 3; pp 368-374

**[Daniel et al, 1986]** Daniel, F.; Weill, R.; Bourdet, P.; "Computer Aided Tolerancing and Dimensioning in Process Planning"; *Annals of the CIRP*; v35; n1, pp.381-386

# Tolerance Synthesis of Higher Kinematic Pairs

**M.-H. Kyung, E. Sacks**
*Computer Science, Purdue University, USA*
*eps@cs.purdue.edu*

**Abstract**: We present a tolerance synthesis algorithm for mechanical systems comprised of higher kinematic pairs. The input is a parametric model of a mechanical system (part profiles and system configuration) with initial tolerance intervals for the parameters. The output is revised tolerances that guarantee correct kinematic function for all system variations. Nominal parameter values are changed when possible and tolerance intervals are shrunken as a last resort. The algorithm consists of a three-step cycle that detects and eliminates incorrect system variations. The first step finds vectors of parameters values whose kinematic variation is maximal. The maximums of the higher pairs are derived by contact zone construction then are combined into system maximums. The second step tests the vectors for correct kinematic function using configuration space matching and kinematic simulation. The third step adjusts the tolerances to exclude the vectors with incorrect functions. The cycle repeats until every vector exhibits correct function. We demonstrate the algorithm on common mechanical systems.

## 1.  INTRODUCTION

We present a tolerance synthesis algorithm for mechanical systems comprised of higher kinematic pairs. Tolerance synthesis is a central part of kinematic synthesis, which is the task of designing a mechanical system that performs a specified kinematic function. Kinematic synthesis is an iterative process in which designers select a design concept, construct a parametric model, and assign tolerance intervals to the parameters. The goal of tolerance synthesis is to derive tolerances that guarantee correct kinematic function at a minimal cost. Overly tight tolerances can necessitate expensive manufacturing processes, whereas overly loose tolerances can lead to unreliable products.

The kinematic function of a system is the coupling between its part motions due to contacts between pairs of parts. A lower pair has a fixed coupling that can be modeled as a permanent contact between two surfaces. For example, a revolute pair is modeled as a cylinder that rotates in a cylindrical shaft. A higher pair imposes multiple couplings due to contacts between pairs of part features. For example, gear teeth consist of involute patches whose contacts change as the gears rotate. The system transforms driving motions into outputs via sequences of part contacts. Small system variations can produce large motion variations, can alter contact sequences, and can introduce failure modes, such as jamming, due to changes in kinematic function.

We ensure correct kinematic function by synthesizing tolerances that preclude failure modes and that limit motion variation. Nominal parameter values are changed when possible and tolerance intervals are shrunken as a last resort. We cannot search the entire



*J.K. Davidson (ed.), Models for Computer Aided Tolerancing in Design and Manufacturing*, 289–299.
© 2007 *Springer*.

parameter space for bad parameter values. Its dimension is prohibitively high because mechanical systems have tens to hundreds of shape and configuration parameters. Tiny steps are required because the kinematic function can vary suddenly or even discontinuously. We limit the search to parameter values that maximize the variation of one or two contacts. We explain the mathematical and empirical rationale for this heuristic below.

The input to our algorithm is a parametric model of a mechanical system (part profiles and system configuration) with initial tolerance intervals for the parameters. The output is revised tolerances that guarantee correct kinematic function for all system variations. The algorithm consists of a three-step cycle that detects and eliminates incorrect system variations. The first step finds vectors of parameter values whose kinematic variation is maximal. The second step tests the vectors for correct kinematic function. The third step adjusts the tolerances to exclude the vectors with incorrect functions. The cycle repeats until every vector exhibits correct function.

The algorithm builds upon our prior work in mechanical design with configuration spaces. The first step uses generalized configuration spaces, called contact zones [Kyung and Sacks, 2003], to find parameter vectors that maximize individual contacts. The second step uses configuration space matching [Kyung and Sacks, 2003a] to identify failure modes and uses kinematic simulation [Sacks and Joskowicz, 1993] to compute motion variation. We review prior work on kinematic tolerancing elsewhere [Kyung and Sacks, 2003]. This work has evolved beyond what is reported here. A full description of the current algorithm with additional examples will appear in [Kyung and Sacks, 2006].

## 2.   CONFIGURATION SPACE

We use configuration spaces [Sacks and Joskowicz, 1995] to represent higher pair kinematic function. The configuration space of a pair is a manifold with one coordinate per part degree of freedom (rotation or translation). Points in configuration space correspond to configurations of the pair. The configuration space partitions into blocked space where the parts overlap, free space where they are separate, and contact space where they touch. Free and blocked space are open sets whose common boundary is contact space. Contact space is a closed set comprised of subsets that represent contacts between part features.

We illustrate these concepts for a Geneva pair comprised of a driver and a wheel (Figure 1). The driver consists of a driving pin and a locking arc segment mounted on a cylindrical base. The wheel consists of four locking arc segments and four slots. The wheel rotates around axis $A$ and the driver rotates around axis $B$. Each driver rotation causes an intermittent wheel motion with four drive periods where the driver pin engages the wheel slots and with four dwell periods where the driver locking arc engages the wheel locking arcs.

The configuration space coordinates are the part orientations $\theta$ and $\omega$ in radians. The pair is displayed in configuration $(0, 0)$, which is marked with a dot. Blocked space is the gray region, contact space is the black curves, and free space is the channel between the curves. Free space forms a single channel that wraps around the horizontal and vertical boundaries, since the configuration at $\pm\pi$ coincide. The defining equations of the chan-

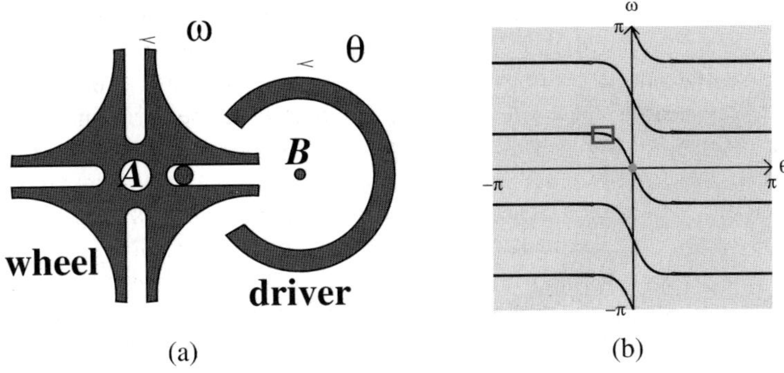



*Figure 1; (a) Geneva pair; (b) configuration space.*

nel boundary curves express the coupling between the part orientations. The horizontal segments represent contacts between the locking arcs, which hold the wheel stationary. The diagonal segments represent contacts between the pin and the slots, which rotate the wheel. The contact sequences of the pair are the configuration space paths in free and contact space. In a typical sequence, the driver rotates clockwise (decreasing $\theta$) and alternately drives the wheel counterclockwise with the pin (increasing $\omega$) and locks it with the arcs (contact $\omega$).

We [Kyung and Sacks, 2003a] model kinematic variation by generalizing configuration space to parametric parts with tolerances. Kinematic variation occurs in contact space. As the parameters vary, the part shapes and motion axes vary, which causes the contact curves to vary. The union of the varying contact curves over the parameter ranges defines a band around the nominal contact space, called a contact zone, that bounds the worst-case kinematic variation of the pair. The contact zone is the subset of the configuration space where contacts can occur for some parameter variation.

Figure 8a shows the contact zone that our algorithm generates for a 26 parameter model of the Geneva pair shown in Figure 1 with parameter tolerances of $\pm 0.02$mm and $\pm 1°$. The zone is a detail of the portion of the configuration space in the box in Figure 1b. This portion is the interface between a horizontal and a diagonal channel where the driver pin leaves a wheel slot and the locking arcs engage. The two dark gray bands that surround the channel boundary curves are the contact zone. The white region between the bands is the subset of the nominal free space that is free for all parameter variations.

The contact zone reveals a possible failure mode. The lower and upper bands overlap near where the horizontal and diagonal channels meet. Some parameter vector might yield a configuration space in which the lower and upper contact curves intersect, block the channel, and cause the mechanism to jam. Figure 2 illustrates this failure mode. But contact zone overlap does not guarantee a faulty parameter vector. The contact zone is a conservative estimate of kinematic variation that ignores dependencies between contacts

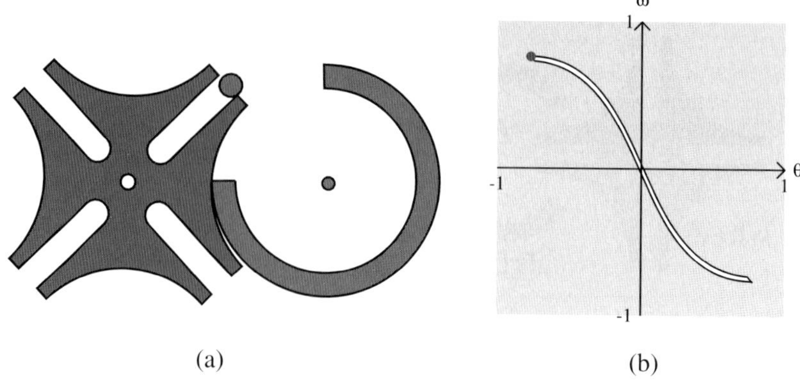

(a)            (b)

*Figure 2; Geneva failure: (a) jamming configuration; (b) configuration space.*

due to shared tolerance parameters. Hence, there might be no parameter vector that moves both curves into the overlap region. The tolerance synthesis algorithm resolves this issue by analyzing vectors that maximize the variation of the contacts whose zones overlap.

## 3.  TOLERANCE SYNTHESIS ALGORITHM

This section describes the tolerance synthesis algorithm, which is summarized in Figure 3. The input system consists of planar higher pairs with parametric tolerances. The part profiles are simple loops of line and circle segments. Line segments are represented by their endpoints and circle segments are represented by their endpoints and radii. Each part has one degree of freedom: translation along a fixed axis or rotation around an orthogonal axis. The segment endpoints, circle radii, and motion axes are represented with algebraic expressions whose variables are tolerance parameters. The parameters have nominal values and range limits, which are collectively called tolerances. This class of higher pair system covers 90% of engineering applications based on our survey of 2,500 mechanisms in an engineering encyclopedia [Sacks and Joskowicz, 1991] and on our industrial experience.

### 3.1.  Candidate selection

The candidate parameter vectors are selected in two steps. The first step finds sets of parameter values that generate points on the boundaries of the system contact zones. Each set contains the parameters that determine the shape and motion axes of two contacting part features. The set specifies parameter values that maximize the kinematic variation of the contact in one configuration of the nominal work cycle. Two sets are called *compatible* when they agree on their common parameters, for example $\{x = 1, y = 2\}$ and

**Input:** system, initial tolerances.
1. Select candidate parameter vectors.
2. Test kinematic function at candidates.
3. If correct, return tolerances.
4. Revise tolerances.
5. Go to step 1.
**Output:** revised tolerances.

*Figure 3; Kinematic tolerance synthesis algorithm.*

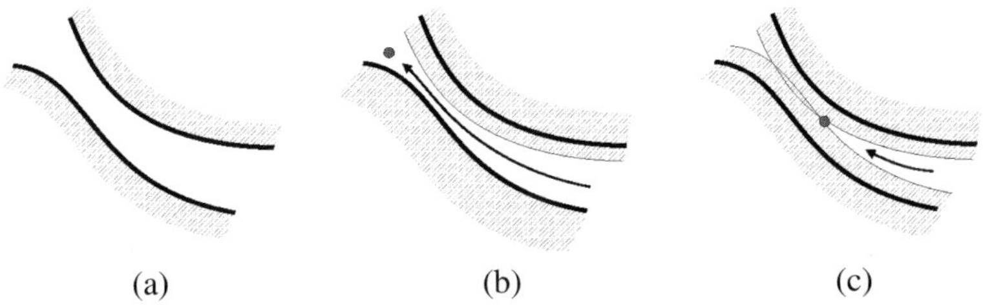

(a)               (b)               (c)

*Figure 4; Candidate computation: (a) nominal kinematics; (b) one maximal variation; (c) two maximal variations.*

$\{y = 2, z = 3\}$, and in particular if they are disjoint. The union of $k$ compatible sets simultaneously maximizes the kinematic variation of $k$ contacts in $k$ nominal configurations. The second step forms the candidate parameter vectors from unions of compatible sets. These candidates represent limiting cases of contact interactions, which is where failures are most likely to occur and hardest to detect.

Figure 4 illustrates the two steps. Part a shows the configuration space of a nominal pair with a free space channel. Part b shows the configuration space for a step 1 parameter set that maximizes the variation of the upper channel boundary; the thin black curve is the new upper boundary and the lower boundary is unchanged. The channel remains open, so the kinematic function is qualitatively correct. Part c shows the configuration space for a step 2 parameter vector that maximizes the variation of both channel boundaries. The boundaries overlap, which indicates incorrect kinematic function.

**Step 1**   Each higher pair is processed separately. The contact space of a pair consists of contact curves that represent contact between feature pairs. Each curve is processed separately. The computation is described in our kinematic tolerance analysis paper [Kyung and Sacks, 2003]. We summarize it here for completeness.

A curve has a parametric equation $C(\mathbf{p}, \mathbf{u}) = 0$ where $\mathbf{p}$ denotes the configuration space coordinates, for example $\mathbf{p} = (\theta, \omega)$ in the Geneva pair. There is one type of

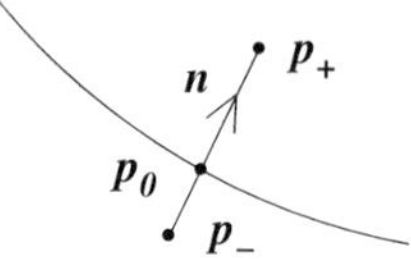



*Figure 5; Contact zone computation: nominal point* $\mathbf{p}_0$ *generates contact zone boundary points* $\mathbf{p}_1$ *and* $\mathbf{p}_2$.

equation for every combination of features and motions, such as rotating circle/translating line. For example, the driver/wheel locking arc equation is $(B + R_\theta m - A - R_\omega n)^2 = (r - s)^2$ where $B, A$ are the centers of rotation, $m, n$ are the arc centers in part coordinates, $R_\theta, R_\omega$ are rotation operators, and $r, s$ are the arc radii. The equation states that the distance between the arc centers equals the difference of their radii.

The kinematic variation at a nominal contact configuration, $\mathbf{p}_0$, occurs along the normal vector, $\mathbf{n}$, to the contact space (Figure 5). It has the form $\mathbf{p}_0 + k\mathbf{n}$ with $k$ a function of $\mathbf{p}$. The $\mathbf{p}$ values that maximize/minimize $k$ yield points on the upper/lower boundaries of the contact zone. They are computed by solving a nonstandard optimization problem with a custom algorithm. Parameter value sets are computed by discretizing the nominal curve, $C(\mathbf{p}, \mathbf{u}_0)$, to an input accuracy ($10^{-5}$ in the paper) and applying the algorithm to the resulting points.

**Step 2** We construct a parameter vector for each pair of compatible parameter sets. The elements of the vector that appear in the sets are assigned their set values and the remaining elements are assigned their nominal values. For example parameters $(w, x, y, z)$ with nominal values of zero and compatible sets $\{x = 1, y = 2\}$ and $\{y = 2, z = 3\}$ yield the vector $(0, 1, 2, 3)$. We also construct a vector for each parameter set that is not compatible with any other set. A set $a$ that is compatible with some $b$ does not need its own vector, since its maximal variation is achieved in the $a \cup b$ vector. We reduce the number of candidates by culling vectors that are within $\epsilon$ of other vectors. We employ Manhattan distance and set $\epsilon$ to 1% of the initial tolerance interval. In our examples, culling eliminates 75% of the candidates without missing any failures.

We do not construct parameter vectors for larger compatible sets because they are expensive to compute and provide little benefit. The expense is clear: $n$ contacts yield $O(n^3)$ triples, $O(n^4)$ quadruples, and so on. The lack of benefit is a mathematical fact when we consider a single higher pair with two degrees of freedom: three curves cannot intersect in a two-dimensional manifold, except for rare degenerate cases. Larger compatible sets might reveal failure modes due to combinations of maximal kinematic variation among several higher pairs. Selective generation of these sets is a topic for future research.

### 3.2. Testing

The candidate parameter vectors are tested for failure modes and for excessive motion variation. The vectors that fail either test are passed to the tolerance revision module.

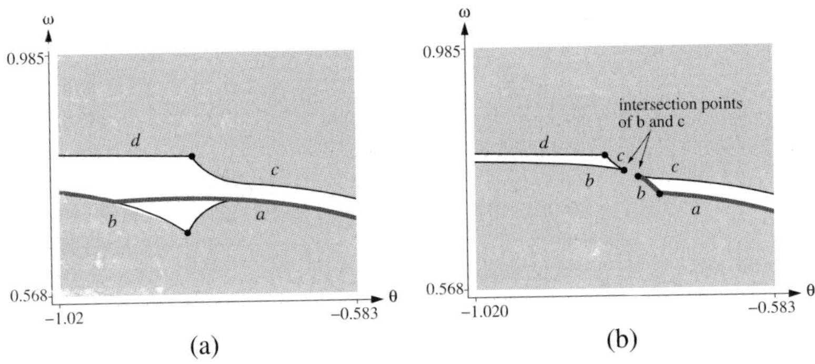


*Figure 6; Mismatch between (a) nominal and (b) jammed Geneva spaces.*

The failure mode test matches the nominal and candidate contact spaces of the higher pairs. The test succeeds when the two spaces have the same structure: they have the same number of components and each component in the first space matches a unique component in the second space. Two components match when they consist of equivalent curves in the same cyclic order. Two curves are equivalent when they are generated by the same pair of part features. The matching algorithm is described in our kinematic synthesis paper [Kyung and Sacks, 2003a].

Figure 6 illustrates matching on the nominal (Figure 1) and jammed (Figure 2) Geneva configuration spaces. The match fails because the nominal contact space has two components, whereas the jammed space has one. The structural mismatch is that curves $b$ and $c$ are disjoint in the nominal space, but intersect in the jammed space.

The motion variation test compares the nominal and candidate motion paths of the driven parts using kinematic simulation [Sacks and Joskowicz, 1993]. The simulator steps the driving part through its work cycle and propagates the motion through the higher pairs of the system. The driving motion and the step are specified as input. The output is a nominal and a candidate configuration for each driven part at each driver configuration. The motion variation of a candidate configuration is the minimal distance to a nominal configuration. The motion variation of the candidate is the maximum over these minimal distances. A candidate passes the motion test when the variation is below a specified limit.

### 3.3. Tolerance revision

The tolerance revision step revises the current tolerances to exclude the failed parameter vectors. The tolerances define an axis-aligned box in parameter space: the box is centered at $u_0$ and its width in the $k$th dimension is the tolerance interval of the $k$th parameter. The failed vectors, $u_i$, lie in this box. The revision excludes them by modifying $u_0$ when possible and by shrinking the box width otherwise.

Each failed vector has a neighborhood of vectors that exhibit the same failure, since

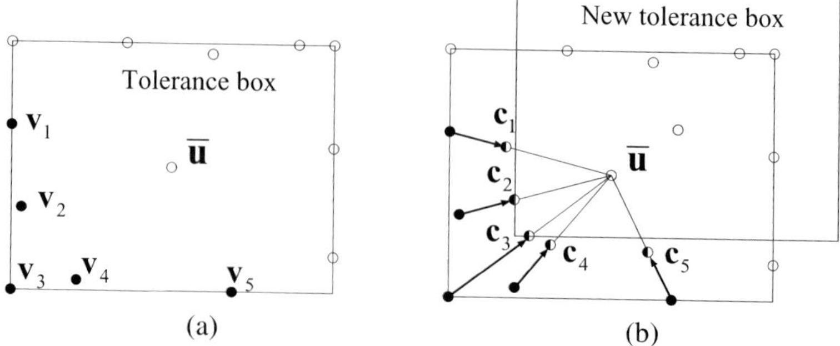


*Figure 7; Tolerance revision: (a) before; (b) after. Unfilled circles mark correct candidates.*

the system depends continuously on the parameters. In one dimension, consider a system with parameter $u$, nominal value $u_0$, and failed value $u_1$. Continuity implies the existence of a critical parameter value $u_{c,1}$ where the failure first occurs: the kinematic function is correct on $[u_0, u_{c,1})$ and is incorrect on $(u_{c,1}, u_1]$. In higher dimensions, $\mathbf{u}_{c,i}$ is the first failure point on the line segment $[\mathbf{u}_0, \mathbf{u}_i]$, as shown in Figure 7. The revision algorithm excludes the critical points, rather than the failed candidates, from the tolerance box.

The critical point $\mathbf{u}_{c,i}$ is found by bisection search. The initial interval is $[\mathbf{u}_0, \mathbf{u}_i]$. At each iteration, the midpoint parameter vector is tested for failure using configuration space matching or kinematic simulation based on the failure type. If failure occurs, the upper interval limit is replaced with the midpoint; otherwise, the lower limit is replaced. The iteration ends when the interval width falls below a user specified accuracy $(10^{-5})$.

The next task is to compute a minimal update, $\mathbf{u}_0' = \mathbf{u}_0 + \Delta\mathbf{u}$, that excludes the critical points from the parameter box. The minimal update for a single point is translation perpendicular to the closest box face by the distance to the face. In our example (Figure 7b), the closest face to $\mathbf{u}_{c,1}$ is the left side of the box, so $\mathbf{u}_0$ moves right along the horizontal axis. The closest face to $\mathbf{u}_{c,5}$ is the bottom of the box, so $\mathbf{u}_0$ moves up along the vertical axis.

The system update is computed as follows. Initialize $\Delta\mathbf{u}$ to zero. Sort the minimal updates by magnitude in decreasing order. Combine each update with $\Delta\mathbf{u}$. Let the current update change $u_i$ by $d_i$. If $\Delta u_i$ is zero, set $\Delta u_i$ to $d_i$. If $\Delta u_i$ and $d_i$ have the same sign, set $\Delta u_i$ to $\text{sign}(d_i) \max\{|d_i|, |\Delta u_i|\}$. If they have opposite signs, the current update is incompatible with a previous update. It is replaced with an update perpendicular to the second closest parameter box face, and so on through the parameters. If no parameter works, the $p_i$ tolerance interval is shrunk minimally to exclude $d_i$.

The initial sorting is a heuristic that reduces the magnitude of $\Delta\mathbf{u}$. For example, $\mathbf{u}_{c,1}$ and $\mathbf{u}_{c,2}$ have distance vectors $(3, 2)$ and $(3, 3)$ from the tolerance box in Figure 7. Processing $\mathbf{u}_{c,2}$ first yields $\Delta\mathbf{u} = (3, 0)$, whereas processing it second yields $\Delta\mathbf{u} = (3, 2)$.

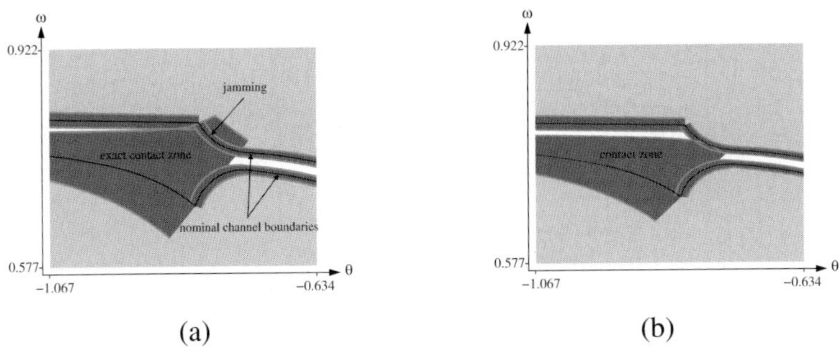


(a)                                                    (b)

*Figure 8; Detail of Geneva pair contact zone: (a) initial; (b) final.*

## 4.  RESULTS

We have tested the tolerance synthesis algorithm on common higher pair systems from the engineering literature. We present two representative examples where it corrects kinematic problems in the initial tolerances by changing the nominal parameter values.

The first example is the 26 parameter model of the Geneva pair (Figure 1) whose contact zone shows possible jamming (Figure 8a). In the first synthesis iteration, step 1 of candidate selection finds 44 maximal parameter sets and step 2 generates 137 parameter vectors. Candidate testing finds 21 vectors with incorrect kinematic function. There are three more iterations in which five, two, and one incorrect vectors are found. The maximum change is 0.00322 which is 8% of the tolerance interval. The Geneva example demonstrates that contact zones can be overly conservative. The lower and upper channel boundaries overlap slightly in the contact zone of the synthesized tolerances (Figure 8b), but there is no parameter vector that realizes both boundary variations.

The second example is a camera shutter mechanism comprised of a driver, a shutter, and a shutter lock (Figure 9a). The user advances the film (not shown), which engages the driver film wheel and rotates the driver counterclockwise. The shutter tip follows the driver cam profile, which rotates the shutter clockwise, which extracts the shutter pin from the shutter lock slot (b). When the pin leaves the slot, a torsional spring rotates the shutter lock clockwise until its tip engages the driver slotted wheel (c).

The mechanism is parametrized by 22 parameters with tolerances of $\pm0.09$mm. In the first synthesis iteration, candidate selection generates 1673 parameter vectors. All the vectors pass the failure mode test, but 40 fail the motion test with a threshold of 0.1 radians. The problem is that the shutter does not move far enough left to clear the shutter lock. Figure 9 shows a failed motion path in the shutter/lock (d) and driver/lock (e) configuration spaces. The problem is fixed in seven iterations. The maximum change is in the $y$ coordinate of the cam axis, which is moved 0.1577mm.

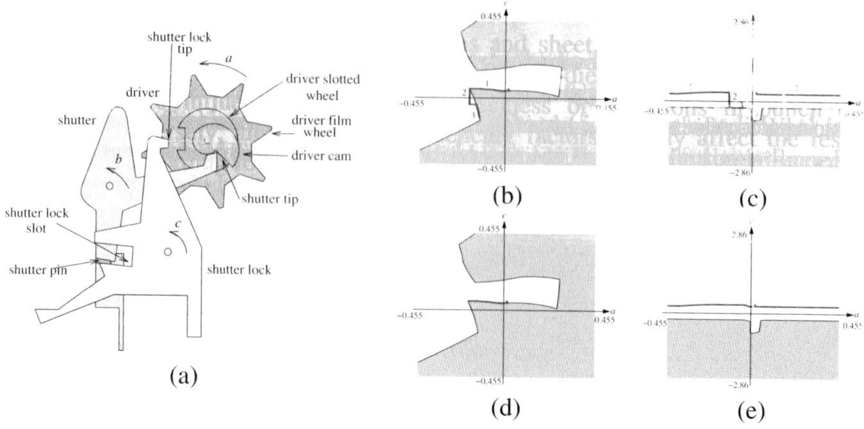

*Figure 9; (a) Camera shutter mechanism; (b–c) nominal configuration spaces of shutter/lock and driver/lock; (d–e) incorrect configuration spaces.*

## ACKNOWLEDGEMENTS

This research was supported by NSF grants IIS-0082339 and CCR-0306214, and by the MIC (Ministry of Information and Communication), Korea, under the ITRC (Information Technology Research Center) support program supervised by the IITA (Institute of Information Technology Assessment). The Figures are reprinted from *Computer-Aided Design*, Kyung and Sacks, *Robust parameter synthesis for planar higher pair mechanical systems*, 38(5), 2006 with permission from Elsevier.

## REFERENCES

[**Joskowicz and Sacks, 1991**] Leo Joskowicz and Elisha Sacks. Computational kinematics. *Artificial Intelligence*, 51(1-3):381–416, October 1991.

[**Kyung and Sacks, 2003**] Min-Ho Kyung and Elisha Sacks. Nonlinear kinematic tolerance analysis of planar mechanical systems. *Computer-Aided Design*, 35(10):901–911, 2003.

[**Kyung and Sacks, 2003a**] Min-Ho Kyung and Elisha Sacks. Parameter synthesis of higher kinematic pairs. *Computer-Aided Design*, 35:567–575, 2003.

[**Kyung and Sacks, 2006**] Min-Ho Kyung and Elisha Sacks. Robust parameter synthesis for planar higher pair mechanical systems. *Computer-Aided Design*, 38(5), 2006.

[**Sacks and Joskowicz, 1993**] Elisha Sacks and Leo Joskowicz. Automated modeling and kinematic simulation of mechanisms. *Computer-Aided Design*, 25(2):106–118, 1993.

[**Sacks and Joskowicz, 1995**] Elisha Sacks and Leo Joskowicz. Computational kinematic analysis of higher pairs with multiple contacts. *Journal of Mechanical Design*, 117(2(A)):269–277, June 1995.

# Geometrical Study of Assembly Behaviour, Taking Into Accounts Rigid Components' Deviations, Actual Geometric Variations and Deformations

**G. Cid*, F. Thiebaut**, P. Bourdet, H. Falgarone*****

**LURPA, ENS de Cachan, 61 avenue du président WILSON 94235 Cachan cedex France*
***LURPA, IUT de Cachan, 9, av de la Division Leclerc, 94235 Cachan cedex France*
**** EADS CCR, 12 r Pasteur 92150 Suresnes France*
*gregory.cid@lurpa.ens-cachan.fr*


**Abstract:** Simulations of mechanisms' geometrical behaviour is a topical industrial need. Thanks to these simulations, gain of time during design and assembly stages can be important. We study the influence of components' geometrical variations on functional conditions. It corresponds to an extension of work realised with rigid body hypothesis, and allows study of mechanisms made by some flexible component. Majority of works dealing with components' deformation on assembly are focused on welded, riveted or bolded assembly. Contrary to numerical results generally obtained, our objective is to determine coefficients of influence of the studied parameters in a symbolic writing, thanks to the study of geometrical behaviour of mechanism, considering geometrical pairs and actual geometry variations.

This method is illustrated through an industrial study, based on an aeronautic example.


**Keywords**: actual geometric deviation, tolerancing, deformation, assembly

## 1 INTRODUCTION

In order to reduce design costs and delays, simulation of mechanisms' geometrical behaviour becomes more and more important in industry. We present in this paper the influence of components' geometrical variations on functional conditions. It corresponds to an extension of work realised with rigid body hypothesis [Thiebaut 2001], [Bourdet and Ballot, 1995], and allows simulation of mechanisms made by some flexible components.

Works dealing with components' deformation on assembly are focused on welded, bolded or riveted assembly and propose variations propagation model [Camelio et al, 2001] or solution to study multi level assembly systems [Hu, 1997]. It sometimes applied to tolerance analysis method [Merkley, 1998], [Samper and Giordano, 1998], to main assembly parameters and their associated criteria [Dahlström and Söderberg, 2001], [Sellem et al, 1999], or even to tolerance allocation method [Shiu et al, 2003].

301

*J.K. Davidson (ed.), Models for Computer Aided Tolerancing in Design and Manufacturing*, 301–310.
© 2007 *Springer.*

Our work belongs to a different approach that consists in determining geometrical behaviour of a mechanism in a symbolic writing. Moreover, in this paper we present the way we include actual geometry variation in our model.

After a presentation of this integration of actual geometrical variations, we present how we take deformation into account in our simulation.

Then, this method is illustrated through an industrial study.


## 2 INTEGRATION OF ACTUAL GEOMETRIC DEVIATION IN 3D EQUATIONS

Geometrical definition of a mechanism is defined through numerical models of its components using CAD system. The geometry of actual components does not exactly match the numerical geometry. The simulation of the behaviour of the mechanism that includes geometrical deviations is necessary.

Geometrical deviations belong to two types. The first type concerns the surfaces that belong to geometrical links and the second type concerns form deviations of the surfaces that do not belong to geometrical links. In order to set up the nominal situation of the components, a geometrical frame is associated to each component from given surface situations.

Let us consider the assembly presented figure 1. Surfaces that belong to links are represented as points and arrows. Nominal assembly is presented on figure 1a. Figure 1b shows deviations associated to component 2 (C2) when no constrain is applied.

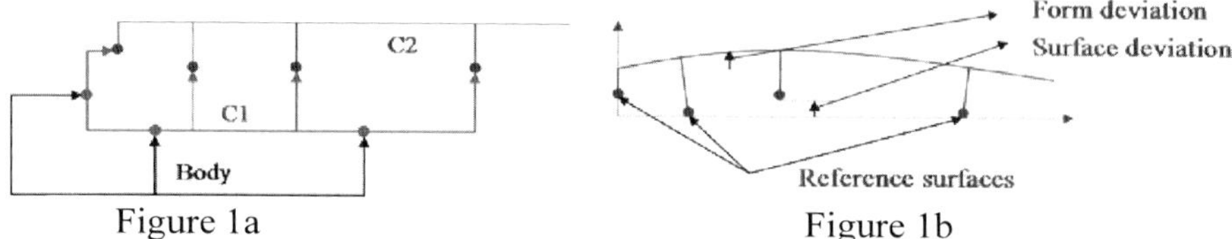


Figure 1a                                    Figure 1b

*Figure 1: Two types of deviations*

Section 2.1 presents how surface deviations are taken into account when 3D geometrical pairs are considered and section 2.2 presents how form deviations are determined.

This model uses Small Displacement Torsor: SDT presented in [Bourdet et al, 1995] and applied in this paper to serial component assembly.


### 2.1. Geometrical deviation model of a geometrical pair

As it is necessary to take component actual geometry into account, we propose to substitute actual surfaces that belongs to geometrical pairs by ideal surfaces that are outward tangent to actual surface. We call these surfaces substitution surfaces.

The model of deviations is given using a SDT called deviation torsor, noted $E[A_{S_i}/A]$ and presented figure 2. The six components of the deviation torsor represent position and orientation of the substitution surface relative to the nominal model. The components determined according to invariance degrees of the surface class are noted $O_{RX}, O_{RY}, O_{RZ}$ for the rotations and $O_{TX}, O_{TY}, O_{TZ}$ for the translations. Components corresponding to actual rotation deviations are noted $\alpha_M, \beta_M, \gamma_M$ and $u_M, v_M, w_M$ for translation deviations.

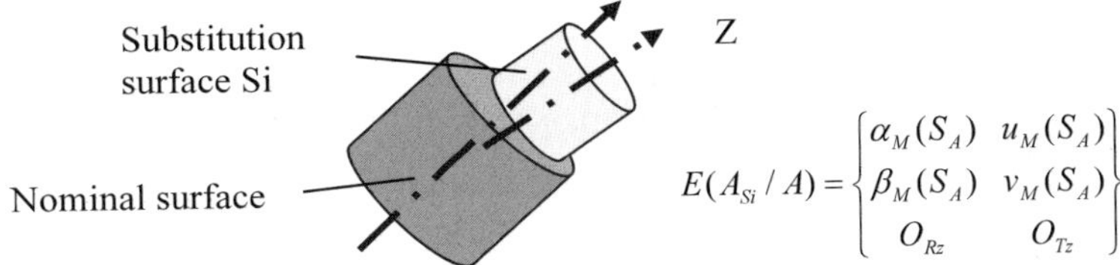


$$E(A_{Si}/A) = \left\{ \begin{matrix} \alpha_M(S_A) & u_M(S_A) \\ \beta_M(S_A) & v_M(S_A) \\ O_{Rz} & O_{Tz} \end{matrix} \right\}$$

*Figure 2: Deviation torsor*

Each actual component of the mechanism is represented thanks to its nominal model and thanks to the components of the deviation torsors that represents effective deviations. The components corresponding to invariance degrees are chosen null for our application.

## 2.2. Actual geometry of a component

Due to process uncertainties, actual form deviations exist between nominal and actual component geometry. Manufactured component is measured to determine its actual form deviations thus, we have to work on discrete geometry, where studied points correspond to measured ones.

Measured actual deviations are expressed in a specific frame, associated with the measured process noted $R_m$. Seeing registration of components, we cut out this frame in order to obtain a null actual geometry deviation on datum surfaces, as presented on figure 3.

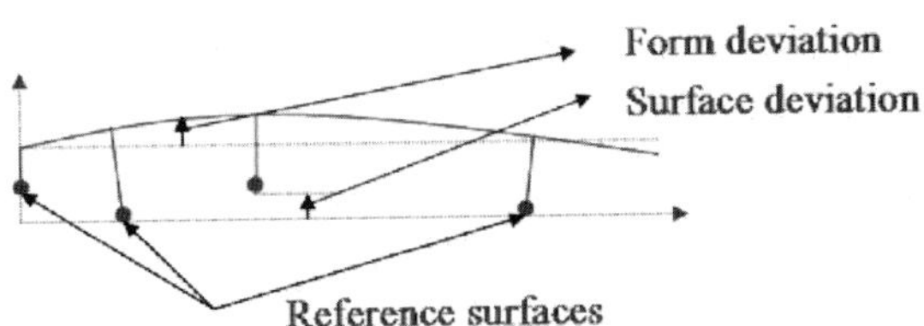


*Figure 3: Form deviations from actual geometry*

## 2.3. Geometrical pair and geometrical relations

Our work is based on a perfect geometrical pair hypothesis. Moreover we consider that all geometrical pairs are defined without clearance, a model of clearance torsor is presented on [Cid et al, 2004].

We consider that each component is set in position in a non over constrained configuration. If the assembly is over constrained, one can either consider flexibility of components as shown on the third part of this paper, or referred to [Thiebaut, 2001] if rigid component hypothesis is to be kept.

Thanks to the serial component assembly assumption, we use an iterative resolution to determine the situation deviations of all the components from their nominal situations. If component (j) situation is known, we find the cut out torsor on the non over constrain registration. By this way, we deduce all new pair's position with deviation torsor of this component. Component (j+1) position is then known and we use the iterative resolution, till the end of mechanism.


## 3 TAKING INTO ACCOUNT COMPONENTS' FLEXIBILITY


### 3.1. Deformation in nominal conditions

A Finite Element Model (FEM) of the component is used to perform the deformation analysis. At each node of the mesh, exerted force is noted $F$ and associated displacement is noted $U_D$. $U_D$ and $F$ are linked by a matrix presented on equation 1. I represents the identity matrix

$$F = KU_D \Leftrightarrow KU_D - F = 0 \Leftrightarrow [K \quad -I]\begin{bmatrix} U_D \\ F \end{bmatrix} = [0] \tag{1}$$

The linear relations (equation 1) that link forces and displacements are always checked and have to be completed with boundary conditions for each node of the mesh. Components' deformations may come from seals, forces or other geometrical pairs. Figure 4 proposes a simple example, where a component is linked with another one thanks to 2 pairs and where an external force is exerted.

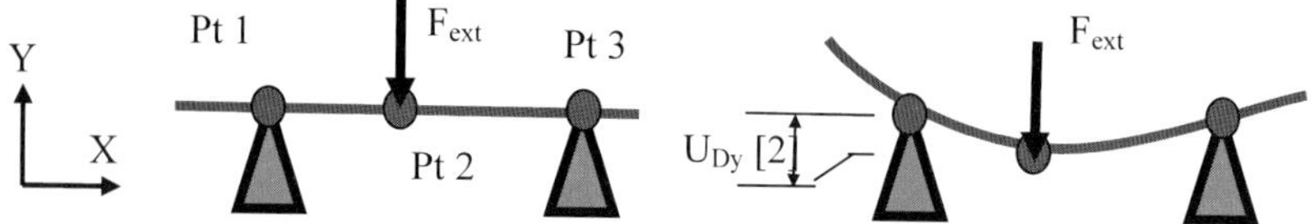


*Figure 4: Classical deformation calculation*

In linear deformation calculation, we can impose either forces $F_{ext}$ and /or displacements values for each point:

$$\begin{cases} F[i] = F_{imposed} \\ U_D[i] = U_{Dimposed} \end{cases} \tag{2}$$

This set of equations is called boundary conditions.

In our example, boundary conditions are presented equation 3 and written using a matrix form where V vector components correspond to imposed values.

$Fx[1] = 0$,     $U_{Dy}[1] = 0$,

$Fx[2] = 0$,     $Fy[2] = F_{ext}$,         $[BC]\begin{bmatrix} U_D \\ F \end{bmatrix} = [V] \tag{3}$

$Fx[3] = 0$,     $U_{Dy}[3] = 0$,

The linear system of equation obtained while merging equations 1 and 3 is presented equation 4.

$$\begin{bmatrix} K & -I \\ BC & F \end{bmatrix} \begin{bmatrix} U_D \\ F \end{bmatrix} = \begin{bmatrix} 0 \\ V \end{bmatrix} \tag{4}$$

This square system of equation composed is then solved in order to find out values of forces and displacements at each node of the mesh.

### 3.2. Combined approach

Compared to classical calculation, boundary conditions depend on position of components after rigid movement.

Figure 5 presented a simple example, where two components are separated by a seal, modelized by springs located at studied points. Both components are set in position with body component thanks to two pairs: L1 and L2 for component 1 and L3 and L4 for component 2.

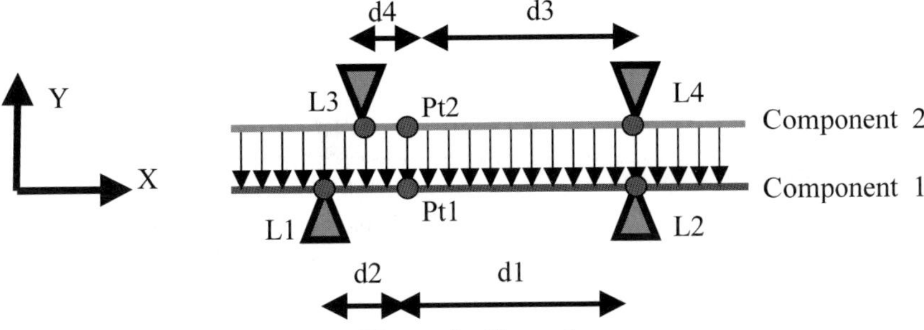


*Figure 5: Example*

Force exerted by this spring on point Pt1 depends on it stiffness noted S but also on actual distance l(1) between point Pt1 and point Pt2. A geometrical pair variation of position introduces a variation of distance between these points, hence a variation of force exerted by the spring, as shown figure 6.

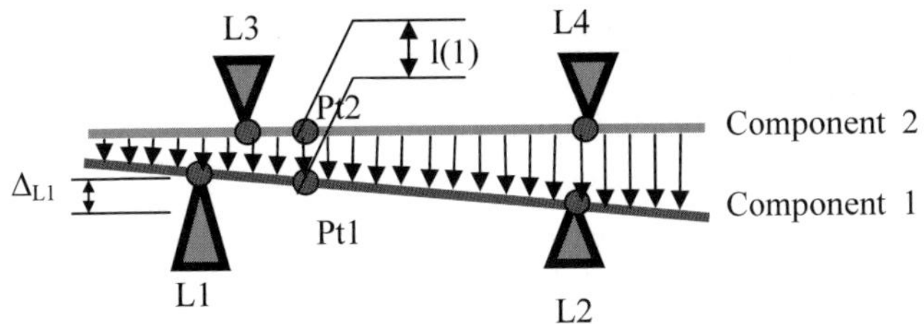


*Figure 6: Rigid body movement influence*

We are able to find equation linking parameters of the geometrical pairs' position and components' point position: equation 5.

$$F_x(1) = S.(L_{ini} - l(1)) \tag{5}$$

$L_{ini}$ corresponds to the spring length when it is free of forces, and $l(1)$ is actual distance between the two points.

G. Cid et al.

$$l(1) = l_{nom}(1) + \frac{d1}{d1+d2}.v_M(1) + \frac{d2}{d1+d2}.v_M(2) - \frac{d3}{d3+d4}.v_M(3) - \frac{d4}{d3+d4}.v_M(4)$$

In our example, $v_M(2) = 0$, $v_M(3) = 0$, $v_M(4) = 0$ and $l_{nom}(1)$ has a numerical determined value. So we are able to write boundary conditions associated to this point:

$$\begin{cases} Fx(1) = (S.l_{ini} - S.l_{nom}(1)) - \dfrac{S.d1}{d1+d2}.v_M(1) \\ Fy(1) = 0 \\ Fz(1) = 0 \end{cases}$$

In this example, we realise that pair positions have an influence on boundary conditions applied on the component, in particular on seal forces values.

By the same way, actual geometric deviation modify distance between the 2 components hence the boundary conditions values. Our model takes into account this influence. These vector form deviation noted ($u_F(i)$, $v_F(i)$, $w_F(i)$) can be found in equations associated to the mechanism.

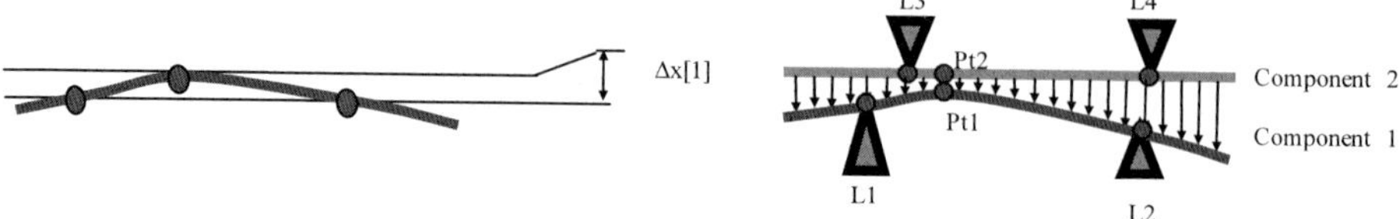


Figure 7: Actual geometry deviations and there influence on boundary conditions

If we consider component 1 actual geometry deviation, we can write equations expressing this influence:

$$F_x(1) = S.(L_{ini} - l_{nom}(1) - \frac{d1}{d1+d2}.v_M(1) - v_F(1))$$

with Δx[1] given by experimentation, for example: $v_F(1) = 0{,}5$

$$F_x(1) = S.(L_{ini} - l_{nom}(1) - 0{,}5) - \frac{S.d1}{d1+d2}.v_M(1)$$

Where $F_x(1)$ belongs to $\begin{bmatrix} forces \\ displacement \end{bmatrix}$ matrix and $v_M(1)$ belongs to [V] matrix.

Once establishing all these boundary conditions, we solve system made of these equations and equation 3.

Obtained results are relations linking actual position of components and geometrical parameters. Symbolic writing of the relations gives the possibility to have influence factor of each parameter on point positions.

## 4 INDUSTRIAL STUDY

### 4.1. Presentation

The above method has been applied to an industrial case that consists in assembling a large compliant door on a structure of helicopter. The door shown figure 8 is to be open and closed many times. Four fixation points (P1,P2,P3,P4) of the door on the structure have been chosen. A set of seven adjusting devices (X2, Y1, Y2, Y3, Y4, Z1, Z2) is available behind the panel the value of whose we search.

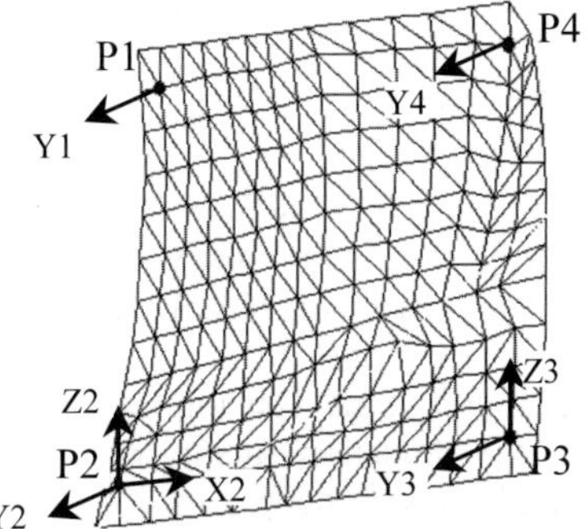


*Figure 8: Nominal panel*

The actual geometry of the panel is measured and does not exactly match the nominal geometry, we present in this paper the method for a given actual geometry. All around this panel a seal assures the waterproofness, its compression has to be controlled, the flush between the panel and the structure has also to be controlled. Eventually, the forces applied to the four fixation points should not exceed a given value.

The study consists in simulating the behaviour of the assembly and in determining the values of the adjusting devices that permit to respect all the requirements.

### 4.2. Method

The method is three steps method.

First, the panel is assembled on a reference structure using an equally constrained positioning. The fixation is only effective on points P1, P2 and P3 and point P4 is free. This first positioning permits to measure a panel and to evaluate the actual form of the actual panel. The results of this step are obtained from measure and are presented on figure 9. The edges of the nominal panel are represented and the deviations between the nominal mesh and the mesh that represents the actual panel are obviously amplified.

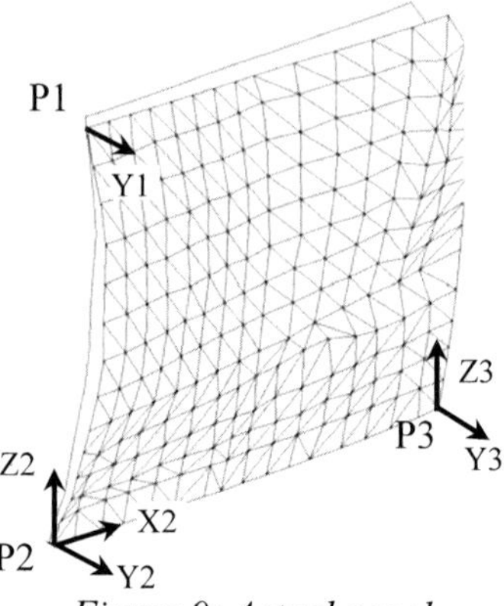


*Figure 9: Actual panel*

The second step of the method consists in simulating the assembly of the panel on the structure in order to determine the relations that link the actual form of the panel, the equally constrained positioning of the panel on the structure and the deformation that provide from the fixation of the fourth point. The application of the method presented in section 2 allows us to write the deviation torsors of surfaces that belong to the geometrical links between the body and the door.

$$E(P1/Body) = \begin{Bmatrix} 0 & 0 \\ 0 & Y1 \\ 0 & 0 \end{Bmatrix}_{P1} \qquad E(P1/Door) = \begin{Bmatrix} 0 & 0 \\ 0 & 0 \\ 0 & 0 \end{Bmatrix}_{P1}$$

$$E(P2/Body) = \begin{Bmatrix} 0 & X2 \\ 0 & Y2 \\ 0 & Z2 \end{Bmatrix}_{P2} \qquad E(P2/Door) = \begin{Bmatrix} 0 & 0 \\ 0 & 0 \\ 0 & 0 \end{Bmatrix}_{P2}$$

$$E(P3/Body) = \begin{Bmatrix} 0 & 0 \\ 0 & Y3 \\ 0 & Z3 \end{Bmatrix}_{P3} \qquad E(P3/Door) = \begin{Bmatrix} 0 & 0 \\ 0 & 0 \\ 0 & 0 \end{Bmatrix}_{P3}$$

$$E(P4/Body) = \begin{Bmatrix} 0 & 0 \\ 0 & Y4 \\ 0 & 0 \end{Bmatrix}_{P4} \qquad E(P4/Door) = \begin{Bmatrix} 0 & 0 \\ 0 & v_M(P4_D) \\ 0 & 0 \end{Bmatrix}_{P4}$$

When the door is measured on the reference structure, surface deviations are null since it is a reference structure since the components of the deviation torsors relative to the body make only appear the adjusting device values. Concerning the deviation torsors relative to the door, the component are null at points P1, P2 and P3 since these points are reference points. A unique deviation component appears on point 4 the value of which is measured.

At this stage of the method, the situation of all the points of the mesh is known as a function of the adjusting device values while considering a non over constrained positioning. The distance between the points of the door and the corresponding points of the body are known through symbolic equations.

The whole of these distances is translated to boundary conditions as shown in section 3 and the resulting set of equations is solved. For each node of the mesh, the deviation between the actual geometry including deformations is then known as a symbolic equation.

The third step of the method consists in determining the values of the adjusting devices. At this stage of the simulation, we know the linear relations that link the values attributed to the adjusting devices to the functional requirements. If we consider the side points of the panel, the values of the flush corresponding on these points depends on the actual form of the panel and on the values given to the adjusting devices along Y-axis. The influence of the values of the adjustments on the position of each point of the mesh is known from the resolution. The initial values of the flush are determined considering the actual form of the panel and the values of all the adjustments set to zero. An illustration of this configuration is shown in figure 10. An optimization of the values of the adjustments can be done to obtain the minima values of the flush all around the door. The illustration that corresponds to the result of the simulation is given figure 10.

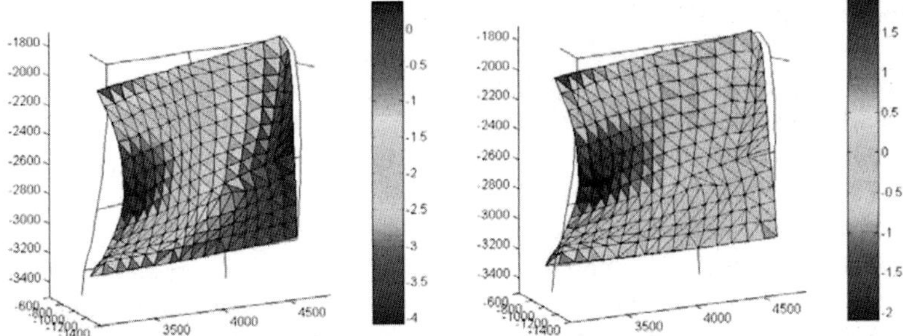

*Figure 10: Initial and optimized deviation of the panel from the structure*


5 CONCLUSION


We have presented our model of simulation, taking into account geometrical variations, deformations and actual geometry variations. Giving results in a symbolic writing allows the study the influence of components' geometrical variations on functional conditions.

This extension of the classical model of 3D link chain allows us to take into account the actual form deviation, and provide the symbolic equations of the assembly.

The application of the method to an industrial study has given some quite interesting results. These results let us envisage the confrontation of the results of simulation to the experiment and the generalization of the approach to other similar cases.

REFERENCES

**[Cid et al, 2004]** Cid G., Thiébaut F., Bourdet P, Taking the deformation into account for components' tolerancing, In *5th conference on Integrated Design on Manufacturing and Mechanical Engineering*, CD Rom., 2004

**[Thiebaut 2001]** Thiébaut F, "Contribution à la définition d'un moyen unifié de gestion de la géométrie réaliste basé sur le calcul des lois de comportement des mécanismes", *PhD Thesis*, Ecole Normale Supérieure de Cachan., 2001

**[Bourdet and Ballot, 1995]** Bourdet P. and Ballot E, "Geometrical behavior laws for computer aided tolerancing". In *4th CIRP Seminar on Computer Aided Tolerancing*, Tokyo, April 5-6, 1995. edited by Fumihiko Kimura, Chapman & Hill, pp. 119-131., 1995

**[Camelio et al, 2001]** Camelio J., Hu S.J., Ceglarek D.J., 2001, "Modeling variation propagation of multi station assembly systems with compliant parts", In *Proceeding of DETC'01, ASME 2001*, Pittsburgh, Pennsylvania, 2001

**[Hu, 1997]** Hu S.J., 1997, "Stream-of-Variation Theory for Automotive Body Assembly", In *Annals of the CIRP*, Volume 46/1, pp1-6, 1997

**[Merkley, 1998]** Merkley K.G, "Tolerance analysis of compliant assemblies", *Ph.D. Thesis*, Department of Mechanical Engineering, Brigham Young University., 1998

**[Samper and Giordano, 1998]** Samper S., Giordano M., "Taking into account elastic displacements in 3D tolerancing. Models and application", *Journal of Materials Processing Technology, n° 78* pp 156-162, 1998

**[Dahlström and Söderberg, 2001]** Dahlström S., Söderberg R, "Towards the method for early evaluations of sheet metal assemblies", In *Proceedings of the 7 CIRP Seminar on Computer Aided Tolerancing*, Ecole Normale Supérieure de Cachan, April 25-25, 2001, pp. 141-150., 2001

**[Sellem et al, 1999]** Sellem E., de Hillerina C. A., Clément A., Rivière A, "Modelling-Simulation-Testing of Compliant Assemblies", In *Global consistency of tolerance, proceedings of the 6th International Seminar on Computer Aided Tolerancing,* pp 355-364, Enschede, The Netherlands, Kluwer Academic Publishers, 1999

**[Shiu et al, 2003]** Shiu B., Apley D.W, Ceglarek D., Shi J, "Tolerance allocation for compliant beam structure assemblies", *In IIE Transactions, n°35*, pp 329-342 ,2003

# Practical Implications in Tolerance Analysis of Sheet Metal Assemblies:
## Experiences from an Automotive Application

**S. Dahlström**[*], **L. Lindkvist, R. Söderberg**[**]

[*]*Volvo Car Corporation, BIW Structure Engineering Dept. 93740, Loc. PV2A,
SE-405 31 Göteborg, Sweden
sdahlstr@volvocars.com*
[**]*Chalmers University of Technology, Product and Production Development
Engineering and Industrial Design,
SE-412 96 Göteborg, Sweden
lali@chalmers.se
rikard.soderberg@me.chalmers.se*

**Abstract:** Over the past years, several approaches for variation simulation of sheet metal assemblies have been presented. However, there are few reports on validation of the methods compared to real production inspection data. Often, the validation consists of comparing the simulation result from different methods against each other and not to real inspection data. In practice, variation simulation methods are difficult to use due to practical implications. This paper illustrates some of the implications and modeling aspects related to sheet metal assembly variation analysis. The tolerance analysis and validation is done by analyzing a real sheet metal assembly with the production inspection data that is used to monitor the process. The aim is to analyze how consistent the result from the tolerance analysis is with the inspection data. Finally, some modeling aspects and techniques will be discussed.
**Keywords**: sheet metal assembly, variation simulation, tolerance analysis, evaluation


## 1. INTRODUCTION

Sheet metal assemblies are commonly used in fabricating various complex products such as automobiles and airplanes. During the assembly process, parts and process variation will influence the final geometry and variation of the assembly. The fulfillment of dimensional specifications is fundamental to assure that the final product will comply with functional, aesthetical and assembly requirements, which are important for the overall quality of the product. To manage this, geometry related production problems must be avoided during early design stages. This is done by analyzing the assembly variation by using different variation simulation techniques.

### 1.1. Sheet metal assembly process

The sheet metal assembly process is often modeled as four sequential steps:
1. Parts are placed in a fixture.
2. Parts are clamped.
3. Parts are joined together.



*J.K. Davidson (ed.), Models for Computer Aided Tolerancing in Design and Manufacturing, 311–320.*

4.       The assembly is released from the fixture.

The most common welding method for joining sheet metal parts is resistance spot welding. There are generally two different weld-guns used for spot welding; the position gun and the balanced gun. The position gun has a fixed lower weld electrode. During welding, sheet metal parts are pushed against the lower electrode despite the initial deviation. The balanced gun applies an equal magnitude of force over both sides of the weld electrodes. This means that the flanges of the sheet metal parts will meet in an equilibrium position. When the welding is complete, the assembly is released from the fixture. Then, the assembly springbacks due to forces introduced during welding and to imperfections such as part and fixture variation. It is this springback deflection that needs to be controlled and analyzed during the development stages.

### 1.2. Dimensional Variation Modeling

Over the past years, different methods have been presented to predict dimensional variation on sheet metal assemblies. Most of the methods are based on the Finite Element Analysis (FEA). [Liu and Hu, 1997], presented a model to analyze the effect of component deviations and assembly springback on assembly variation by applying linear mechanics and statistics. Using FEA, they constructed a sensitivity matrix for compliant parts of complex shapes. The sensitivity matrix established a linear relationship between the incoming part deviation and the output assembly deviation. [Camelio *et al.*, 2001], extended this methodology to multi-station assembly systems and [Lee *et al.*, 2000], complemented the method with a robustness evaluation method. In addition, [Chang and Gossard, 1997], presented the transformation vectors to describe variation and displacement of features. The method represented the interaction between parts and tooling by contact chains. [Ceglarek and Shi, 1997], presented a Beam-based model for tolerance analysis for sheet metal assemblies. [Sellem and Riviére, 1998], developed a methodology based on influence matrices taking into account three different kinds of variation in the simulation: positioning, conformity and shape variability. The effect of weld configurations such as weld sequence and weld gun type on the final geometry has been presented by [Dahlström and Söderberg, 2002]. Weld pattern optimization for sheet metal assemblies was presented by [Liao, 2003]. A fixture design methodology for sheet metal assemblies by using computer experiments has been presented by [Dahlström and Camelio, 2003]. Finally, a contact modeling algorithm implemented into the *Method of Influence Coefficient* to prevent penetrations between parts has been presented by [Dahlström and Lindkvist, 2004].

### 1.3. Validation of variation simulation methods for sheet metal assemblies

Despite that many methods and tools have been developed for variation simulation in the past years, there are few papers on validation of the methods compared to real production inspection data. The reason for this could be that it is difficult to access inspection data and models from industry. Often, the validation consists of comparing the simulation result from different methods against each other and not to real inspection data. [Cai *et al.*, 2003] compared two methods, the Monte Carlo Simulation and the Taylor Series Expansion that are provided from the method EAVS (Elastic

Assembly Variation Simulation). [Liu *et al.*, 1995] performed a validation of *Method of Influence Coefficient* with a lift gate door assembly for two inspection points. They showed that a mechanistic model with PCA (Principle Component Analysis) is a better estimation of the assembly standard deviation than RSS (Root Sum Square). Furthermore, [Liu and Hu, 1997] presented experimental verification of the springback deflection of 16 lap boxes that were made with one type of part variation. The result from the simulation was consistent with the inspection data. [Hu *et al.*, 2001] presented a simulation method, NVAM (New Variation Analysis Method) that were verified for one instrument panel consisting of two parts, assembled with different weld sequences without part and process variation. Finally, a validation of a more complex assembly with relatively few inspection points, with involvement with real production inspection data was presented by [Sellem *et al.*, 1999]. [Sellem *et al.*, 2001] concluded also for the same assembly case, that modeling aspects such as type and size of elements, in- and compatible meshes and the use of contact points have minor impact on the simulation result.

### 1.4. Scope of the paper

In practice, variation simulation methods are difficult to use due to practical implications. The aim of this paper is to illustrate some of these implications and modeling aspects related to sheet metal assembly variation analysis. The tolerance analysis and validation will be done by analyzing a real sheet metal assembly with the accessible production inspection data that is used to monitor the process. The aim is to analyze how consistent the result from the tolerance analysis can be with the inspection data.

## 2.  VARIATION SIMULATION OF SHEET METAL ASSEMBLIES

One way of performing variation simulation of sheet metal assemblies is to use Direct Monte Carlo simulation. This method uses inspection data or a random number generator that produces the desired distributions for the position and part variation. Part variation is simulated by applying a displacement in the inspection points. New part geometry is calculated by FEA. The parts are then joined together by either applying a prescribed displacement or a force in the weld points depending on what type of weld gun to be used. The assembly is then released from the fixture and the springback is calculated by FEA. The simulation starts all over again with new variation. A minimum of 500 to 1000 iterations is necessary to receive a satisfactory distribution of the assembly variation [Liu and Hu, 1997]. This method is considered the most accurate, since each iteration is calculated with the actual geometry. Furthermore, it is relatively easy to implement contact modeling and other parameters such as weld and clamping sequence, weld gun type etc. Direct Monte Carlo is therefore used in the case study.

## 3.  CASE STUDY

The dash board panel consists of 4 sheet metal parts that are spot welded together in two assembly stations (see figure 1). All welding is performed with a balanced gun. In the first station, part 3 is joined to part 2 and then to part 1 on contact surface 1. This sub-assembly is then released from the fixture and repositioned at the second station where it is joined to part 4 on contact surfaces 2, 3 and 4. The weld points are marked with a small x in the figure. The locating scheme is illustrated by boxes that indicated which directions the locators constrain the assembly. Figure 1 also shows the complete assembly and a section cut of the complete assembly at contact surface 4. The simulation case has been analyzed using the FEA software ABAQUS Standard and MATLAB has been used to analyze and run the simulations. Material for all parts is steel with an elastic modulus E = 210 GPa and Poisson ratio $\upsilon$ = 0.3. The parts have been modeled by triangular and square shell elements S3 and S4 (ABAQUS). The complete assembly consists of approximately 12500 elements.

### 3.1. Inspection data

Inspection data for the assembly was collected over a time period of six months. During this period no process changes were made. The number of inspection points for single part 1, 2 and 4 are 52, 59 and 92 respectively. The inspection points of the parts are mainly concentrated on the flanges of the parts and the rest are distributed over the whole part geometry. The complete assembly consists of 64 inspection points. Approximately 50 single parts and 80 complete assemblies have been measured. One important aspect when using inspection data is to analyze the correlation between the inspection points in order to avoid producing part geometry in the simulation model that is unlikely to occur. The correlation coefficients in matrix $R$ are calculated by equation 1 and are calculated from an input matrix $M$ whose rows are observations and whose columns are inspection points. The correlation coefficients in matrix $R$ give the correlation -1 to 1 between the inspection points. In general, points that are located close to each other on the same normal surface should have a strong correlation. In this application the inspection data was analyzed with the correlation coefficient and showed a reasonable correlation for the inspection points located close together. When simulating different part geometries, the actual measured inspection data can be used. However, often the number of observations is not sufficient when variance is simulated. Instead, one can generate random part geometry $M_s$ with maintained correlation by using equation 2.

$$R(i, j) = \frac{C(i, j)}{\sqrt{C(i,i)C(j,j)}} \, , \quad \text{where} \quad C = \text{cov}\,(M) \tag{1}$$

$$M_s = \sqrt{\text{cov}(M)} \cdot N(0,1) + \overline{M} \tag{2}$$

$N(0,1)$ is random generated numbers with mean of zero and a standard deviation of one and $\overline{M}$ is the mean of the inspection points. In this simulation equation 2 is used to generate correlated part geometry.

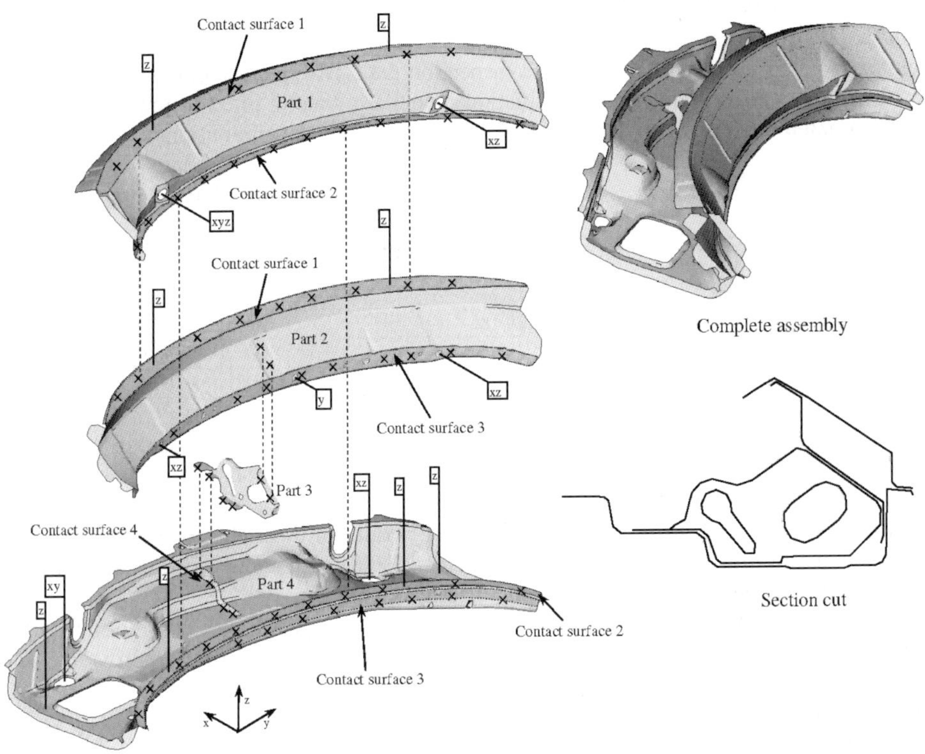


*Figure 1: Assembly model*

### 3.2. Simulation model

FEA is used to produce part variation by applying a displacement corresponding to the inspection data in the nodes closest to the inspection point. The parts are joined together by applying an equal magnitude of force in the spot welds. After a spot weld has been joined, a rigid beam is activated in the weld to prevent the weld to open during the rest of the simulation. When the welding is complete, the assembly is released from the fixture and the assembly springback deflection are calculated with FEA. A total of 500 iterations are conducted for each simulation.

## 4.  RESULT

Five different simulations have been conducted in order to investigate how different modeling aspects influence the result. The result from the different simulations is summarized in table 1. After each simulation the mean and variance ($6\sigma$) is compared to the inspection data for the complete assembly. Correlation coefficients $R_{mean}$ and $R_{6\sigma}$ are calculated between the simulated and measured inspection points, by using equation 2.

$R_{mean}$ is a measure of how consistent the simulated mean for the inspection points is with the measured and $R_{6\sigma}$ is the consistency for the variance. A low $R$ value indicates that the result is not consistent with the inspection data. Different modeling aspects are considered and refinements of the simulation model are done in order to improve the $R$ values. Each modeling technique and aspect is kept for the next simulation.

*Table 1: Correlation coefficient $R_{mean}$ and $R_{6\sigma}$ for the different simulations*

| # | Modeling aspect | $R_{mean}$ | $R_{6\sigma}$ |
|---|---|---|---|
| 1 | Simultaneously joining | 0.08 | - |
| 2 | Springback for part 1 and 2 | 0.11 | - |
| 3 | Welding performed in clusters | 0.23 | 0.11 |
| 4 | Removal of inspection points | 0.43 | 0.21 |
| 5 | Mating surfaces are modeled with contact elements | 0.79 | 0.45 |

1)      In the first simulation the part variation is applied to all single parts. Part 1, 2 and 3 are joined together and then to part 4. All joining to part 4 is done simultaneously. Since a balanced gun is used, an influence matrix is calculated in order to apply the correct forces in the weld points to prevent over closure of the weld points. The result from the first simulation ($R_{mean}$=0.08) is not consistent with the measured data. The comparison with variance is therefore not performed.

2)      The assembly is divided into two assemblies so that part 1 and 2 can springback before they are joined to part 4. The result from this simulation does not improve $R_{mean}$ significantly, however, this is more similar to the real assembly process and is therefore needed in order refine the simulation model.

3)      The welding of the assembly is divided into sequence. Each contact surface is welded in sequence. Within the contact surfaces the welding is performed simultaneously. $R_{mean}$ improved to 0.23.

4)      Surface 3 on part 2 was crumpled in the previous steps after applying part variation, see figure 2. Applying a displacement to the inspection points parallel to the part surface such as center position of holes, embossing and re-enforcements etc. and calculate new part geometry with FEA may cause the surface to crumple. These types of variation are probably caused during the stamping process of the holes. These inspection points are instead applied as a translation without any deformation. The crumpling will also have an impact on the magnitude of the weld force and consequently on the springback deflection. The result shows an improvement of both the $R_{mean}$ and $R_{6\sigma}$.

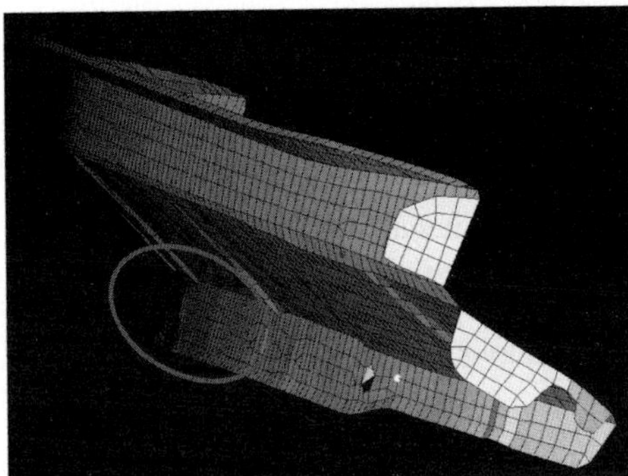

*Figure 2: Crumpling of the contact surface3*

*Figure 3: Penetration in the corners of contact surface 1*

5)      There is penetration between part 1 and 2 in both corners after they are joined together (see figure 3). This can have a major impact on the springback deflection [Dahlström and Lindkvist, 2004]. All mating surfaces are therefore modeled with contact elements in order to prevent the parts from penetrating each other. This results in $R_{mean}$=0.79 and $R_{6\sigma}$=0.45. The values for each inspection point for mean and variance is presented in figure 4.

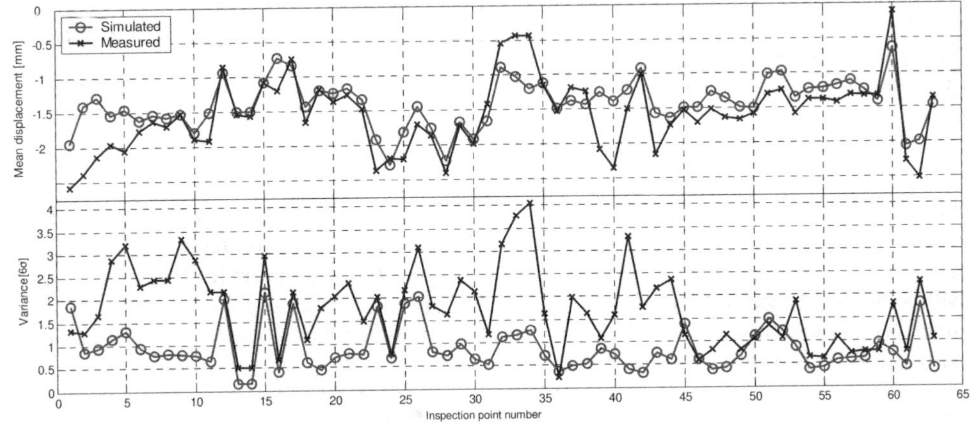

*Figure 4: Comparison mean deviation (upper) and 6σ (lower)*

## 5. DISCUSSION

In this application, we have incorporated modeling aspects such as full springback between the assembly steps, welding of the contact surfaces separately in sequence, removal of inspection points that were parallel to the surface and using contact elements in order to receive a satisfactory correlation between the inspection data. The weld sequence and contact modeling was required since this assembly has curved contact surfaces which the springback is strongly influenced by. The incorporation with contact elements on several surfaces and the number of assembly steps increases the computational time significantly. The total time for one iteration in the last simulation was approximately 40 minutes. The use of a linear model would be useful in this case to reduce the computational time. However the linear model must manage contact modeling and weld sequence. [Dahlström and Lindkvist, 2004], presented a contact algorithm that can be used in linear models.

The correlation coefficient for the comparison in mean for the last simulation is satisfactory. The inspection points 1 to 5 and 39 and 40 are measured in the x-direction of part 4 (see figure 4). The weak correlation in these points could be caused by some translation in the x-direction that is not captured in the model.

The correlation coefficient for the comparison in $6\sigma$ is not satisfactory. There are several of reasons for this. The number of measured single parts may not be sufficient in order to simulate variance. The inspection data collected for the single parts is measured in an inspection fixture. This means that neither positioning variations nor the repeatability index for the assembly fixture are captured. In general, the simulated $6\sigma$ is lower than measured, which indicates that the assembly is probably subjected to other types of variations than simulated. The variation could be weld gun variation such as wear and position accuracy. Balanced weld guns are designed to be balanced in the direction where they perform the welding. In practice, this is hard to achieve since one weld gun often performs welding in different directions (horizontal and vertical welding of flanges), and will therefore always have some inertia in some directions. These variations are very difficult to include in the simulation model.

One other important modeling aspect when part variation is applied by giving the inspection points a described displacement can cause the part to deflect in an unnatural way. This could occur if the inspection points are widely distributed or that the part is of a complex geometry. Assume that the geometry illustrated in figure 5a is the nominal geometry. The part has one inspection point that is measured in the arrow direction. The part variation $d$ is measured and the figure 5b shows the actual part geometry. If a described displacement with a distance $d$ is applied and new part geometry is calculated with FEA, the part would deform as in figure 5c. In this case, the simulated part variation does not correspond to the actual part geometry. This phenomenon can occur when using FEA to produce part variation.

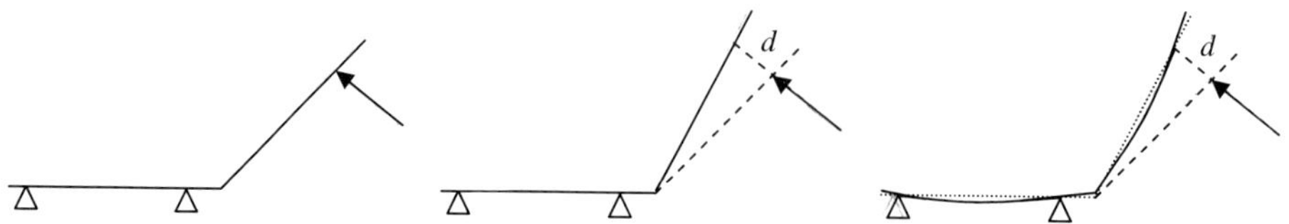

*Figure 6: a) nominal geometry, b) actual part variation, c) simulated part variation*

It is therefore important to analyze the inspection data regarding this effect, especially, between contact surfaces since it is via these surfaces the imperfections of one part is transferred to the other. To fully verify variation simulation methods for complex sheet metal assemblies it is probably necessary to involve scanned single parts.


## 6. CONCLUSIONS

In practice, variation simulation of sheet metal assemblies is difficult to conduct due to practical implications. In this paper an automotive application with the accessible inspection data used to monitor the process has been analyzed. This application illustrates some of the implications in variation simulation of sheet metal assemblies. It shows that correlation between simulation result and inspection data is depended on a number of modeling aspects. It was required to incorporate assembly sequence, contact modeling, weld sequence and analysis of inspection data in order to receive a satisfactory correlation between simulated variation in mean and inspection data.

The correlation between simulated variance and measured was not satisfactory. The probable cause to this is that other types of variations are not captured in the model due to that it is very difficult to estimate or measure these variations.

It is also important to be aware of that simulating part variation by applying a displacement and calculating new part geometry with FEA could result in a geometry that is unlikely to occur in real assembly. It is therefore important to analyze inspection data to confirm that the amount of inspection points is adequate. In order to fully verify variation simulation methods it is probably required to have scanned single parts. More verification of other types of assemblies is needed in order to derive more general conclusions.


## REFERENCES

[Cai *et al.*, 2003] Cai, W., Long, Y., and Hsieh, C., 2003, "VARIATION SIMULATION FOR DIGITAL PANEL ASSEMBLY", 2003 ASME International Mechanical Engineering Congress and Exposition, Washington, D.C., November 15-21.

[Camelio *et al.*, 2001] Camelio, J. A., Hu, J. S., and Ceglarek, D., 2001, "Modeling Variation Propagation of Multi-Station Assembly Systems With Compliant Parts", ASME 2001 Design Engineering Technical Conference and Computers & Information in Engineering Conference, Pittsburgh, Pennsylvania, September 9-12.

[Ceglarek and Shi, 1997] Ceglarek, D., and Shi, J., 1997, "Tolerance Analysis for Sheet Metal Assembly Using a Beam-Based Model", Concurrent Product Design and Environmentally Conscious Manufacturing, Dallas, Texas, USA, November.

**[Chang and Gossard, 1997]** Chang, M., and Gossard, D. C., 1997, "Modelling the Assembly of Compliant, Non-Ideal Parts", Computer Aided Design, Vol. 29, pp. 701-708.

**[Dahlström and Camelio, 2003]** Dahlström, S., and Camelio, J. A., 2003, "Fixture Design Methodology for Sheet Metal Assembly Using Computer Simulations", 2003 ASME International Mechanical Engineering Congress and Exposition, Washington, D.C., November 15-21.

**[Dahlström and Lindkvist, 2004]** Dahlström, S., and Lindkvist, L., 2004, "Contact Modeling in Method of Influence Coefficient for Variation Simulation of Sheet Metal Assemblies", 2004 ASME International Mechanical Engineering Congress and Exposition, Anaheim, California, November 13-19.

**[Dahlström and Söderberg, 2002]** Dahlström, S., and Söderberg, R., 2002, "Analysis of the Final Geometry Due to Weld Process Effects in Sheet Metal Assemblies", The 4th International Symposium on Tools and Methods for Competitive Engineering, Wuhan, Hubei, China, April 22-26.

**[Hu et al., 2001]** Hu, M., Lin, Z., Lai, X., and Ni, J., 2001, "Simulation and Analysis of Assembly Processes Considering Compliant, Non-Ideal Parts and Tooling Variations", International Journal of Machine Tools & Manufacture, Vol. 41, pp. 2233-2243.

**[Lee et al., 2000]** Lee, J., Long, Y., and Hu, J. S., 2000, "Robustness Evaluation for Compliant Assembly Systems", The 2002 ASME International DETC and CIE / Design for Manufacturing Conference, Baltimore, Maryland, USA, September 10-13.

**[Liao, 2003]** Liao, G. Y., 2003, "A Generic Algorithm Approach to Weld Pattern Optimization in Sheet Metal Assembly", 2003 ASME International Mechanical Engineering Congress and Exposition, Washington D.C., November 15-21.

**[Liu and Hu, 1997]** Liu, S. C., and Hu, J. S., 1997, "A Paremetric Study of Joint Performance in Sheet Metal Assmebly", International Journal of Machine Tools & Manufacturing, Vol. 37, pp. 873-884.

**[Liu and Hu, 1997]** Liu, S. C., and Hu, J. S., 1997, "Variation Simulation for Deformable Sheet Metal Assemblies Using Finite Element Methods", Journal of Manufacturing Science and Engineering, Vol. 119, pp. 368-374.

**[Liu et al., 1995]** Liu, S. C., Lee, H.-W., and Hu, J. S., 1995, "Variation Simulation for Deformable Sheet Metal Assemblies Using Mechanistic Models", Transactions of NARMI/SME, Vol. XXIII, pp. 235-240.

**[Sellem and Riviére, 1998]** Sellem, E., and Riviére, A., 1998, "Tolerance Analysis of Deformable Assemblies", 1998 ASME Design Engineering Technical Conference, Atlanta, GA, USA, September 13-16.

**[Sellem et al., 1999]** Sellem, E., Riviére, A., Hillerin, C. A. D., and Clement, A., 1999, "Validation of the Tolerance Analysis of Compliant Assemblies", 1999 ASME Design Engineering Technical Conference, Las Vegas, Nevada, Sept. 12-15.

**[Sellem et al., 2001]** Sellem, E., Sellakh, R., and Riviére, A., 2001, "Testing of Tolerance Analysis Module for Industrial Intrest", 7th Cirp International Seminar on Computer Aided Tolerancing, ENS de Cachan, France, 24-25 April.

# Predicting Deformation of Compliant Assemblies Using Covariant Statistical Tolerance Analysis

**M. R. Tonks, K. W. Chase, C. C. Smith**
*Brigham Young University, 4351 CTB, Provo, UT 84602*
*chasek@byu.edu*

**Abstract:** In assemblies with compliant parts, dimensional variation causes misalignment between mating parts. To correct the misalignment, the compliant parts are deformed before being fastened. The resultant springback and residual stress can hinder performance. A new method uses statistical tolerance analysis and stochastic finite element analysis to predict the probable range of deformation caused by dimensional variation. To account for surface variation, a hybrid method models the surface covariance in which Legendre polynomials are used to model long wavelengths and the frequency spectrum is used to model shorter wavelengths. The hybrid covariance model accurately predicts the covariance of simulated parts and the covariance calculated from sheet-metal part data. The hybrid covariance method is an important part of an effective system for statistical analysis of variation in compliant assemblies.
**Keywords**: Assembly Variation; Tolerance Analysis; Stochastic Finite Element Method

## 1. INTRODUCTION

Thin, compliant parts, whether sheet metal, plastic, or composite, are subject to warping, distortion, misalignment and surface waviness due to manufacturing variation. The variation can cause assembled parts to spring back to a new equilibrium position, with attendant residual stresses. The final shape may have objectionable appearance or aerodynamics and the residual stresses could shorten the fatigue life. A tool that predicts the springback and residual stress in the assembled parts would be of great value to industries that rely on flexible assemblies: aerospace, automotive, electronics, etc.

Tolerance and variational analysis methods predict the amount of misalignment that will occur between parts, but cannot predict the amount of springback and stress. Compliant Statistical Tolerance Analysis (CSTA) methods have been developed to predict the amount of springback and residual stress that will occur within assemblies containing compliant parts due to manufacturing variation.

Many CSTA methods exist in the literature, employing various variational methods but universally employing finite element analysis (FEA) to account for part compliance. [Liu and Hu, 1997] combine FEA with Monte Carlo simulation and lower the



*J.K. Davidson (ed.), Models for Computer Aided Tolerancing in Design and Manufacturing*, 321–330.

computational cost using influence coefficients. [Chang and Gossard, 1997] simulate the assembly and measurement processes, along with the part stiffness matrix, to find the variation introduced into the assembly. [Sellem and Riviere, 1998] combine the results of Liu and Hu and Chang and Gossard to create a linear method that uses influence coefficients to find the part variation. [Camelio *et al.*, 2003] use the method of Liu and Hu in a multi-station assembly method that accurately represents the assembly process and [Camelio *et al.*, 2004 ] express the part covariance matrix in terms of its eigenvectors to identify the critical variation modes. [Merkley and Chase, 1996] develop a set of linear equations that account for part covariance in conjunction with part stiffness matrices with only two finite element solutions. [Bihlmaier, 1999] includes surface variation using spectral analysis in Merkley and Chase's method. In a similar method, [Huang and Ceglarek, 2002] use a discrete-cosine-transformation to model the part form error. [Soman, 1999] shows that the surface variation of compliant parts often have significant variation with wavelengths longer than the part-length, which cannot be accurately modeled using spectral analysis, and [Stout, 2002] presents a polynomial-based method to model such variation. [Tonks and Chase, 2004] develop a method to model long wavelength variation using a series of orthogonal polynomials.

In this work, the CSTA method of [Merkley and Chase, 1996] is summarized, the need to model the effects of surface variation is explained, and typical surface variation is investigated. A hybrid geometric covariance model that combines the work of [Bihlmaier, 1999] and [Tonks and Chase, 2004] is presented. The hybrid geometric covariance model is used to predict the geometric covariance of a set of simulated parts and the covariance calculated from measured data taken from a set of sheet-metal parts.


## 2. LINEAR CSTA METHOD

The CSTA method, first developed by [Merkley and Chase, 1996], finds the residual stress and springback after assembly due to dimensional and surface variation. The solution can be divided into three sections, solving for the total misalignment in the assembly; finding the covariance of the misalignment of each part, and determining the mean and standard deviation of the springback and stress in each assembled compliant part.

### 2.1.  Calculating the Total Misalignment

The total misalignment in the assembly is due to dimensional variations from each part, as well as fixture and tooling error. To account for the sources of dimensional variation, all parts are considered rigid and statistical tolerance analysis (STA) is used to find the mean and variance of the misalignment for each set of mating parts. Many STA methods exist and are summarized in [Chase and Parkinson, 1991]. An additional STA method is presented in [Shen *et al.*, 2004]. A description of this process using a vectorized STA method is found in [Mortensen, 2002].

### 2.2.  Calculating the Misalignment Covariance

STA gives the mean and variance of the misalignment between each set of mating parts, but to solve for the springback and stress in the compliant parts, the covariance of the misalignment at the closure points (rivet/spot weld locations) is needed as is the portion of the total misalignment absorbed by each part on assembly. In the misalignment covariance matrix $\Sigma_{\delta_0}$ the diagonal terms are the variance of the variation at each point and the off-diagonal terms are a measure of the interaction between the variations at two points. The off-diagonal terms are obtained using part data or a covariance model that accounts for part surface variation. Because $\Sigma_{\delta_0}$ is affected only by geometric variation, it is called the *geometric covariance*. The methods used to model the geometric covariance are presented later in this work.

When two compliant parts are joined together and released, the joint moves to a force equilibrium position (see Fig 1). To find the mean and covariance of the equilibrium position for each part a stochastic finite element method (SFEM) is used, as shown in [Tonks and Chase, 2004]. This problem varies from typical SFEM, because the displacements applied to assemble the model are analyzed rather than an external load applied to a fully assembled model. Various SFEM methods exist and are summarized in [Tonks and Chase, 2004]. The method used by [Merkley and Chase, 1996] requires that each part be meshed and an equivalent stiffness matrix be created according to

$$\mathbf{K}_{eq,a} = \left(\mathbf{K}_a + \mathbf{K}_b\right)^{-1}\mathbf{K}_b$$
$$\mathbf{K}_{eq,b} = \left(\mathbf{K}_a + \mathbf{K}_b\right)^{-1}\mathbf{K}_a$$

$$(1)$$

With the equivalent stiffness matrices and the mean of the misalignment between the parts $\mu_{\delta_0}$ , the means of the misalignment of the individual parts are found from

$$\mu_{\delta_a} = \mathbf{K}_{eq,a}\mu_{\delta_0}$$
$$\mu_{\delta_b} = \mathbf{K}_{eq,b}\mu_{\delta_0}$$

$$(2)$$

where $\mu_{\delta_0} = \mu_{\delta_a} + \mu_{\delta_b}$. For a linear relationship such as in Eq. (2), the covariance of the part misalignment is found from the equivalent stiffness matrix and $\Sigma_{\delta_0}$ according to

$$\Sigma_{\delta_a} = \mathbf{K}_{eq,a}\Sigma_{\delta_0}\mathbf{K}_{eq,a}^T$$
$$\Sigma_{\delta_b} = \mathbf{K}_{eq,b}\Sigma_{\delta_0}\mathbf{K}_{eq,b}^T$$

$$(3)$$

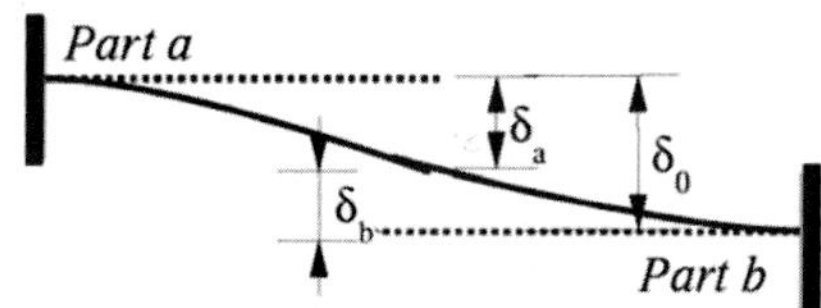


*Figure 1; Equilibrium position of assembled parts*

as shown in [Johnson and Wichern, 2002]. The covariance of the part misalignment depends on the material stiffness and is therefore called the *material covariance*.

### 2.3. Stress and Springback Solution

The springback and residual stresses throughout each compliant part due to assembly are obtained from the mean and covariance of the part misalignment. An individual FE solution is performed for each compliant part, where the part has prescribed displacements at all mating edges. Two FE solutions provide maximum and minimum values for the springback and stress or, if a statistical solution is required, SFEM is used. Typically, the misalignment is small and the problem is treated as linear, elastic with small variations. If the variations are too large to be treated in this way, the resultant springback and residual stress would likely render the assembly unfit for use.

After the residual stresses and springback are found for each part, the performance of the assembly can be evaluated. If the stresses and springback violate design constraints, the critical tolerances are iterated until the performance is within design limits. To identify the critical tolerances, the critical closure points are identified with an FE sensitivity analysis and the critical tolerances are then identified with an STA sensitivity analysis. The design iterations are efficient, as they do not require repeated simulations or remeshing.

## 3. GEOMETRIC COVARIANCE

As explained above, when part data is not available, *i.e.* the parts are not in production, the geometric covariance is modeled because STA does not provide any covariant information. [Bihlmaier, 1999] shows that to accurately model the geometric covariance, the surface variation of the mating parts must be accounted for. It is useful to divide the surface variation into three frequency domains;

1. Warping – Wavelengths longer than the part-length

2. Waviness – From one to five wavelengths over the part-length

3. Roughness – More than five wavelengths over the part-length

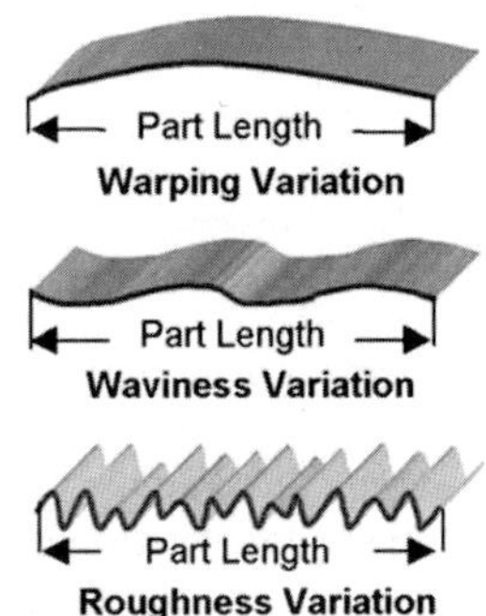



*Figure 2: Examples of types of surface variation*

where Figure 2 shows examples of each type of variation. The surface variation on any surface is composed of a combination of variation from the three domains and therefore an effective geometric covariance model should be capable of quantifying variation from all three domains. Often short wavelength variation also has small amplitude and need not be accounted for, but the amplitude that can be neglected varies with application and should be decided by the engineer.

Several covariance models have been proposed, each with different modeling capabilities. Promising models are presented in [Bihlmaier, 1999] and [Tonks and Chase, 2004]. These models require two sets of inputs, the variance of the misalignment at each closure point and descriptors that define typical surface variation caused by the manufacturing process. The methods are summarized below.

### 3.1. Frequency Spectrum Model

The geometric covariance model of [Bihlmaier, 1999] uses the frequency spectrum to model the surface variation. In the frequency spectrum model, the average autospectrum of the mating part surfaces, found by multiplying the frequency spectrum by its complex conjugate, describes the surface variation. To find the geometric covariance from the average autospectrum, **a,** the autocorrelation **c** is first found using the inverse discrete Fourier transform

$$c_i = \frac{1}{N}\sum_{0}^{N-1} a_j e^{\frac{2\pi j i}{N}}.$$  (4)

The autocorrelation function describes the correlation of two points on the surface and the center value corresponds to the normalized variance. To construct the geometric covariance matrix, **c** is placed along each row of the covariance matrix so that the peak value falls along the diagonal. Each row is then scaled such that the diagonal values equal the input variances at each node of the mating surfaces. Figure 3 depicts this shifting of the autocorrelation function.

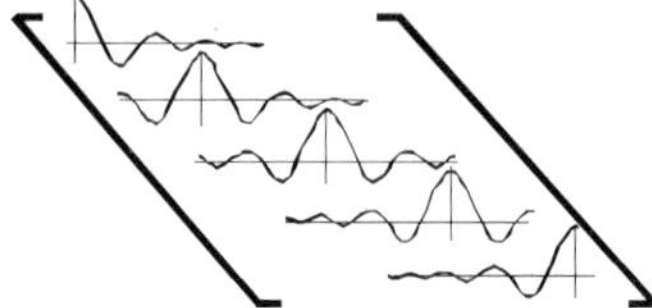

*Figure 3; Finding the covariance matrix from the autocorrelation, [Bihlmaier, 1999]*

The frequency spectrum model accurately predicts the covariance from waviness and roughness variation, but cannot provide information about warping variation. This is a serious shortcoming because warping is often the dominant type of surface variation in thin, compliant parts.

### 3.2.  Orthogonal Polynomial Model

[Soman, 1999] shows that the warping variation can be separated from the waviness and roughness variation by fitting and subtracting polynomials from the variation data. [Tonks, 2002] finds that a series of orthogonal polynomials are an effective means of modeling the warping variation. He develops a model that uses a series of Legendre polynomials to model the covariance. Typical surface variation resulting from a manufacturing process is defined by the average polynomial coefficient vector **a**.

In the orthogonal polynomial model, an uncorrelated covariance matrix $\Sigma_0$ is created by placing the closure point variances down the diagonal of a diagonal matrix. A correlation matrix is created according to

$$S_{ij} = \sum_{l=o}^{M-1} a_l \sqrt{\frac{2l+1}{N}} P_l(x_i) P_l(x_j) \tag{5}$$

where $P_l(x_i)$ is the $l$th order Legendre polynomial at point $x_i$. The geometric covariance, $\Sigma_\delta$, is found using $\Sigma_0$ and **S** according to

$$\Sigma_\delta = S\Sigma_0 S \tag{6}$$

Orthogonal polynomials were found to accurately model warping variation, but the accuracy quickly decreased for variation with more than one wavelength over the part-length (see Table 1). Above four wavelengths, the polynomials could not accurately model the variation.

### 3.3.  Hybrid Covariance Model

The frequency spectrum model accurately predicts the geometric covariance of waviness and roughness variation and the orthogonal polynomial model accurately predicts the geometric covariance of warping variation, but neither method models the entire range of variation. [Tonks and Chase, 2004] show that when the surface variation is dominated by warping variation the orthogonal polynomial method accurately models

*Table 1; Number of polynomials needed to accurately model surface variation*

| Wavelengths over part-length: | 1 | 2 | 3 | 4 |
|---|---|---|---|---|
| Number of polynomials: | 7 | 11 | 15 | *inaccurate* |

the geometric covariance, but when this is not the case, neither model is sufficient. [Soman, 1999] showed that after subtracting the warping variation with a polynomial fit, the remaining higher frequency variation could be modeled using the frequency spectrum method. Following this approach, a hybrid model has been developed that combines the warping variation covariance found from the orthogonal polynomial model with the waviness and roughness variation covariance found from the frequency spectrum model to obtain the full surface variation covariance.

Given a vector $\mathbf{a}$ that defines the variation of a surface, a linear combination of a vectors representing the normalized warping variation, $\mathbf{a}_w$, and the normalized waviness and roughness variation, $\mathbf{a}_{wr}$, can be obtained such that

$$\mathbf{a} = c_w \mathbf{a}_w + c_{wr} \mathbf{a}_{wr} \tag{7}$$

where $c_w$ and $c_{wr}$ scale the normalized vectors. Assuming any dependence between $\mathbf{a}_w$ and $\mathbf{a}_{wr}$ is negligible, the covariance of $\mathbf{a}$ is found from the covariances of $\mathbf{a}_w$ and $\mathbf{a}_{wr}$,

$$\Sigma_{\mathbf{a}} = c_w^2 \Sigma_w + c_{wr}^2 \Sigma_{wr} \tag{8}$$

where $\Sigma_w$ is found from Eq. (6) and $\Sigma_{wr}$ is found using the frequency spectrum method. The hybrid covariance model uses three variation descriptors to accurately model the geometric covariance for a general surface. The descriptors are the coefficients $c_w$ and $c_{wr}$, the frequency spectrum of the waviness and/or roughness variation, and the polynomial coefficient matrix of the warping variation.

The hybrid method is carried out according to the following steps:
1. Model the geometric covariance of the waviness and roughness variation using the frequency spectrum method. Inputs: frequency spectrum
2. Model the geometric covariance of the warping variation using the orthogonal polynomial method. Inputs: polynomial coefficient vector
3. Combine the two covariances according to Eq. (8) to model the total covariance. Inputs: $c_w$ and $c_{wr}$

To develop the descriptors for the hybrid model, typical variation from a manufacturing process is required. The descriptors are obtained by first obtaining the warping portion of the variation data by fitting and subtracting successive polynomials; the remaining variation is the waviness and roughness portion. The warping variation and the waviness and roughness variation are then normalized such that their average values have a maximum of one. Finally, the average polynomial coefficient matrix is found from the normalized waviness variation and the average autospectrum is found from the normalized waviness and roughness variation.

## 4. HYBRID COVARIANCE MODEL EVALUATION

To evaluate the modeling capability of the hybrid covariance model, two comparisons are shown; first, with simulated surface variation data and second, with measured surface data.

### 4.1. Simulated Data

A set of surfaces were simulated as if being manufactured using a process that gives surface variation of all three types and for which descriptors of typical variation have been identified. 10,000 surfaces were generated and the covariance was calculated directly from the simulated data. The covariance was also calculated using the hybrid method according to the known variation descriptors. Figure 4 shows four plots comparing the simulated covariance to the covariance generated with the hybrid model, a 3D bar graph of the geometric covariance matrix generated by each method (Fig. 4 a. and 4 b.), a plot of the error in the hybrid covariance matrix (Fig. 4 c.), and a plot of the variances (the diagonal of the two covariance matrices) (Fig. 4 d.). The hybrid model accurately reproduced the covariance calculated from the simulated data

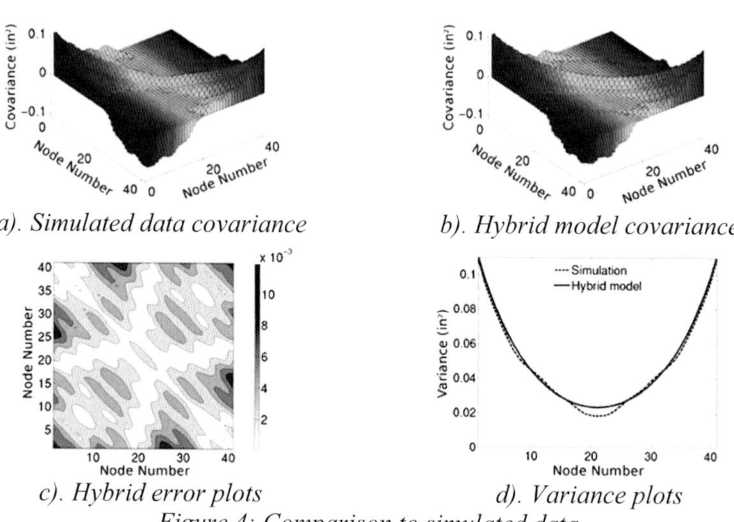

a). Simulated data covariance

b). Hybrid model covariance

c). Hybrid error plots

d). Variance plots

Figure 4; Comparison to simulated data

### 4.2. Measured Part Data

The hybrid model was also used to model the geometric covariance from a set of 6 sheet-metal parts for which the surface variation was obtained using a coordinate measure machine (CMM) and characterized using the three variation descriptors. The covariance was calculated directly from the CMM data and was also modeled using the hybrid method. Similar plots to those in Fig. 4 are shown in Fig. 5 and demonstrate that the hybrid model effectively modeled the geometric covariance of the part data.

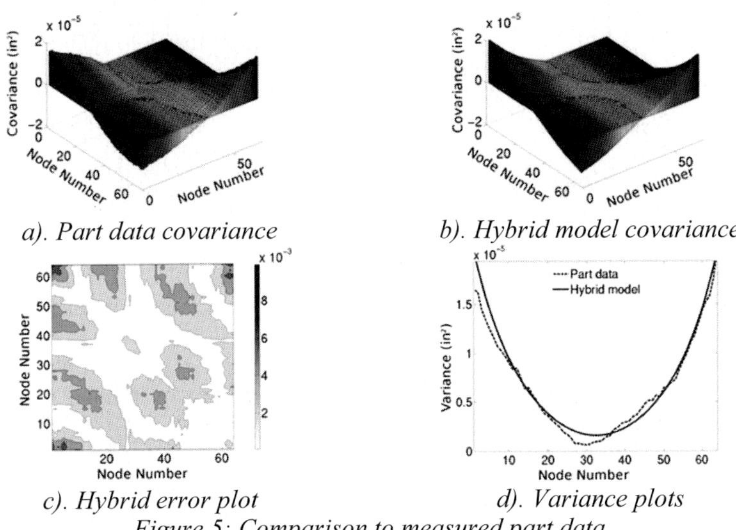

a). Part data covariance          b). Hybrid model covariance

c). Hybrid error plot             d). Variance plots

*Figure 5; Comparison to measured part data*

## 5. CONCLUSIONS

The hybrid covariance model provides a robust means of modeling the geometric covariance of surface variation in assemblies with compliant parts. Three descriptors of typical manufacturing process variation are required, but databases of these descriptors can be gathered for different processes to make this information readily available for design. The hybrid model uses the orthogonal polynomial model, requiring the polynomial coefficient matrix, for the warping variation and the frequency spectrum model, requiring the average autospectrum, for waviness and roughness variation. The two covariances are combined using the coefficients $c_w$ and $c_{wr}$. The hybrid model has been shown to efficiently and accurately predict the geometric covariance from simulated and real data.

Combining STA with FEA provides a tool for performing variation analysis on compliant assemblies. The CSTA method, using the hybrid covariance model, can efficiently estimate the residual stresses and springback in each compliant part in an assembly due to the dimensional and surface variation of the parts. The analysis may be repeated with different tolerances without requiring remeshing, providing an efficient means of experimenting with tolerances to maintain the springback and residual stress within design limits. Thus, this tool provides an effective means for analysis and design of robust assemblies using rigid and compliant parts and for the simulation of current assembly processes and evaluation of product quality.

## REFERENCES

**[Liu and Hu, 1997]** Liu, S.; Hu, S.; "Variation Simulation for Deformable Sheet Metal Assemblies using Finite Element Methods"; *Manufacturing. Sci. and Eng.*, Trans. of the ASME, 119(3), pp. 368-374; 1997.

**[Chang and Gossard, 1997]** Chang, M.; Gossard, D.C.; "Modeling the Assemblies of Compliant, Non-ideal Parts"; Computer Aided Design, 29(10), pp. 701-708; 1997

**[Sellem and Riviere, 1998]** Sellem, E.; Riviere, A.; "Tolerance Analysis of Deformable Assemblies"; Design Automation Conference, *ASME Design Eng. Tech. Conf*, Atlanta, GA, 1998; DETC98 – DAC4471.

**[Camelio *et al.*, 2003]** Camelio, J.; Hu, S. J.; Ceglarek, D.; "Modeling Variation Propagation of Multi-Station Assembly Systems with Compliant Parts," *Journal of Mechanical Design*, 125, pp 673-681; 2003.

**[Camelio *et al.*, 2004]** Camelio, J.; Hu, S. J.; Marin, S. P.; "Compliant Assembly Variation Analysis Using Component Geometric Covariance," *Journal of Manufacturing Science and Engineering*, 126, pp 355-360; 2004.

**[Merkley *et al.*, 1996]** Merkley, K., Chase; K.W., Perry, E.; "An Introduction to Tolerance Analysis of Flexible Systems"; *MSC World Users Conference; 1996.*

**[Bihlmaier, 1999]** Bihlmaier, B.; *Tolerance Analysis of Flexible Assemblies Using Finite Element and Spectral Analysis*; MS. Thesis. BYU, Provo, UT; 1999.

**[Huang and Ceglarek, 2002]** Huang, W.; Ceglarek, D.; "Mod-based Decomposition of Part Form Error by Discrete-Cosine-Transform with Implementation to Assembly and Stamping System with Compliant Parts"; *Annals of CIRP*, 51, pp. 21-26; 2002

**[Soman, 1999]** Soman, S., *Functional Surface Characterization for Tolerance Analysis of Flexible Assemblies*; MS. Thesis. BYU, Provo, UT; 1999.

**[Stout, 2000]** Stout, J. B; *Geometric Covariance in Compliant Assembly Tolerance Analysis*; MS Thesis. BYU, Provo, UT; 2000.

**[Tonks and Chase, 2004]** Tonks, M. R.; Chase, K. W.; "Covariance Modeling Method for Use in Compliant Assembly Tolerance Analysis"; In: *Proceedings of DETC'04*; DETC2004-57066, SLC, UT; 2004.

**[Chase and Parkinson, 1991]** Chase, K.W.; and Parkinson, A.; *"Survey of Research in the Application of Tolerance Analysis to the Design of Mechanical Assemblies"; Research in Engineering*, 3, pp. 23-27.

**[Shen *et al.*, 2004]** Shen, Z.; Ameta, G.; Shah, J. J.; Davidson, J. K.; "A Comparative Study of Tolerance Analysis Methods"; In: *Proceedings of DETC'04*; DETC 2004-57699, SLC, UT; 2004.

**[Mortensen, 2002]** Mortensen, A. J.; *An Integrated Methodology for Statistical Tolerance Analysis of Flexible Assemblies*; MS. Thesis, BYU, Provo, UT; 2002.

**[Johnson and Wichern, 2002]** Johnson, R.A.; Wichern, D.W.; *Applied Multivariate Statistical Analysis*; Prentice Hall, Upper Saddle River, N.J., p. 77; 2002;

**[Tonks, 2002]** Tonks, M. R.; *A Robust Geometric Covariance Method for Flexible Assembly Tolerance Analysis,* BYU, Provo, UT; 2002.

# Elastic Clearance Domain and Use Rate Concept Applications to Ball Bearings and Gears

*S. Samper, J.-P. Petit, M. Giordano*
*LMECA ESIA BP 806,*
*74016 ANNECY Cedex FRANCE*
*serge.samper@esia.univ-savoie.fr*

**Abstract:** In an assembly, we model joints by a clearance domain. This domain represents the set of configurations for a surface from the other of the joint. The clearance is most of the time associated to contact conditions, but here we extend this concept to flexible joints. For this purpose, we have built the concept of Use Rate (UR). This rate has no dimension and ranges from 0 to 1 the component use. When UR equals 0, then the component has not been used (or is like unused), and UR equals 1 when the component has reached its limit (or maximum). The relevance (or importance) of UR is demonstrated through ball bearings and gears applications. It is then possible to make the tolerance analysis of a mechanism by taking into account the flexibility of joints and the technological limits. The domain model is presented and the
**Keywords**: flexible mechanism, clearance domain, elastic joint, ball bearing, gear

## 1.  INTRODUCTION

The concept of clearance domain and deviation domain [Giordano *et al.*], [Petit *et al.* 2004] helps in tolerance analysis. It is based on the Small Displacement Torsor model [Bourdet *et al.* 1988]. With the domains theory, the assembly assessment and functional requirements are computed. In order to be used by the designer, the domain model result is projected in a zone [Samper *et al.* 2006]. When we study some components like those presented here, we see that the elastic displacements are linked to technological limits. The Use Rate (UR) concept can be built for each of them in order to include those technological assessments into geometrical specifications. We present, here, two examples of joint analysis in an assembly process with possible assembly forces.

## 2.  CLEARANCE DOMAIN MODEL

A clearance is defined by the gap between the two surfaces of the two parts linked by a joint. But what is a clearance? We can define it as the set of possible displacements between the two linked parts. Those are limited by contact conditions for UR=0, and we



*J.K. Davidson (ed.), Models for Computer Aided Tolerancing in Design and Manufacturing*, 331–340.
© 2007 *Springer*.

call it (the set) Rigid Clearance Domain. The displacements are the components of the torsor of small displacements of a frame attached to one surface from the other one.

## 2.1. Substitution surface

The model we use is based on the concept of associated surfaces. This one is attached to the measured surface (skin model) with a specific criterion (mean squares, …). In the following, we assume that all surfaces are associated surfaces. We build a frame on each joint (associated) surface of the system as shown in figure 1. Thus, a joint will have two frames, one for each part attached to the joint.

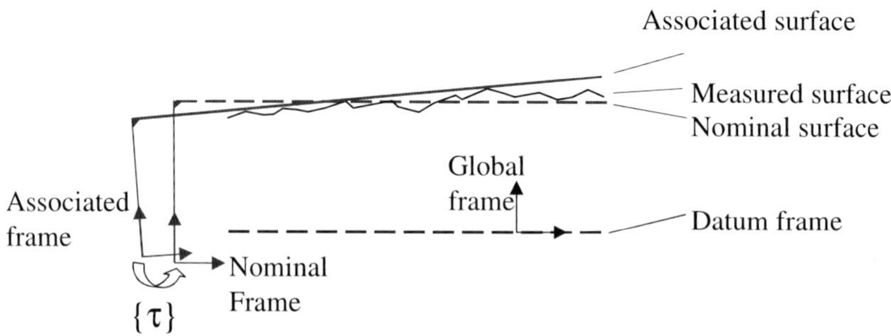


*Figure 1 Associated Surfaces*

## 2.2. Clearance Torsor and domain

In figure 1, $\{\tau\}$ represents the small displacements torsor of the associated surface from the nominal surface. $\{C_{1A}\}$ is the small displacement torsor of the A surface of the part 1 from the ideal A surface and $\{C_{2A}\}$ is the small displacements torsor of A belonging to 2. $\{C_{1A2}\}$ is the small displacement torsor of A.

In tolerance analysis, the knowledge of clearance in joints is useful in order to determine the corresponding deviations. The mechanism is defined as an assembly of parts linked by joints as shown in figure 2.

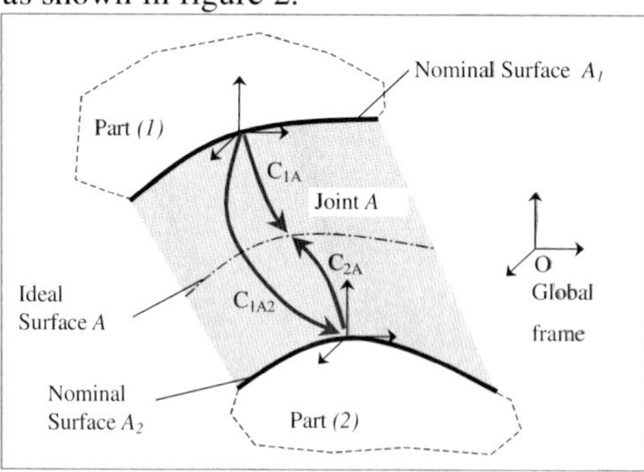


*Figure 2 Joint and Frames*

The corresponding domain is the set of values of the coordinates of $\{C_{2A}\}$. This domain is defined as a list of inequalities (Half space representation) that gives in the 6D space of components (3 translations and 3 rotations) a list of vertices (V representation). In the following, we only show the V representations of the domains.

### 2.3. Zone and domain

As we shown in the previous section, the domain model is the set of small displacements torsor components corresponding to the specified domain. Thus a domain is a "torsor view" of a given zone. In the other way, we can compute [Samper *et al.* 2006] a zone from the domain definition. In the general case, the bijection is not true, thus we add a fitting criterion (the smallest, nearest or biggest zone).

## 3.   USE RATE DEFINITION

Use **R**ate (UR) is a dimensionless limiting criterion. It allows to input functional limits in the tolerancing process. UR is used as a metric and corresponds to the inverse of the safety factor.

As the allowable displacements of the surfaces are limited by a technological criterion, we characterise the UR concept by a dimensionless number that ranges from 0 to 1. This UR can be a limiting label for either joints or for parts. That is to say that UR can be used for limiting displacements of deviations in parts or for limiting clearances in joints. When UR equals 0 then forces are nil and when UR equals 1 then the object is at its limit. In a tolerancing simulation, UR can be higher than 1 then the limit is overloaded.

In this paper, we only present UR for joints, but the aim of this concept is to be used in a tolerance analysis with different kinds of technological limits such as life span, static forces, or stresses. Then it will be possible to compare different kinds of limits, because the 0 and 1 have the same signification for all.

We define a set of domains parameterized by UR ranging from 0 (rigid clearance domain) to 1 (upper limit clearance domain). When UR is 0 the corresponding clearance can be nil or positive, depending on the mounting conditions (pre-stressed or not).

Here, the analysis is made in an assembly process, thus forces are not external forces but assembly forces.

A joint is a model, and we can call joint a component like ball bearings or gears as shown bellow. As UR is presented here for joints, we can also define UR for deviations.

## 4.   BALL BEARINGS JOINT

A ball bearing is described by its geometrical parameters and by the material properties of the corresponding steel. In literature [Harris 1991], [Hernot *et al.* 2000], [Houpert

1997], we can find analysis of bearings behaviour in order to know their stresses and we had to build the 5D analysis of displacements corresponding to Hertz stresses and force torsor. We made the assumption that each of its parts were perfect. We set the internal clearance of the ball bearing. Then, we compute a set of extreme displacements with contact conditions, and next under stress.

A ball bearing has one kinematic degree of freedom. Thus, the clearance domain might be represented in 6D, or at least in 5D. We have built it for the general case, but we can show it in 3D by choosing the good components of the clearance torsor. If we affix the outer ring and move the inner ring, like in figure 3, we can observe the displacement in a specific plane and reduce the components to 3 real numbers. In the assembly analysis, we always use the 6D domains. If the domain is built in 3D and the other screw parameters are undefined (kinematic,…) the corresponding axis is like a 5D cylinder axis in 6D (any cut on this axis gives the same 5D feature).

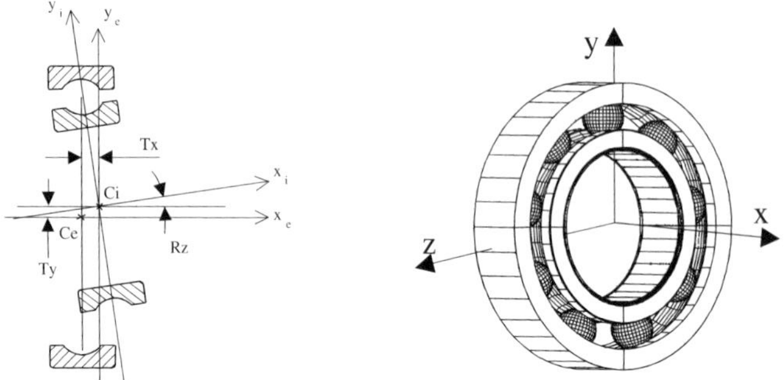


*Figure 3 Ball Bearing Model*

Then the clearance domain can be displayed in 3D figure 4. The method used here consists in testing the interference between balls and rings by moving the inner ring against the outer one. This program [Samper *et al.* 2001] was written in the Mathematica software. It is based on the Hertz Theory of Contact and the kinematics model of the parameterized ball bearing. The elastic clearance domain in figure 4 is defined for the static limit (4200 MPa).

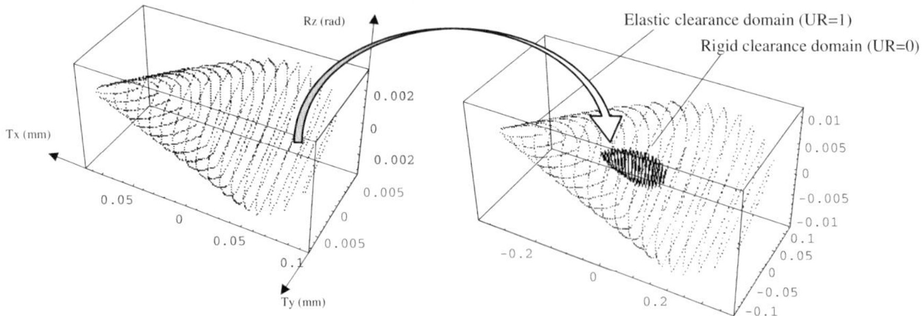


*Rigid Clearance Domain*              *Rigid Compared to Elastic Domain*
*Figure 4 Rigid and Clearance Domain of a Ball Bearing*

The results give the evolution of the clearance domain depending on the UR value but we also compute the corresponding forces domain. Those results give the same values than the standards of design of ball bearings. This UR parameterized clearance domain can then be used in a Computer Aided Tolerancing software (mock up in [Fukuda *et al.*2003]) in order to take into account the use of the system.

As the domain concept is not a geometric object and the zone concept is easy to manage for a designer we propose to show on 3D the equivalent zone to each domain.

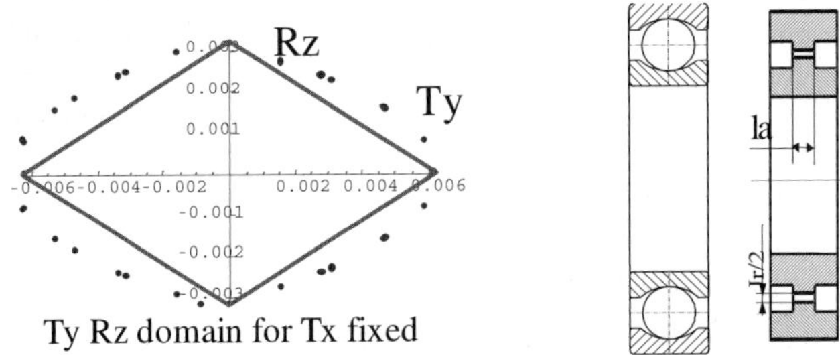

Ty Rz domain for Tx fixed

*Figure 5 Clearance Domain and fitted zone of a Ball Bearing*

In figure 5, the computed domain is shown in 2D (dots) and is fitted (lines). This approximate domain can be shown in the drawing like a plain bearing with a *jr* radial clearance equal to $Ty_{MAX}$ and a *la* length equal to $Ty_{MAX}/Rz_{MAX}$ . Then, in a tolerancing analysis, the designer can substitute a ball bearing to a parameterised plain bearing where *Jr* and *la* are calculated by a given UR and a corresponding domain.

In figure 6, the force domain is linked to the clearance domain of a joint in order to show that a small displacements torsor is a force torsor out of the nil UR domain. A UR>0 domain has two equivalent representations (displacements and forces). A zero UR domain is the rigid one.

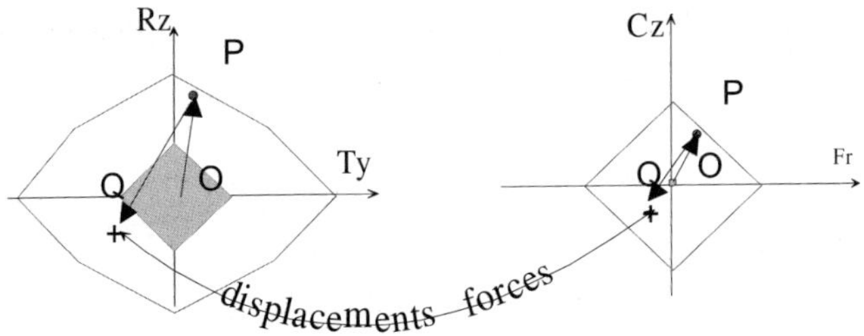


*Figure 6 Clearance and force domain of the joint*

Forces can be assembly forces (internal forces of the system) or external forces. The internal forces are depending of the assembly positions and the external forces are given by the using of the system. Most of time, the external forces are fixed. In this case, we can separate forces and compute the "central point of use" *P* if loaded *O* otherwise. An assembly torsor should be PQ measured in the force domain or in the displacement domain. The designer can by this way analyse loads and assembly in the same process.

## 5.   CYLINDRICAL GEARS JOINT

The gear joint is built with two gears as shown in figure 7. The components of the clearance torsor are Tx, Ry and Rx. They are limited by the contact conditions, and we obtain the set of contact torsors with a finite element analysis. This clearance domain is shown in figure 9. As in the ball bearings analysis, we assume that surfaces have no form defects.

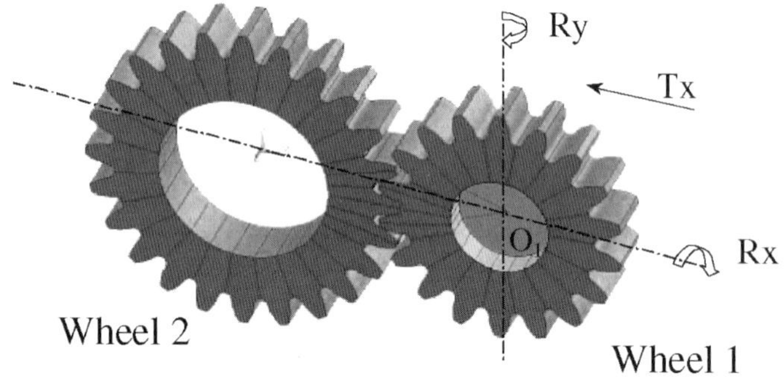


*Figure 7 Clearance Torsor Components*

We could add Rz to this analysis as a backslash parameter but Rz results from an assembly process. Tx, Rx and Ry are the small displacements torsor components used in the assembly analysis. When this analysis is made, we can obtain Rz easily.

### 1.4. Rigid domain of a gear (UR=0)

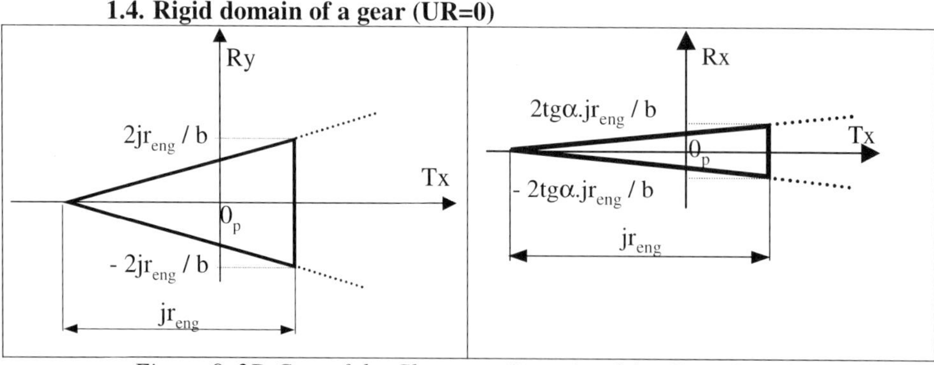


*Figure 8  2D Cuts of the Clearance Domain of the Gear (UR=0)*

In the figure 8, we show the two cuts of the gear domain. If we input the Rz component, we should obtain a 4D domain and cuts would be 3D polytopes. $Jr_{eng}$ is the radial clearance and $\alpha$ is the pressure angle.

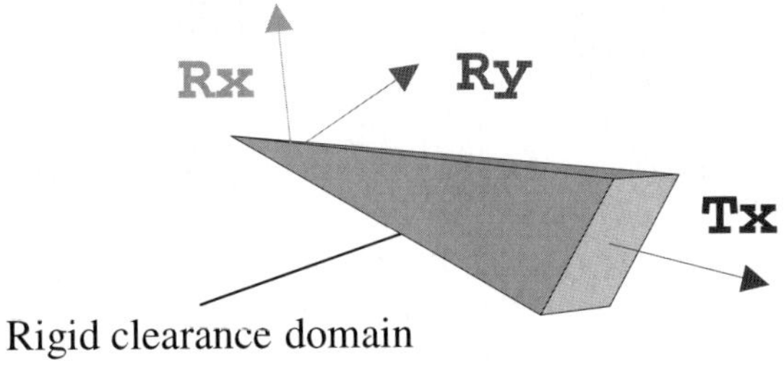


*Figure 9 Clearance of the Gear Joint*

In the figure 9, the rigid clearance domain is shown as a limited one but the rhomb corresponds to a technological limit value for Tx. This domain can be computed by the using of the simple formulas of figure 8.

### 5.1. Elastic domain of the gear joint

The finite element model was built on ANSYS software by a specific automatic parameterized program. Any conceivable classical cylindrical gear joint can be defined. Next, displacements are introduced in the centre of one wheel, with the other wheel fixed. The maximum Von Mises stress in the assembly is then measured. As we wanted to limit this one, UR is 1 at this limit.

The FEM is made of brick elements for the wheels and contact elements between them. We made some optimisations in our model in order to minimise the computation time and obtain a satisfying accuracy. The model presented here is build automatically

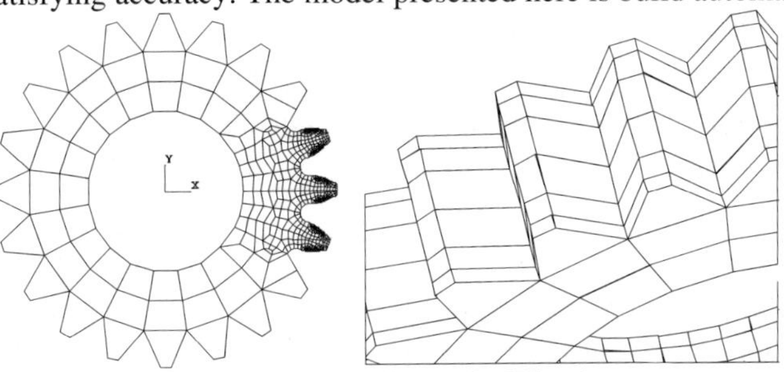


*Figure 10 Meshing of a Wheel*

*Figure 11 FEM of the Gear Joint*

In the figure 11 above, a rotation Rx and a translation Tx  are performed and the resultant stresses may be observed. A wheel is set (the left one) and the other is moved step by step in the program. Three loops are needed to measure all the position sets of the gears. Each position provides a Von Mises maximum stress, the displacements torsor, and the corresponding forces (13 elements in general but 7 here). This set of torsors can be shown as a clearance domain or as a force domain.  This set of 7 real numbers (three are independent) can then be used in an assembly analysis.

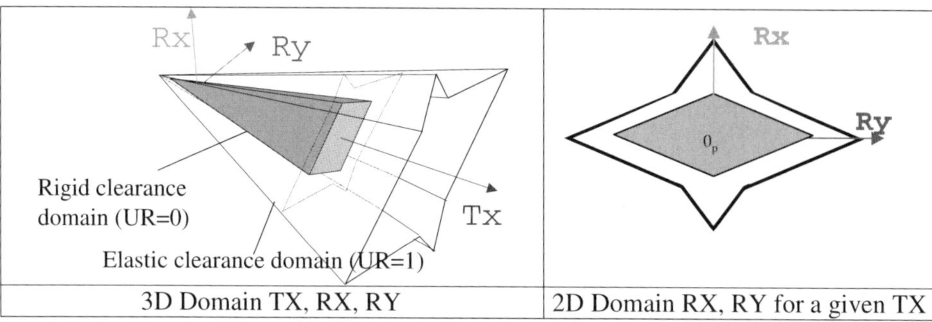


*Figure 12 Clearance domains of the gear joint*

In this figure, the clearance domain of the rigid gear can be compared to the maximum flexible one (UR=1). As one can see, the gain in the rotation Rx is bigger than the one associated to Ry. It is the representation of a flexure of the teeth. In order to use those results in a tolerancing analysis, we can use an equivalent clearance domain (proportional transformation of rigid clearance domain) inside of the elastic non convex one or outside depending on the requirement. As we shown in the case of a ball bearing, it is possible to compute a zone (corresponding to a fitted domain) for each value of UR, than we can show on a CAD system, the different values of the clearance zone for a gear joint. Those results are then used in order to make the tolerancing analysis of the assembly, taking into account elastic component behaviour.

## 6.  CONCLUSION

The clearance domain provides the set of possible displacements of a specified joint. The Use Rate concept (UR) allows us to redefine the limits of the surfaces displacements. As for joint components, this concept has been translated into a parameterized clearance domain (with $0 \leq UR \leq 1$). Then this domain can be used in a tolerancing analysis software with UR as input for each component. Another way is to output UR from the analysis of an assembly. Assembly forces are depending on tolerancing, they must remain in the UR forces domain that can be computed in the same way than the clearance domain. In order to use those results we show how parameterized domains give zones on joints. The designer can then use a tolerancing process method in order to take into account elastic behaviour of components.


## REFERENCES

**[Giordano *et al.* 1992]** M. Giordano, D. Duret., S. Tichadou, «Clearance space in volumic dimensioning», Annals of the CIRP, vol. 40 (1), 1992

**[Petit *et al.*, 2004]** J-Ph. Petit, S. Samper. "Tolerancing analysis and functional requirement", In: *Proceedings of the 5th International Conference on Integrating Design and Manufacturing in Mechanical Engineering, paper n°205* Bath(UK), April 5-7 2004.

**[Harris 1991]** HARRIS Tedric A. *"Rolling Bearing Analysis."* John Wiley & Sons, Third Edition, 1991.

**[Fukuda *et al.*2003]** K. Fukuda and J.-P. Petit. *"Optimal tolerancing in mechanical design using polyhedral computation tools"*, 2003. 19th European Workshop of Computational Geometry, March 24-26, Bonn.

**[Hernot *et al.* 2000]** Hernot X., Sartor G., Guillot J. *"Calculation of the Stiffness Matrix of Angular Contact Ball Bearings by Using the Analytical Approach"* Journal of Mechanical Design, ISSN 1050-0472, ASME 2000, Vol. 122, 83-90.

**[Houpert 1997]** HOUPERT Luc *"A uniform analytical approach for ball and roller bearings calculations"*. Journal of Tribology ; Vol n° 119 ; p 851-858 ; October1997.

**[Samper *et al.* 2001]** Samper S., Giordano M., Perroto S. *"Fiabilité et tolérancement d'un assemblage élastique"* proceedings of the 4th Multidisciplinary International Conference Quality and Dependability, Qualita2001  May 22 & 23 2001 Annecy France

**[Bourdet *et al.* 1988]** P. BOURDET, A. CLEMENT *"A Study Of Optimal-Criteria Identification Based On Small- Displacement Screw Model"*, CIRP, Annals 1988, Manufacturing Technology, Volume 37, janv.1988

**[Samper *et al.* 2006]** S. SAMPER, J-P. PETIT *"Computer Aided Tolerancing - solver and post processor analysis"* Advances in Design pp. 487-498, Springer Ed., 2006, ISBN: 1-84628-004-4

# Tolerance Verification for Sheet Metal Bending: Factors Influencing Dimensional Accuracy of Bent Parts

**T. H. M. Nguyen*,***, J. R. Duflou*, J.-P. Kruth*,
I. Stouten**, J. Van Hecke**, A. Van Bael****
**Katholieke Universiteit Leuven, Department of Mechanical Engineering,
Celestijnenlaan 300B, B-3001 Leuven, Belgium
**XIOS Hogeschool Limburg, Department of Industrial Sciences and Technology-Engg.,
Universitaire Campus – Gebouw H, B-3590 Diepenbeek, Belgium
***NguyenThiHongMinh@mail.hut.edu.vn*

**Abstract:** The dimensional accuracy of bent sheet metal parts is influenced by many factors and possible sources of inaccuracy such as the sheet material, machine, and material handling. This paper addresses the issue of tolerance verification for sheet metal bending by analytically and experimentally exploring the associations between each of these factors and the achievable dimensional accuracy of bending operations. Making use of the GUM method for quality assessment, in a first step, the influencing factors on the angular and linear dimensions of the parts are listed. The influences of these factors on each type of dimension are subsequently determined by geometry analysis. Secondly, special experimental setups were designed and experiments were conducted with industrial machines to establish the statistical characteristics of these factors. Therefore, the dominant factors determining the dimensional variations of the bending operations were fully identified and quantified. The result of this study can be used to predict the error range and thus the scrap ratio for the process. The developed methodology allows to point out possible improvements in the process plan, and most importantly, to predict the achievable dimensional accuracy of complex bent parts.
**Keywords**: tolerance verification, sheet metal bending, quality analysis

## 1.  INTRODUCTION

Bent parts are produced on press brake machines by folding flat patterns cut out from metal sheets. During the process, workpieces are positioned against backgauges before being bent linearly by means of a punch penetrating in a V-die. While tolerances have been a traditional concern in manufacturing processes and aspects related to tolerance verification have been rather well defined for conventional processes, studies dedicated to tolerance issues in process planning for non-conventional processes such as sheet metal bending have only been initiated in recent years

Similar to conventional manufacturing processes, tolerances have been used as a drive for operation planning, often referred to as bend sequencing, in sheet metal bending. In the first instance, rules helped to drive the search for a bending sequence



*J.K. Davidson (ed.), Models for Computer Aided Tolerancing in Design and Manufacturing,* 341–350.
© 2007 *Springer.*

resulting in sufficiently accurate realisation of critical dimensions for a given part, without requiring the time consuming tasks of geometric calculations or simulations [Shpitalni *et al.*, 1999]. Other authors used tolerance verification as a tool to generate or validate the (partial) bend sequences [De Vin, 1996], [Inui *et al.*, 1998]. These contributions consisted of a deterministic approach in which accumulated errors in a worst-case scenario were compared to preset tolerances. In contrast, 3D Monte-Carlo simulations have been used to estimate the total achievable accuracy, which is used in turn as a tool to select an appropriate process plan from a list of prepared solutions [Hagenah, 2003]. However, due to the computational restrictions, mainly simplified physical models and limited influencing factors have been covered in this approach. Other contributions focussed on investigating the factors leading to inaccuracy in sheet metal bending. Among those, an intensive sensitivity analysis of the influencing factors was provided, with focus on material properties [Streppel *et al.*, 1993]. FEM analysis was used for both machine and material factors influencing the dimensional accuracy [Singh *et al.*, 2004]. However, the stochastic nature of the process is typically not included due to the fast increase in the complexity of the model.

Despite a large number of studies related to tolerance aspects, a quantitative analysis on statistical characteristics of the influencing factors and their impacts on the dimensional accuracy of parts produced are not yet available. This fact hinders successful applications of the methods mentioned above on computer-aided tolerancing for sheet metal bending.

In this paper, tolerance verification for sheet metal bending is addressed by providing a complete method to investigate the actual sources of errors in sheet metal bending and the resulting dimensional errors. Firstly, the crucial factors influencing the dimensional accuracy of the product are analysed. Analytical formulas are derived to model the magnitude of the impacts that specific factors have on the angular and linear dimensions of bent parts. Secondly, the uncertainties of these influencing factors are quantified by means of dedicated experimental setups. A series of tests has been carried out using industrial machines. The result of this exploration provides the statistical characteristics of each influencing factor. Based on this analysis, dominant factors determining the dimensional variations of the bending operations have been fully identified and quantified.

## 2. FACTOR ANALYSIS

### 2.1. Sources of errors

To be able to identify the possible sources of errors influencing the dimensional accuracy of parts produced by sheet bending using press brakes, the sequence of the process, including workpiece preparation and part processing phases was investigated.

To prepare the workpiece for sheet metal bending, at first the envisaged part is modelled and the corresponding flat pattern is calculated based on specific bend models. Such models apply geometric approximations of the bend features to be produced to provide the bend allowance for each bend line, depending on the characteristics of the bend features, such as bend angle, inner radius, sheet thickness, and material properties. Since there are always discrepancies between the bend model and the actual material behaviour, there exists an error in the calculation of the unfolding. The flat pattern calculated is then cut out from a sheet by a cutting process such as shearing, punching, nibbling, or laser cutting, which can provide various quality of cutting. The precision of the processes selected for workpiece preparation directly affects the dimensional accuracy of the workpiece.

After being prepared, the workpiece is positioned on top of the die by placing one of its edges or flanges against a backgauge of the machine before being bent linearly by a punch penetrating into the die cavity. In this phase, the positioning accuracy is determined by factors such as the repeatability of the gauging system itself, the gauging method, and the estimation of the gauging edge according to the process plan. Moreover, the material handling method, such as robot or manual assistance, also has an effect. Similarly, in the bending phase, the accuracy of the punch positioning also directly influences the dimensional quality of the resulting part. The main influences to the dimensional accuracy of bent parts are depicted in a fishbone diagram as shown in Figure 1.

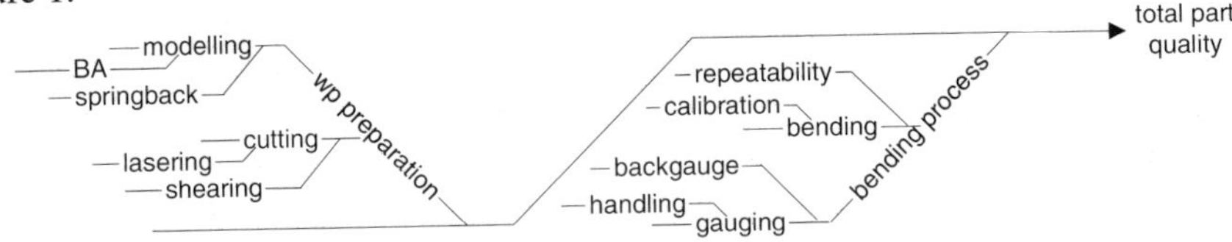


*Figure 1: Sources of errors in sheet metal bending*

## 2.2. Analytical model

In order to thoroughly investigate the magnitude of the impacts these factors have on the dimensional accuracy of bent parts, a foil model as shown in Figure 2 is considered.

**a- Bend angle**. Due to elastic springback, the final bend angle is achieved after retraction of the punch. Therefore, besides the V-width, the actual bend angle resulting from a bending operation depends on two main factors. The first factor is *the relative position of the punch tip* to the die, defined by the X and Y coordinates. Such positioning accuracy is in turn influenced by the positioning accuracy of the ram and that of the punch on the ram. The second factor is *the actual springback* after releasing of the bending force and relaxation of the part after production.

As depicted in Figure 2.a, assuming that (1) the sheet is bent perfectly under three contact point condition with optimal holding time, (2) a complete springback occurs immediately after unloading, (3) the same punch penetration is attained for all points along the ram, the achievable bend angle can be calculated by the following formula:

$$\alpha = \alpha_F + \alpha_B - \zeta \tag{1}$$

Where, $\alpha$ is the actual achievable angle, the two angles $\alpha_F$ and $\alpha_B$ are formed at the lowest punch penetration point, $\zeta$ is the actual springback angle. $V$ is the effective V-die width; $\Delta X_P$ is the X coordinate of the punch tip in machine coordinate system, which is actually the positioning error from the correct centre line; $Y_p$ is the Y coordinate of the punch tip, with $\Delta Y_p$ as the error of punch displacement in Y direction as compared to the intended value $Y_p^0$; and $s$ is the sheet thickness, with $\Delta s$ as the error of sheet thickness as compared to the ideal thickness $s^0$. All angles are measured in radians, and all lengths are measure in mm. Therefore, (1) can be rewritten as:

$$\alpha = \tan^{-1}\left[\frac{\frac{V}{2} + \Delta X_P}{Y_P + s}\right] + \tan^{-1}\left[\frac{\frac{V}{2} - \Delta X_P}{Y_P + s}\right] - \zeta \tag{2}$$

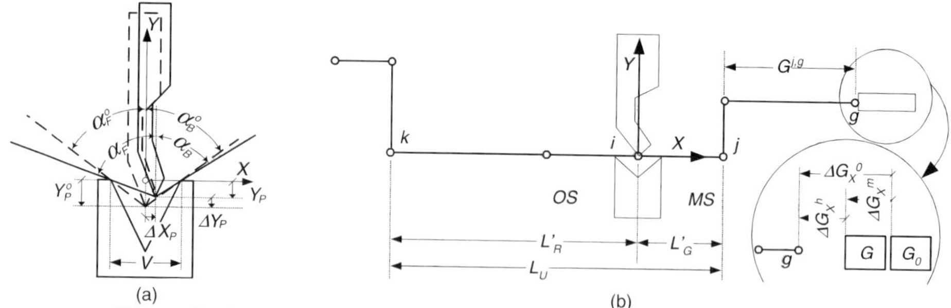

Figure 2: *Analytical models for (a) bend angle and (b) bend lengths*

**b- Bend length.** Each bending operation results in two bend lengths. The first length $L_G$ is formed between the bend line and the backgauge, the other length $L_R$ is the residual length of the unfolding $L_U$ at the other side of the bend line, as seen in Figure 2.b. The relation between the dimensions before and after bending is expressed by (3), where $L_U$ is the unfolded length, and $BA$ is the bend allowance:

$$L_U = L_R + L_G - BA, \text{ or } L_R = L_U - L_G + BA \tag{3}$$

According to the analysis on the possible sources of errors, the following factors influencing the accuracy of bend lengths are accounted for. Each factor can contain both random and systematic errors.

1- *Workpiece preparation error* $\Delta L_U$ takes into account the inaccuracy of the workpiece preparation process for $L_U$, which is either cutting or previous bending steps.

2- *Gauging error* $\Delta G$ is introduced in the gauging phase, representing the dislocation of the actual gauging position from the ideal one. Considering a single bend, the two main factors influencing the total gauging error are the machine factor $\Delta G^m$ and the human factor $\Delta G^h$. For complex parts, if direct gauging is not possible, the accuracy of gauging is also affected by the cumulative error $\Delta G^{j,g}$ of the geometries lying between the current bend line and the gauging line, where $j$ is the first performed bend line adjacent to the current bend line $i$ and $g$ is the effective gauging line, as illustrated in Figure 2.b. Therefore, the total gauging error in X direction, which linearly influences the bend length, can be estimated by:

$$\Delta G_X = \Delta G_X^m + \Delta G_X^h + \Delta G_X^{j,g} .\tag{4}$$

3- *Bend allowance error* $\Delta BA$ represents the discrepancy between the bend model used and the actual material behaviour.

4- *Tool positioning error* in X direction $\Delta X_P$.

Thus, taking into account all the errors, the achievable lengths of the bend legs can be expressed as a function of nominal dimensions and the influencing errors as follows:

$$\begin{cases} L_G = L_G^0 + \left(\Delta G_X + 1/2\,\Delta BA - \Delta X_P\right) \\ L_R = L_R^0 + \left(\Delta L_U - \Delta G_X + 1/2\,\Delta BA + \Delta X_P\right) \end{cases}\tag{5}$$

### 2.3. Sensitivity analysis

Equations (2) and (5) represent the quantities of interest, namely the bend angle and the bend length, as functions of various factors, which can be rewritten as:

$$\begin{cases} \alpha = f_\alpha\left(V, X_P, Y_P, s, \zeta\right) \\ L_j = f_{L,j}(\Delta L_U, \Delta G_X, \Delta BA, \Delta X_P) \end{cases}\tag{6}$$

According to [GUM, 1992], these quantities can be estimated by:

$$\begin{cases} \alpha = \alpha^0 + \Delta\alpha \pm 2u_c(\alpha) \\ L = L^0 + \Delta L \pm 2u_c(L) \end{cases}\tag{7}$$

where $\Delta\alpha$ and $\Delta L$ are the correctable systematic errors of $\alpha$ and $L$; $u_c(\alpha)$ and $u_c(L)$ are the combined uncertainties of quantities $\alpha$ and $L$ respectively, which are estimated by $u_c^2(f) = \sum_{i=1}^{N} c_i^2 u^2(x_i)$. With $i$ ranging from 1 to N, $c_i$ and $u(x_i)$ are respectively the sensitivity coefficients and the standard uncertainty of the influencing factors $x_i$. According to [GUM, 1992], with $i = 1$ to $N$, $c_i$'s are calculated by:

$$c_i = \partial f / \partial x_i\tag{8}$$

In other words, the sensitivity coefficients are determined based on the function $f$ expressing the quantity of interest. Substituting equations (2) and (5) into (8) allows estimating the sensitivity of the uncertainty of the bend angle and bend lengths to those of the influencing factors, as expressed in (9) and (10).

$$\begin{cases} c_V(\alpha) = 1/2(1/A + 1/B)(Y_P + s) & c_{Xp}(\alpha) = (1/A - 1/B)(Y_P + s) & c_\zeta(\alpha) = -1 \\ c_{Yp}(\alpha) = -\left((V/2 + \Delta X_P)/A - (V/2 - \Delta X_P)/B\right) & c_s(\alpha) = -\left((V/2 + \Delta X_P)/A + (V/2 - \Delta X_P)/B\right) \\ \text{where } A = (Y_P + s)^2 + (V/2 + \Delta X_P)^2 & \text{and } B = (Y_P + s)^2 + (V/2 - \Delta X_P)^2 \end{cases}\tag{9}$$

$$\begin{cases} c_{\Delta L_U}(L_G) = 0 & c_{\Delta Gx}(L_G) = 1 & c_{\Delta BA}(L_G) = 1/2 & c_{\Delta Xp}(L_G) = -1 \\ c_{\Delta L_U}(L_R) = 1 & c_{\Delta Gx}(L_R) = -1 & c_{\Delta BA}(L_R) = 1/2 & c_{\Delta Xp}(L_R) = 1 \end{cases}\tag{10}$$

If $V = 12mm$, $Y_P = 6mm$, $s = 1mm$, and $\Delta X_P = 0$, the values in (9) can be estimated as $c_V(\alpha) = 0.08$, $c_{Xp}(\alpha) = 0$, $c_\zeta(\alpha) = -1$, $c_{Yp}(\alpha) = -0.14$, and $c_s(\alpha) = -0.14$. (9')

Note that since the sensitivity coefficients are combined with the uncertainties based on their squared values, only the absolute values of the coefficients are important. From (9) and (9'), it can be seen that for the uncertainty of the resulting angles, the variations of the springback angles have the most pronounced effect. The influences of variations in punch displacement Y directions and sheet thickness are of the second biggest magnitudes. While an incorrect width of the V-die only causes a limited effect, it is much more influential than the incorrectness or variations in punch centre alignment. Conversely, almost all the influencing factors equally affect the resulting bend lengths, according to equation (10). It could be noted also from this equation that the errors of cutting do not have impact on the bend leg produced at the gauging side. Since the factors influencing gauging accuracy are linearly combined in equation (4), they also produce the same impact on the total gauging error.

## 3. EXPERIMENTAL MEASUREMENT

### 3.1. Sample preparation and experimental setups

Based on the analysis of the factors influencing the accuracy of the elementary bend dimensions, samples were prepared and experiments were designed with special setups to study the statistical characteristics of these factors.

All samples used in this study were prepared by laser cutting from seven sheets of stainless steel 304 with dimensions of 2000×1000×2 mm in rolling, transverse and thickness direction, respectively. In order to minimise the fluctuations in material properties, the sheets were taken from the same coil that was cold rolled, annealed and skin-passed. The commercial thickness tolerance was ± 0.09 mm. The actual bend allowance has been determined as described in 3.1.d so that the flat pattern could be calculated. After laser cutting, a code number was engraved in each sample to keep track of its precise location within the initial sheets. The samples were oriented so that all subsequent bend lines were parallel to the rolling direction.

**a. Sheet thickness.** For each sample, the thickness has been measured at the four corners with a digital micrometer as described in 3.2.a.

**b- Workpiece dimension.** The lengths have been measured at both sides of the bend lines for each sample with a digital calliper as described in 3.2.b.

**c- Springback angle determination.** Due to the high magnitude of influence, springback angles should be determined to eliminate a large systematic error caused by this factor. A press brake that allows for adaptive bending has been used to impose a single bend angle of 90° before unloading and to measure the angle after unloading by a device as described in 3.2.c. The difference between both angles is the springback angle. In total, 33 samples have been used for this purpose.

**d- Bend allowances.** The same press brake as for 3.1.c was used to impose a bend angle of 90° after unloading in another separate set of 36 samples. The lengths of the two bend legs have been measured using a digital calliper as described in 3.2.d near both ends of the bend line to calculate the actual bend allowance.

**e- Gauging.** Two main factors causing gauging errors were investigated. For the *machine factor*, the experimental setup to analyse the repeatability of the positioning of the back gauges is shown in Figure 3.a. It consists of a dial gauge as described in 3.2.e mounted on the die using a magnetic dial gauge stand. Such a setup is used for each of the two back gauges of a given press brake. The back gauges are instructed to move first backward and then forward to their original position, and the maximum deflections of the dial gauges are recorded. Another possible error source concerns the positioning of the sheet against the back gauges by the operator and is called the *human factor*. Therefore, a particular bend line was made in 28 samples without repositioning the back gauges. The lengths of the bend legs were measured on a CMM machine as described in 3.2.f at both ends of the bend lines to analyse this factor.

**f- Punch positioning.** The positioning of the punch in both X and Y directions was explored. The experimental setup to measure the repeatability of the horizontal position of the punch with respect to the die in this direction is shown in Figure 3.b. It uses a dial gauge, as described in 3.2.e, in contact with the punch and a magnetic dial gauge stand that is mounted on the die. The values of the dial gauge were recorded for both the upper and lower positions of the punch when making an angle of 90° for 28 samples. Meanwhile, the repeatability of the vertical punch movement was monitored using the experimental setup shown in Figure 3.c. Two magnetic dial gauge stands are mounted on the punch so that, when the punch moves down, the two dial gauges make contact with the die surface. The dial gauge values are recorded for the lowest punch position when making a bend angle of 90° for 28 samples.

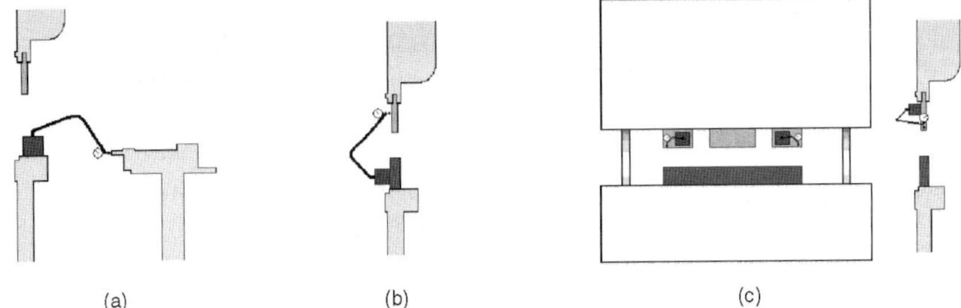


*Figure 3: Experimental setups for measurement of the position of*
*(a) the back gauges, (b) the punch centre line, and (c) the vertical punch movement.*

## 3.2. Data retrieval and processing method

Eleven press brakes, named A to K and installed in different companies, have been used for this study, from which two machines J and K were equipped with devices for in-process controlling of bend angles. A die with V-opening of 16 mm has been used on all press brakes except on machine A, for which such a die was not available. A V-opening of 12 mm on a die block with several V-openings has been used in this case.

To investigate the process-induced errors, a systematic mix of 34 samples, taken from different positions within all initial sheets, has been used for each press brake. The samples were numbered in random order. The first six samples were provided for setting up the production parameters, and the remaining 28 were the actual test samples.

The following measurement equipment was used:

a- Digital micrometer with ± 0.016 mm uncertainty for test 3.1.a;

b- Digital calliper with ± 0.03 mm uncertainty for test 3.1.b;

c- Angle measurement device: LVD Easy-form® with ±0.1° uncertainty for 3.1.c.

d- Digital calliper with ± 0.02 mm uncertainty for test 3.1.d;

e- Digital dial gauges with 0.001 mm resolution and ± 0.002 mm repeatability for tests 3.1.e/f;

f- 3D-CMM with U3 = 3.00 + L/350 uncertainty (VDI/VDE 2617) for test 3.1.e;

The micrometer, calliper and dial gauges were all equipped with a RS232C interface for a direct registration of the measured values to a spreadsheet on PC. All measurement results have been analysed using the statistical software package Minitab for the calculation of averages, standard deviations, tests for equal variances and the construction of boxplots and control charts with individual values. These control charts reveal whether production is under control or not, and all observed out-of-control values have been investigated in detail to decide whether they represent outliers.

### 3.3. Error analysis of the influencing factors

The statistical characteristics of the factors related to material input, including sheet thickness, initial length, springback, and bend allowance, are presented in Table 1, with the nominal value, the range, the mean, and the standard deviation of the samples.

*Table 1: Factors related to input material.*

|  | Thickness [mm] | Length [mm] | Springback [o] | Bend allowance [mm] |
|---|---|---|---|---|
| Nominal | 2 | 240.96 | | |
| Max | 1.948 | 241.03 | 4.92 | 3.96 |
| Min | 1.935 | 240.81 | 4.46 | 4.07 |
| Mean | 1.942 | 240.89 | 4.69 | 4.01 |
| Stdev | 0.015 | 0.041 | 0.09 | 0.02 |
| Sample size | 30 | 30 | 33 | 36 |

Since the machine and process related factors, such as gauging and punch positioning errors, are strongly related to the individual production environment, the possible shift of the average values, the range of variations and the standard deviations are comparatively presented for all the tested machines in Figure 4 and 5, except A, D and J in Figure 4.a since the machine configurations did not allow the particular setup.

### 3.4. Discussion

Considering *bend angles*, the factor with the highest sensitivity coefficient, being the springback angle, was determined for the material used in the experiments, and a noticeable uncertainty represented by a standard deviation of 0.09° was found. A procedure as described in 3.1.c, in which accurate springback values are obtained for every processed material, is recommended to minimise the systematic errors in bend angle predictions. The use of an updatable database or in process control of the angle can also be applied. While a database cannot help compensating for random errors occurring due to material anisotropy and thickness variations throughout the sheet, adaptive bending techniques in principle can be applied to minimise the range of variations. For most machines, the repeatability of punch vertical positioning was rather good, characterising by standard deviations ranging barely from 0.002mm to 0.008mm. However, the sheet thickness showed a standard deviation of 0.015mm, which caused a higher order impact to the accuracy of the resulting angle. Thereby, though having a rather low sensitivity coefficient, the contribution of the uncertainty of the sheet thickness represented approximately half of the total uncertainty of the resulting bend angle. For machines J and K, equipped with adaptive control of bend angles, the deviations were high due to the adjustment of the punch positioning to the variations of individual sheet properties in order to obtain accurate angles. Though having a small sensitivity coefficient, the measured accuracies in punch alignment were rather low as compared to the accuracy of ±0.01mm typically announced in machine specifications [Kroeze *et al.*, 1994]. Poor calibration and low quality of tooling system were the problems encountered at machines B, E, and F respectively, as seen from Figure 4.a.

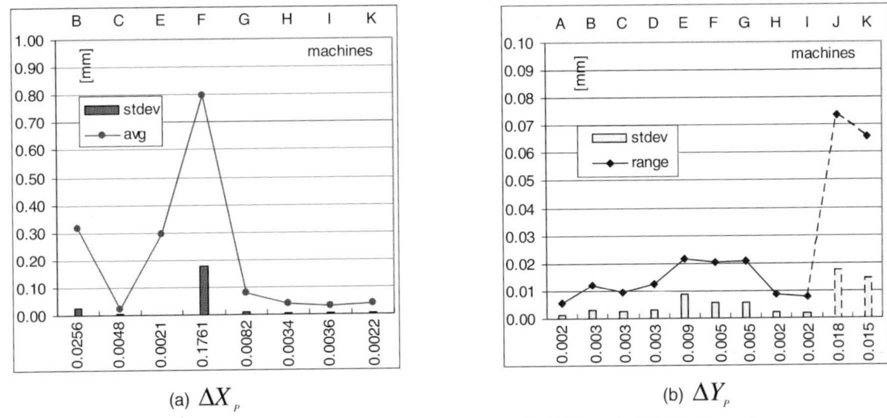

(a) $\Delta X_P$                                     (b) $\Delta Y_P$

*Figure 4: Punch positioning errors in (a) X and (b) Y direction*

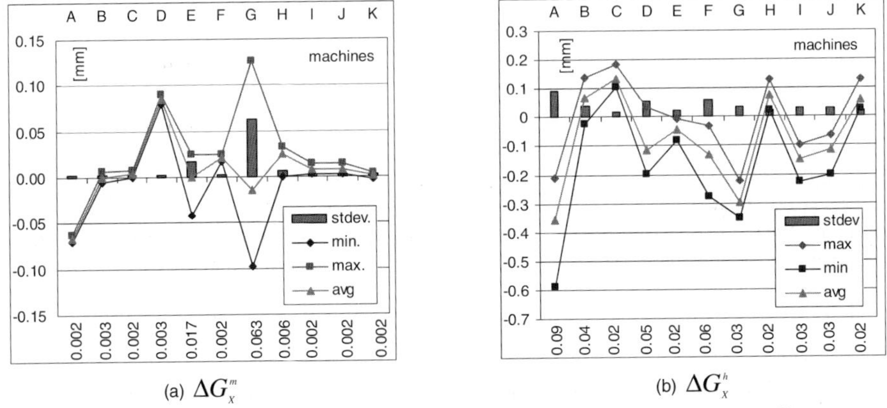

(a) $\Delta G_X^m$                                     (b) $\Delta G_X^h$

*Figure 5: Gauging errors due to (a) machine and (b) material handling*

For *bend lengths*, material handling was found to be the dominant factor, with the magnitude of standard deviations ranging from 0.02 to 0.09mm. However, poor calibration seen in many gauging systems caused a systematic error of more than 0.1mm. Next to this factor, the residual bend length also suffered from a remarkable uncertainty of the workpiece preparation process, even when an advanced process, i.e. laser cutting, was utilised. The contributions by other factors were rather negligible.


## 4. CONCLUSIONS

The paper has contributed to tolerance verification for sheet metal bending by identifying the error sources in the process that significantly affect the dimensional accuracy of bent parts. The sensitivity coefficients of these influencing factors have been investigated by analytical models for bend angle and bend lengths, while their uncertainties have been quantified by a series of tests on industrial press brakes. The sensitivity analysis of this study shows the importance of each factor in the total dimensional quality of bent parts. Therefore, the same test procedures can be used in the industry for an adequate estimation of the process capability and possible calibration for an efficient quality improvement. The results of this study provide a basis for a system allowing fast estimation of dimensional accuracy of complex bent parts, where the effect of error propagation can be simulated stepwise in order to evaluate proposed bend sequences.


## 5. ACKNOWLEDGEMENT

The authors acknowledge the financial support by the Institute for the Promotion of Innovation by Science and Technology in Flanders through contract HOBU/20110.


## 6. REFERENCES

**[De Vin, 1996]** de Vin, L.J., Streppel, A.H. and Kals, H.J.J., The accuracy aspect in setup determination for sheet bending, Int. J. Advanced Manuf., Vol. 11, 1996, pp 179-185.

**[Inui *et al.*, 1998]** Inui, M. and Terakado, H., "Fast Evaluation of Geometric Constraints for Bending Sequence Planning", *Proc. of IEEE Int. Conference on Robotics and Automation*, Vol. 3, 1998, Leuven, pp 2446-2451, ISBN0-7803-4300-X.

**[Shpitalni *et al.*, 1999]** Shpitalni, M. and Radin, B., "Critical Tolerance Oriented Process Planning in Sheet Metal Bending", *Trans. of ASME J. of Mech.Design*, Vol. 121, 1999, pp. 136-144.

**[Hagenah, 2003]** Hagenah, H., "Simulation based Evaluation of the Accuracy for Sheet Metal Bending caused by the Bending Stage Plan", *Proc. 36th CIRP Int. Seminar on Manufacturing Systems*, 2003, Saarbrücken, Germany.

**[Streppel *et al.*, 1993]** Streppel, A.H., Vin, L.J. de, Brinkman, J., Kals, H.J.J., "Suitability of Sheet Bending Modelling Techniques in CAPP Applications", *J. Materials Processing Technology*, Vol 36, 1993, pp. 139-156.

**[Singh *et al.*, 2004]** Singh, U. P., Maiti, S. K., Date, P. P. and Narasimhan, K., Numerical simulation of the influence of air bending tool geometry on product quality", *J. Materials Processing Technology*, Vol 145, 2004, pp. 269-275.

**[GUM, 1992]** ISO standard, "Guide to the expression of uncertainty in measurement", *ISO/IEC/OIML/BIPM*, First edition, 1992

**[Kroeze *et al.*, 1994]** Kroeze, B., Streppel, A.H. and Lutters, D., "Tools and Accessories for Press Brakes", *Proc. of the 2nd Int. Conf. on Sheet Metal*, Belfast, April 1994, ISBN: 1 85923 025 3, pp 251-260.

# Keyword Index